SWIFTS

A GUIDE TO THE SWIFTS
AND TREESWIFTS
OF THE WORLD

SWIFTS

A GUIDE TO THE SWIFTS
AND TREESWIFTS
OF THE WORLD

PHIL CHANTLER

AND

GERALD DRIESSENS

PICA PRESS
SUSSEX

RUSSEL FRIEDMAN BOOKS
SOUTH AFRICA

THE NETHERLANDS
AND BELGIUM

Published in Southern Africa by
  Russel Friedman Books CC
  PO Box 73,
  Halfway House
  1685,
  South Africa.

ISBN 0-9583223-9-2
(Southern Africa only)

Published in the Netherlands and Belgium as *Dutch Birding Vogelgids 4* by
  Ger Meesters Boekprodukties
  Vrijheidsweg 86,
  2033 CE Haarlem.

ISBN 90-74345-08-5
(Netherlands and Belgium only)

In the United States of America and Canada this book is distributed by
  Common Ground Distributors
  370 Airport Road,
  Arden
  NC 28704.

and to members of the American Birding Association by
  ABA Sales
  PO Box 6599,
  Colorado Springs
  CO 80934-6599.

Design, computer graphics and typesetting by Fluke Art, Bexhill-on-Sea, E Sussex.
Colour separation by Scan House, Kettering, Northants.
Printed and bound by Hartnolls Limited, Bodmin, Cornwall.

CONTENTS

LIST OF FIGURES

LIST OF TABLES

INTRODUCTION

This book is first and foremost an identification guide. The species accounts that form the bulk of the book provide essential identification points and detailed descriptions as well as notes on distribution and movements, details on the habitats in which the species are likely to occur and their typical habits. The descriptions, although dealing with many features that might not be easily viewable in the field, are intended for both field and in-the-hand purposes. It is worth noting that many of the finer feather details that are not viewable in the field may contribute to the general appearance of the bird and it is therefore important that the reader is aware of them. Some of the identification criteria put forward, especially concerning the notorious swiftlets, need thorough testing in the field and the author welcomes comments on any aspect of swift identification.

For most species detailed studies on breeding behaviour and food are lacking although the introductory chapters provide a general synopsis on swift biology and behaviour. For those wanting to pursue these points further, reference should be made to the several excellent accounts of swift behaviour listed in the bibliography. In particular, studies made by Collins (1967 and 1968a), Lack (1956c), Marin and Stiles (1992), Rowley and Orr (1962 and 1965) and Tarburton (1993) are essential starting points.

Similarly, the taxonomic accounts in this guide are general in comparison to those whose studies I have extensively worked from: Brooke (1970b, 1971d and 1972c), Collins and Brooke (1976), Dickinson (1989a and 1989b), Marin and Stiles (1992), Medway (1966), Navarro *et al.* (1992), Orr (1963), Salomonsen (1983), Sibley and Monroe (1990) and Somadikarta (1967 and 1985). These are some of the most comprehensive reviews of swift taxonomy and are invaluable starting points for anyone who wants to tread the minefield of swift systematics.

ACKNOWLEDGEMENTS

A great many people have helped me in a variety of ways. Without their help I would not have reached this final stage. Special thanks are due in particular to those who commented extensively on specific parts of the text. Edward Dickinson's expert comments on the *Collocalia* caused a vital rethink resulting in many improvements. Similarly, Manuel Marin's great knowledge of Neotropical swifts clarified several areas of confusion, especially with regard to the breeding biology of a number of species. His pioneering work is a must for all students of bird biology. Richard Brooke's unequalled knowledge of African swifts was most apparent in his constructive comments on the accounts for that continent. I am grateful to Robert Clay for sharing his notes and views on field identification of Neotropical swifts, especially the *Chaetura*. Both Guy Kirwan and Günter de Smet reviewed the *Chaetura* accounts and made useful contributions. Peter Kennerley added much to the understanding of *Hirundapus* movements and, together with Martin Hales, made me aware of the interesting series of swiftlet records from Hong Kong and the existence of Edible-nest Swiftlets on Hainan. J S Serrao of the Bombay Natural History Society reviewed species accounts of the Indian subcontinent and brought to my attention a number of sources that I was not aware of. Iain Robertson helped both in his very informative correspondence and also with his comments on the species accounts for the African species. Michael Tarburton helped make the account on Australian Swiftlet, and *chillagoensis* in particular, more complete.

A number of responses were received in reply to pleas published in Dutch Birding and British Birds and also to direct requests. Paul Salaman supplied a great many details regarding swift distribution in Colombia. Robert Ridgely kindly allowed me to see his swift text for the forthcoming Birds of Ecuador. John Ash helped me with some problems regarding Forbes-Watson's and Nyanza Swifts. René Dekker added to the distributional data for Grey-rumped Treeswift and measured Purple Needletail rectrix spines. Both Ralph Browning and Somadikarta replied to questions about Linchi Swiftlet. Craig Robson commented on swiftlet accounts and added in particular to vocalisations of Philippine Grey Swiftlet and Whiskered Treeswift. Anders Anderson kindly supplied a photocopy of his notes of a day's birding at Khao Yai which detailed a movement of White-throated Needletails. Richard Thewlis supplied very detailed distributional data for Laos. Peter Hayman provided details of the first South African record of Scarce Swift.

Several comments were earlier incorporated into other articles which also form part of the current text. Jon Curson kindly helped with several ideas concerning Vaux's and Chimney Swifts. Tony Clarke checked the comments regarding Plain Swift for the Dutch Birding article and C J Hazevoet did the same for Alexander's Swift. Alan Dean shared his views on the variation in the shape of the belly patch in Alpine Swift. Ian Sinclair, Dave Buckingham and Jeff Blincow all corresponded regarding unusually-plumaged swifts in Africa and Asia.

The staff at the Natural History Museum (BMNH), Tring were a constant help, especially Peter Colston, Robert Prys-Jones, Michael Walters and Effie Warr. Similarly René Dekker (National Natuurhistorisch Museum, Leiden, the Netherlands), C Erard (Museum National d'Histoire Naturelle Zoologie, Paris, France), Rene-Marie Lafontaine (Koninklijk Natuurhistorisch Museum, Brussels, Belgium) and Michel Louette (Koninklijk Museum voor Midden-Afrika, Tervuren, Belgium) allowed access to many important specimens.

The quality of the plates has been greatly enhanced by the comments of Jack Chantler, Raf Drijvers, Richard Heading, Christopher Helm, Nigel Redman, Iain Robertson, Günter de Smet and Filip Verbelen. For the loan of photographic material and literature Gerald Driessens wishes to thank Jef de Ridder, Mrs de Roo-de Ridder, Kris de Rouck, Michael Knoll, Chris Steeman and Eduard Vercruysse.

Various trips in recent years were made all the more enjoyable thanks to the company of the following: Jack Chantler, Judy Hodgson and Ian Hodgson in northern Thailand; Wiel Poelman, Anders Anderson and Hedvig Kolboholen in southern Thailand; Nach for his great hospitality and curries in Malaysia; Richard Heading in both the last places and for the unforgettable pilgrimage to Niah; Jenny Hawkes and Jane Lyons in Venezuela; Julian Russell, David Ngala, Patrick Buys and Gerald Driessens in Kenya; Heather Chantler and Nigel Jarman in Ecuador.

A great debt is owed to Julie Reynolds and Marc Dando of Fluke Art for all their hard work, and to Christopher Helm at Pica Press for his encouragement and expertise. In particular I acknowledge the help and encouragement I have received from Nigel Redman, both in the gestation period of this project and in his role as editor.

Finally, thanks to all the friends who have supported and inspired me: Phil Rye for the hours spent in conversation, Rachel Guest and Dave Nash for helping me with several translation problems, Chris Powell for sorting out my computing problems, Luke Downs, Richard Heading, Jenny Hawkes, Sarah Hales, Julian Russell, Nigel Jarman, Lee McRee, Heather, Becky and Brian Godden, and to my parents for putting up with me and for their constant assistance.

For help throughout the entire project Gerald Driessens thanks Patrick Buys, Viki Meeuwis and Gerald Oreel.

STYLE AND LAYOUT OF THE BOOK

A number of short introductory chapters give an overview of swift biology and behaviour. These are inevitably fairly brief but references point to more detailed accounts which the reader can refer to. The plates and species accounts comprise the greater part of the book. There is a strong emphasis on identification and distribution in the species accounts but other aspects of biology and ecology are also covered where known.

Relationships and Taxonomy

This chapter presents a discussion of the taxonomy of the order and highlights some of the taxonomic problems involved. The relationships and species limits of each genus are dealt with individually.

Breeding Behaviour

Specific information on nesting (where known) is included in the species accounts, but this section presents an overview of swift breeding behaviour including mate selection, displays, incubation, fledgling growth rates, special adaptations to environmental extremes and breeding seasons.

Feeding and Ecological Separation

This chapter summarises feeding behaviour and diet. Due to the similarity of the prey of all species, the emphasis in this section is on prey size and selection, and the effect these factors have on ecological separation.

Mortality and Predators

This is a discussion of the main causes of swift mortality, both as a result of direct predation and by indirect factors. Some data on survival rates are included.

Moult

Information on swift moult is far from comprehensive. A brief summary of present knowledge is given but knowledge of moult is incomplete for many genera.

Flight

This section highlights the special adaptations to flight which are so characteristic of the order.

Conservation

The status of the rarer swift species is discussed, as well some of the problems concerning conservation issues, such as the overharvesting of swiftlet nests.

Undescribed Species?

Some ten species of swifts have been described to science in the last fifty years, and it is quite likely that one or two more may still be undiscovered. Some recent claims of possible new species are detailed in the hope that some at least may be substantiated in the future.

Watching Swifts

Useful hints for observing swifts are presented, together with a number of identification pitfalls and topography charts.

SPECIES NUMBERS

For the purpose of this book each species has been given a number for ease of locating plates and text.

PLATES

Within the constraint of the number of plates we have attempted to illustrate distinctive races and age-related plumages, as well as any sexual dimorphism (if relevant) and unusual plumages caused by wear or moult. The nominate form has not always been chosen as the main figure for a species. For example, the nominate form of Glossy Swiftlet occurs solely on New Guinea where there are no confusion species. The main form illustrated is *cyanoptila* which occurs throughout the Malay Peninsula, Borneo and eastern Sumatra where it is is partly sympatric with the very similar Linchi Swiftlet. We have endeavoured throughout to make the plates as useful as possible by careful choice of the forms illustrated and by adding useful comparison species where necessary. In some cases, species have been grouped geographically rather than taxonomically.

The plate captions give, for each species, a summary of world range and the most important or diagnostic features of each form or plumage illustrated.

SPECIES ACCOUNTS

Each genus begins with an introduction summarising key features of the genus and dealing with general structural differences which will enable separation from sympatric swift genera. For some of the more difficult genera, a separate section deals with in-the-hand identification.

Each species account is subdivided into sections as follows:

Identification

These sections detail key features for field identification and should be used in conjunction with the plates. For the most challenging identification problems exclusion by range has to form a major part of the identification process. Cautious comments for those species that have proved unidentifiable in the field have been incorporated and should be noted. Species that fall into this latter category have supplementary details on in-the-hand identification. In particular, the swiftlets generally have a separate paragraph detailing museum characteristics. Some emphasis is also placed on identification by 'jizz', but this should be treated with caution; it is rarely wise to identify a species on jizz alone.

Voice

Voice transcriptions are included where available. In many cases these have been taken directly from the published literature (and referenced as such), but some have been transcribed by the author.

Distribution

These sections commence with a brief summary of world range followed by a more detailed treatment. For several regions information is very scanty (e.g. Indochina and Amazonia). The final paragraph in this section gives, where possible, an indication of the species' status.

Movements

Some species are resident and sedentary, but many others are migratory. This section covers all movements, both local and long distance, including altitudinal movements and vagrancy.

Habitat

Although swifts feed aerially, variations in the land surface ecology affect ecological separation. Many species are faithful to particular habitats but it is important to note that all swifts react to changes in feeding conditions, which are frequently brought on by climatic factors. This can result in species being found in unexpected habitats. Altitude ranges are given where known.

Description

These are detailed descriptions of plumage features and, for ease of use, the section is divided into subsections such as 'Head', 'Body', 'Upperwing', 'Underwing' and 'Tail'. For most polytypic species the nominate, or in some cases the most widespread, race forms the subject of the description. Non-adult plumages or structural differences are described (where relevant) as well as sexual dimorphism (if any). Although many of the features described may be exceptionally hard to view in the field a knowledge of how the features are formed is often essential. Additionally, the depth of most descriptions will serve to add to the usefulness of this guide to those working on museum specimens or researchers trapping swifts.

Measurements

Wherever possible, measurements from the largest samples have been used, and refer to the race described in full under 'Description'; measurements for other races are detailed in 'Geographical Variation'. Measurements taken from the Natural History Museum at Tring (England) are indicated as BMNH. Data on weights are rather infrequent, but the incompleteness of data presented should not hinder identification. Tarsus measurements are only included for the *Cypseloides* swifts as it has been shown that they are useful for identification in this genus. Measurements are frequently abbreviated thus: wing: (13) 97-103 (99.6), i.e. 13 birds were measured, having a range of 97-103mm and a mean of 99.6mm. All species' measurements are in millimetres unless otherwise stated (body length is always given in centimetres); weights are in grams; altitudes are in metres.

Some species of swifts are only known from a handful of specimens (notably Mayr's and Whitehead's Swiftlets, and Schouteden's Swift) and therefore the measurements (and descriptions) should be treated with caution. Individual variation can be considerable and a larger sample is ideally required in order to make meaningful comparisons with similar species.

Geographical Variation

In many swift species geographical variation is very slight. However, we have attempted to describe briefly all recognised races, concentrating on the differences between them. Future taxonomic revisions are likely to subdivide certain polytypic species (or even amalgamate other taxa) and some possible 'splits' and 'lumps' have been noted in these sections. The range of each subspecies is briefly given.

Breeding

The nesting habits of each species, where known, are included in this section. Nest construction is an important aid to identification in some species. For example, in certain caves in South-east Asia up to four species of swiftlet can be identified through a careful examination of their nests.

References

References given in the text and at the end of the species accounts are detailed in full in the Bibliography.

BLACK-AND-WHITE FIGURES

A number of line drawings are included to illustrate specific features mentioned in the text and to supplement the plates.

MAPS

A map is included for each species. Generally, three intensities of shading indicate breeding, wintering and resident ranges. In some cases, hatching indicates an area which is only sparsely inhabited by the species. A cross indicates a vagrant record or an area of irregular occurrence. A question mark indicates an area of uncertain status.

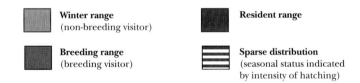

Winter range
(non-breeding visitor)

Resident range

Breeding range
(breeding visitor)

Sparse distribution
(seasonal status indicated
by intensity of hatching)

RELATIONSHIPS AND TAXONOMY

The work on DNA analysis by Sibley and Ahlquist (1990) confirmed the views of many previous researchers that swifts and hummingbirds are closely related. Their place within the taxonomic order is as follows:

Superorder APODIMORPHAE

    Order APODIFORMES
        Family Apodidae: Typical Swifts
        Family Hemiprocnidae: Treeswifts

    Order TROCHILIFORMES
        Family Trochilidae
            Subfamily Phaethornithinae: Hermits
            Subfamily Trochilinae: Typical Hummingbirds

The following structural features described are mainly external. Sibley and Ahlquist (1990) also describe a series of anatomical differences that are largely beyond the scope of this book.

Within the superorder all species are small to tiny birds with ten primaries (with the tenth or ninth being the longest), 6-11 significantly shorter secondaries (in swifts the longest primary is about three times the length of the longest secondary), and ten rectrices. All species are nidicolous and gymnopaedic, and all (except Alexander's Swift) have white eggs.

Within the order Apodiformes the following features apply: bill short and broad with a deeply cleft gape and nostrils that open vertically; non-extensile, short, triangular tongue; 8-11 secondaries and 2-3 alula feathers; all except Hemiprocnidae (and Three-toed Swiftlet) have a claw on the manus; the salivary glands are large and the size increase they undergo during the breeding season is critical for the production of saliva used in nest building (except in the Cypseloidinae and *Hirundapus*).

Within the family Apodidae there has been some considerable discussion as to the correct subdivisions. Two or three subfamilies are generally recognised. Chaeturinae is sometimes recognised as a third subfamily (e.g. Cramp 1985), but Brooke (1970a) noted that no known characters above the generic level, with the exception of foot structure, separated *Chaetura* and *Apus* and therefore the subfamilies based on these genera should not be maintained. Sibley & Ahlquist (1990) were not able to clarify the situation due to lacking the DNAs of many genera. Current thinking tends to divide the subfamilies, tribes and genera of the Apodidae in the following way:

APODIDAE	CYPSELOIDINAE	Cypseloidini	*Cypseloides*
			Streptoprocne
	APODINAE	Collocaliini	*Collocalia*
			Schoutedenapus
		Chaeturini	*Mearnsia*
			Zoonavena
			Telacanthura
			Rhaphidura
			Neafrapus
			Hirundapus
			Chaetura
		Apodini	*Aeronautes*
			Tachornis
			Panyptila
			Cypsiurus
			Tachymarptis
			Apus

Brooke (1970a) discusses evolutionary levels within the order. He considers that the Cypseloidinae are the most primitive surviving group of swifts. This view is based on their possession of two carotid arteries, anisodactyl feet, a primitive palate and their lack of sticky saliva as a nest-building material. They are followed by the more complex Collocaliini which share anisodactyl feet with the Cypseloidinae but have a single carotid artery, a more specialised palate and use saliva as a nest cement. Further specialisation occurs in the Chaeturini as they have spiny rectrix projections and as a consequence more specialised nest sites. The Apodini are the most advanced tribe with their pamprodactyl feet and their tendency to build elaborate nests in trees or under overhanging rocks.

Furthermore he provides an excellent account of the evolutionary path that the New World genus *Chaetura* may have emerged from. In his discussion of the spinetails (*Chaetura, sensu* Lack 1956b) he notes that, as a group, the Old World spinetails are highly dissimilar, seldom have subspecies and have discontinuous to relict ranges, whereas the New World spinetails are very similar in appearance and usually have many subspecies and extensive continuous ranges. Brooke notes that the evidence implies the former are old species in decline with the New World species being at a fairly early stage of speciation and expansion. His suggestion is that 'an Old World stock, the *Zoonavena* group.......invaded the New World fairly recently (say the beginning of the Pleistocene) and radiated afresh'.

The allocation of swift genera has also been a subject of some debate. For example, the genus *Chaetura* formerly included a wide range of species until formally split by Brooke (1970b) into five genera, following the work of Lack (1956b) and Orr (1963). Some authors continue to divide the *Collocalia* into three genera, and *Tachymarptis* is still frequently included in *Apus*. The position of *Schoutedenapus* within the order is perhaps the most contentious problem and it is here placed in the Collocaliini. Further details concerning generic relationships are given in the individual genera accounts below.

In this book, the taxonomy of the order at a species level largely follows Sibley and Monroe (1990 and 1993). Where Sibley and Monroe digress from more widely adopted opinions we have only accepted their views if supported by a mainly unchallenged paper. For example, we have retained the controversial Madagascar Swift *Apus balstoni* within African Swift *A. barbatus* as no reasons are given for its separation. Inevitably, precise species limits in certain genera will remain controversial and unresolved. This is nowhere more apparent than in the swiftlets *Collocalia*, which have long been recognised by many as amongst the worst of taxonomic nightmares. Peters (1940) noted that 'the genus *Collocalia* constitutes one of the most difficult of all groups of birds. The principal recent revisers of this genus have also realized this fact, admitting that their results are purely tentative and their conceptions of relationships are liable to modification'. Although considerable progress has been made since then, the status of many taxa still remains unresolved. The *Chaetura, Cypseloides* and *Apus* are only marginally less problematic. For example, the recently described White-fronted Swift *Cypseloides storeri*, has not met with universal acceptance and may prove to be a subspecies of White-chinned Swift *C. cryptus*. Taxonomy is not an exact science and many of the specific divisions are arbitrary, merely fulfilling the human desire to neatly pigeonhole each taxon. At our current level of understanding this is simply not possible for many swifts and it is hoped that DNA analysis will eventually unravel some of the mysteries.

Recent articles by Hazevoet (1994) and Knox (1994a and 1994b) challenge the traditional use of the biological species concept (BSC), which resulted in a wealth of subspecies. The phylogenetic species concept (PSC) is defined by Cracraft (1983 and 1987, in Hazevoet 1994) who argues that a 'species is the smallest diagnosable cluster of individual organisms within which there is a parental pattern of ancestry'. Hazevoet (1994) notes that 'phylogenetic species are irreducible, terminal (basal) taxa that cannot be subdivided into smaller groups that are themselves discrete, diagnosably distinct taxa'. The implications for avian taxonomy of the acceptance of the PSC would not necessarily mean that all taxa currently considered subspecies would be given full specific status but there is no doubting that many current subspecies would be elevated to species level, while the taxa of many polytypic species which show little more than clinal variation would remain as subspecies. To suggestions that trinomials may be retained to name clinal populations, Hazevoet (1994) argues that it will be 'probably better to abandon the subspecies category altogether and look for other ways of describing intraspecific variation'.

Species limits in this book are based on the biological species concept, but all generally recognised subspecies have been described in the maximum possible detail. Clinal variation within a species can be considerable resulting, in some cases, in a large number of subspecies. However, environmental factors can also affect morphology and not all clinal variations are sufficient to warrant subspecific status. For example, the central New Guinea form of Mountain Swiftlet *Collocalia hirundinacea excelsa* shows a clinal increase in wing length at higher altitudes (table 1).

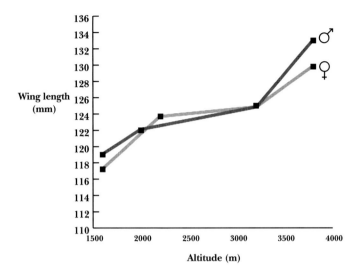

Table 1. Altitudinal increase in wing length in *excelsa* race of Mountain Swiftlet (after Rand 1942).

RELATIONSHIPS AND SPECIES LIMITS OF SWIFT GENERA

Cypseloides

In 1963 Orr argued that the Cypseloidinae should comprise four genera *Streptoprocne, Nephoecetes* (*niger*), *Aerornis* (*senex*) and *Cypseloides* (comprising all remaining species). Brooke (1970b) agreed with this and further split the *Streptoprocne*. However, these views have not persisted. Short (1975) argued for the inclusion of the *niger* superspecies into *Cypseloides,* and *Aerornis* has not been widely supported. More recently Marin and Stiles (1992) have shed much light on the issue. Studies of the plumages and breeding biology of *Cypseloides rutilus* show that, like *Streptoprocne,* it usually lays two eggs, its young develop rapidly and it shares several plumage traits (most notably the collar and the ashy-grey down of the nestling). These features apparently separate *rutilus* (and probably the closely related but poorly studied *phelpsi*) from single egg-laying other members of the genus, all of which show similar plumage characters.

Sibley and Monroe (1990 and 1993) concluded that the genus comprises three superspecies: *rutilus* (comprising *rutilus* and *phelpsi*), *niger* (comprising *niger, lemosi, rothschildi* and *fumigatus*) and *cryptus* (comprising *cryptus* and *storeri*) with *cherriei* and *senex* as simple species. This arrangement has been followed in this book.

At the specific level, debate surrounds the status of the species pairs *fumigatus* and *rothschildi,* and *cryptus* and *storeri.* Eisenmann and Lehmann (1962) noted the difference in tail structure between the former pair, but Marin and Stiles' recent studies questioned the validity of tail structure as a taxonomic determinant. Marin and Stiles (1992) further postulate that *rothschildi* may not be a full species and that *cryptus* might even be conspecific with *fumigatus.* Howell (1993b) has stated that describing *storeri* as a species 'seems premature based on the information available' and this form is duly treated as a race of White-chinned Swift in Howell & Webb (1995). Marin (pers. comm.) agrees with this view.

Streptoprocne

Orr (1963), and later Brooke (1970b), proposed that White-naped Swift was different enough to warrant subgeneric status and Brooke coined the name *Semicollum.* The lack of a nest was thought to be particularly important, but the work of Whitacre (1989) shows that *semicollaris* or *zonaris* will make a nest or not according to the nature of the nest site. Similarly, Orr (1963) put too much emphasis on rectrix shape. Marin and Stiles (1992) note that the presence of tail spines seems to reflect body size and roosting habits rather than phylogenetic affinities.

At the species level this is one of the more straightforward genera but Chestnut-collared Swift and perhaps Tepui Swift may be best placed in this group as a result of their similar plumage sequences and biologies (Marin and Stiles 1992).

Collocalia

All swiftlets are considered here to belong to the genus *Collocalia*. Brooke (1970b) split the genus into three subgenera: *Hydrochous*, a monotypic genus comprising Giant Swiftlet, *Collocalia*, comprising the glossy, non-echolocating species, and *Aerodramus* consisting of the remaining non-glossy, grey species. Later, Brooke (1972c) proposed that *Aerodramus* be given generic status.

Some authors have accepted this division entirely (e.g. Howard and Moore 1980, Medway and Pye 1977), whilst others have only maintained *Hydrochous* as separate from *Collocalia* (e.g. Sibley and Monroe 1990). Salomonsen (1983) considered that the separation of *Hydrochous* was based, at least partly, upon characters that were futile and trivial. For example, he noted that the maximum wing length of Papuan Swiftlet is 141 compared to a minimum of 142.5 for 25 Giant Swiftlets measured by Somadikarta (1968). Furthermore, he noted also that *Aerodramus* was debatable as it was not known whether or not Pygmy Swiftlet could echolocate and that some species in the *Aerodramus* group will only echolocate according to the darkness of the nest site. Considering the importance of nest-type in the swiftlets, it is interesting to note that Pygmy Swiftlet's nest is similar to many *Aerodramus* and different from that of Glossy and Linchi Swiftlets. Similar conclusions were reached by Browning (*in litt.*) who noted that the only morphological character distinguishing *Collocalia* and *Aerodramus* is the glossiness of the plumage which does not hold up.

The species limits of the genus *Collocalia* used in this work are as detailed below. The names in square brackets, placed between the generic and specific names of certain taxa, indicate the senior names of the superspecies as recognised by Sibley and Monroe (1990 and 1993).

Giant Swiftlet *Collocalia gigas* Hartert and Butler 1901
Glossy Swiftlet *Collocalia esculenta* Linnaeus 1758
Linchi Swiftlet *Collocalia linchi* Horsfield and Moore 1854
Pygmy Swiftlet *Collocalia troglodytes* Gray and Mitchell 1845
Seychelles Swiftlet *Collocalia [francica] elaphra* Oberholser 1906
Mascarene Swiftlet *Collocalia [francica] francica* (Gmelin) 1789
Indian Swiftlet *Collocalia unicolor* (Jerdon) 1840
Philippine Grey Swiftlet *Collocalia [spodiopygius] mearnsi* (Oberholser) 1912
Moluccan Swiftlet *Collocalia [spodiopygius] infuscata* Salvadori 1880
Mountain Swiftlet *Collocalia [spodiopygius] hirundinacea* Stresemann 1914
White-rumped Swiftlet *Collocalia [spodiopygius] spodiopygius* (Peale) 1848
Australian Swiftlet *Collocalia [spodiopygius] terraereginae* (Ramsay) 1875
Himalayan Swiftlet *Collocalia brevirostris* (Horsfield) 1840
Whitehead's Swiftlet *Collocalia whiteheadi* (Ogilvie-Grant) 1895
Bare-legged Swiftlet *Collocalia nuditarsus* (Salomonsen) 1963
Mayr's Swiftlet *Collocalia orientalis* (Mayr) 1935
Mossy-nest Swiftlet *Collocalia [vanikorensis] salangana* (Streubel) 1848
Uniform Swiftlet *Collocalia [vanikorensis] vanikorensis* (Quoy and Gaimard) 1830
Palau Swiftlet *Collocalia [vanikorensis] pelewensis* Mayr 1935
Guam Swiftlet *Collocalia [vanikorensis] bartschi* (Mearns) 1909
Caroline Swiftlet *Collocalia [vanikorensis] inquieta* (Kittlitz) 1857
Sawtell's Swiftlet *Collocalia [leucophaeus] sawtelli* Holyoak 1974
Polynesian Swiftlet *Collocalia [leucophaeus] leucophaeus* (Peale) 1848
Black-nest Swiftlet *Collocalia maxima* Hume 1878
Edible-nest Swiftlet *Collocalia fuciphaga* (Thunberg) 1812
Papuan Swiftlet *Collocalia papuensis* Rand 1941

This list largely follows Medway (1966) and Medway and Pye (1977). *Collocalia (esculenta) marginata* was considered a separate species in Medway (1966) but not in Medway and Pye (1977). Separation of *C. linchi* from the *esculenta* group follows Somadikarta (1986). Separation of *C. mearnsi* follows Dickinson (1989a), and the latter's view that it should tentatively be placed in the *spodiopygius* super-species is also followed. Treatment of the other four species within the *spodiopygius* superspecies follows White and Bruce (1986). *C. sawtelli* has been described since Medway's work. *C. salangana* is treated as an allospecies despite being considered conspecific with *C. vanikorensis* by Medway (1975) and Medway and Pye (1977). Salomonsen (1983) noted that this view was precarious and Dickinson (1989a) treated them as separate species.

The species limits of the *Collocalia* adopted in this book differ from Sibley and Monroe (1990 and 1993) in the treatment of the following taxa:

1) Grey-rumped Swiftlet *Collocalia (esculenta) marginata*. Considered conspecific with *esculenta* by Medway and Pye (1977). Sibley and Monroe (1990) treated *marginata* as an allospecies but sub-sequently (1993) considered it to be conspecific with *esculenta*, following Dickinson (1989b).

2) Philippine Grey Swiftlet *Collocalia [spodiopygius] mearnsi*. Sibley and Monroe (1990) placed this form within the *vanikorensis* superspecies, but later (1993) considered it to be part of the *spodiopygius* superspecies following Dickinson's view (1989a) that it is probably more closely related to the Mountain Swiftlet *C. hirundinacea* of New Guinea than to any Bornean taxon.

3) Volcano Swiftlet *Collocalia [brevirostris] vulcanorum*. This isolated form is treated as an allospecies by Sibley and Monroe (1990), and followed by Collar et al. (1993). As its taxonomic status has not been fully reviewed since Medway (1966) such treatment may be unsafe.

4) Indochinese Swiftlet *Collocalia rogersi*. Included in Himalayan Swiftlet *C. brevirostris* by Medway (1966) although with the reservation that it may ultimately be found to be more properly related to *vanikorensis*. Importantly, Deignan (1955) considered that there may be some intergradation between *rogersi* and other forms of Himalayan Swiftlet in the north of the former's range. Treated as a full species by Sibley and Monroe (1990) on the strength of Browning's unpublished opinion that *rogersi* appears closer to Indian Swiftlet *C. unicolor*.

5) Palawan Swiftlet *Collocalia [vanikorensis] palawanensis*. Treated as an allospecies within the *vanikorensis* superspecies by Sibley and Monroe (1990) on the strength of Kennedy's unpublished opinion, but Dickinson (1989a) treated this form as a subspecies of Uniform Swiftlet *C. vanikorensis*.

6) Grey Swiftlet *Collocalia [vanikorensis] amelis*. Sibley and Monroe (1990) placed this form within the *fuciphaga* superspecies but later (1993) considered it to be part of the *vanikorensis* superspecies. Dickinson (1989a) treated this form as a subspecies of Uniform Swiftlet *C. vanikorensis*.

7) Guam Swiftlet *Collocalia [vanikorensis] bartschi*. Sibley and Monroe (1990) treated this form as an allospecies within the *vanikorensis* superspecies, but later (1993) considered it to be conspecific with Caroline Swiftlet *C. inquieta* (using the English name Micronesian Swiftlet for the two forms) following the opinion of H. D. Pratt. This view is not supported by Browning whose review of species limits in Micronesian swiftlets (1993) showed that morphological differences exist between the two forms.

8) Marquesan Swiftlet *Collocalia [leucophaeus] ocista*. Sibley and Monroe (1990) considered this form to be a subspecies of Polynesian Swiftlet *C. leucophaeus*, following Mayr and Vuilleumier (1983), but later (1993) treated it as an allospecies following Holyoak and Thibault (1978).

9) German's Swiftlet *Collocalia [fuciphaga] germani*. Recognised as an allospecies within the *fuciphaga* superspecies by Salomonsen (1983) and White and Bruce (1986). However, as its taxonomic status has not been formally reviewed, Medway's assessment of the *fuciphaga* complex as a ring cline remains uncontested.

Schoutedenapus

The exact position of this genus within the order remains an area of debate. Its superficial similarity to *Apus* ensured that it was originally included in that genus. However de Roo (1968) realised impor-tant differences in foot structure (the feet are anisodactyl as with *Cypseloides*, *Collocalia* and *Chaetura*, as opposed to pamprodactyl in *Apus*) and formed a new genus accordingly. Brooke (1970b) consid-ered that the genus might best be included in the Collocaliini, but placed it at that time in the Apodini, due to its deeply forked tail and the 'increasing emargination of the outermost rectrix between juvenal and adult plumage as in *Apus* and *Cypsiurus*'. Brooke also noted that Collins considered that *Schoutedenapus* might be an aberrant member of the Cypseloidinae as Scarce Swift and Black Swift have similar *Dennyus* Mallophaga (feather lice) with peg-like setae, a character otherwise unknown in swift Mallophaga. Later, Brooke (1972c) postulated that it may prove to be an aberrant member of *Collocalia* and indeed the metallic clicking calls of Scarce Swift are not unlike the echolocating calls of swiftlets. Clearly further studies are needed of this genus. Its true affinities may only be revealed when its nesting habits are known and (possibly) its distinctive calls have been analysed. In this book, the genus has been included with Collocaliini.

Mearnsia

Peters (1940) included the two *Neafrapus* species in *Mearnsia*. *Novaeguineae* was placed alone in the genus *Papuanapus* by Mathews in 1918 on account of its different plumage and longer tail, but this revision received little support. Lack (1956b) paved the way to separation of the *Chaetura*, which at that time included *Mearnsia*, and this view was supported by the reviews of Orr (1963) and Brooke (1970b).

Zoonavena

In 1940 Peters included *sylvatica* and *thomensis* in *Chaetura*, with *grandidieri* as the sole representative of *Zoonavena*. Lack (1956b) looked at the, then, large genus *Chaetura* which he suggested could be subdivided on the basis of plumage coloration and zoogeographical relationships. He grouped *Zoonavena*, *Neafrapus*, *Telacanthura* and *Rhaphidura*, but without naming them, within *Chaetura*. Brooke (1970b) formally split the *Chaetura* along these lines. This genus is thought to be most closely related to *Rhaphidura* and Lack (1956b) postulated that *thomensis* was a well marked race of *sabini*. Brooke (1970b), however, noted that there are great dissimilarities in appearance and that Lack's viewpoint was not well supported.

Telacanthura

This genus was previously included in the *Chaetura* until formally split by Brooke (1970b). Brooke notes that de Roo believed that two species may be present in *ussheri* but this view has not been supported by subsequent researchers.

Rhaphidura

Despite some suggestions that this genus is closely linked to *Zoonavena* no changes have occurred to its taxonomic status since it was split from *Chaetura* by Brooke (1970b).

Neafrapus

The taxonomic status of this genus is not contentious since it was split from *Chaetura* by Brooke (1970b).

Hirundapus

Sibley and Monroe (1990) divide the genus into two superspecies *caudacutus* and *giganteus*, with the former comprising *caudacutus* and *cochinchinensis* and the latter *giganteus* and *celebensis*. Earlier views that the pairs constituted conspecific forms were dismissed by Collins and Brooke (1976).

Chaetura

A confusing genus. Sibley and Monroe distinguish the superspecies *martinica* (comprising *martinica*, *cinereiventris* and *egregia*) with the other six taxa considered as simple species. Lack (1956b) suggested that *vauxi*, *pelagica* and *chapmani* might constitute a single species. Wetmore (1957) argues strongly, however, for the retention of the three as distinct species mainly on morphological grounds. Peters (1940) recognised *richmondi* (including *ochropygia*) and *gaumeri* as separate from *vauxi*, despite the work of Griscom (1932). Sutton (1941) and Sutton and Phelps (1948) reiterated that these taxa are conspecific. Greatest confusion within the genus surrounds the status of some races of *cinereiventris* and in particular the validity of *egregia*. Field observations by Parker and Remsen (1987) proved at least some sympatry between the two and also confirmed plumage and morphological differences.

Aeronautes

Andecolus was included in *Apus* until transferred by Lack (1956b). Brooke (1970b) noted that this treatment was correct but that the other two species were more closely related to each other than to *andecolus*.

Tachornis

Peters (1940) assigned different genera for the three species now included in *Tachornis: Reinarda squamata*, *Micropanyptila furcata* and *Tachornis phoenicobia*. Lack (1956b) noted the great similarity in nesting behaviour and morphology, and consolidated the three; this view has remained unchallenged.

Panyptila

One of the least contentious genera. Lack (1956b) noted that this genus is closely related to *Tachornis*.

Cypsiurus

The two species in the genus comprise the superspecies *parvus*. Brooke (1972a) noted differences in juvenile plumage and tail structure which justify specific status despite previous suggestions of conspecificity. Brooke also considered that the Malagasy *gracilis* might be treated as a separate species, a view that has not been supported (but which may prove to be justified).

Tachymarptis

Not all authors (e.g. Cramp 1985) have followed the separation of *Tachymarptis* from *Apus* which is justified by Brooke (1972c) because of the difference in nestling foot structure, larger size and different Mallophaga (feather lice).

Apus

Alpine and Mottled Swifts were formerly included in this genus.

The taxonomy of the genus continues to be a matter of debate. Sibley and Monroe (1990) suggest that five superspecies exist: *apus*, *pallidus*, *pacificus*, *affinis* and *horus*, with *alexandri*, *caffer* and *batesi* being regarded as simple species.

The species *apus*, *unicolor* and *niansae* are included within the *apus* superspecies; *alexandri* (which has been considered conspecific with *unicolor* e.g. Lack 1956a) was formerly included, but the aberrant egg colour led Brooke (1971d) to state that 'it is pure guesswork to ally it with any form'. Sibley and Monroe (1990) suggest that it may be related to the *pallidu*, group. Some debate still surrounds the status of *A. niansae somalicus* which is considered by some authors to be a race of *pallidus* despite close morphological and biological similarities between both *niansae* and *somalicus* (and indeed between *niansae* and *apus* as pointed out by Fry *et al.* 1988).

Within the *pallidus* superspecies Sibley and Monroe (1990) recognise *pallidus*, *barbatus*, *berliozi*, *bradfieldi* and *balstoni* as distinct species. Some authors also recognise *sladeniae* (e.g. Snow 1978 and Dowsett and Forbes-Watson 1993) but this seems to be rather tenuous and is not universally accepted. Similarly, few other authors support Sibley and Monroe's specific recognition of *balstoni* and it is not treated separately in this book. This superspecies is perhaps the most troublesome in the genus. Morphologically all species are very similar with differences restricted to subtle colour changes; plumage is darker in the wettest areas and paler in the most arid. If the same principles were applied to Little Swift (or Alpine Swift) it is interesting to speculate how many species might be recognised.

Little debate surrounds the status of *acuticauda* and *pacificus* (in the *pacificus* superspecies) despite Lack's (1956a) rather isolated view that they were conspecific.

Affinis has been split into two species, *affinis* and *nipalensis*. The ranges of the two species meet without any sign of intergrading (Snow 1978) and Sibley and Monroe (1990) suggest they are treated as allospecies.

Despite Prigogine's (1985) view that *toulsoni* is a paraspecies it is included in *horus* here following Fry *et al.* (1988), Dowsett and Forbes-Watson (1993) and Brooke (pers. comm.).

Hemiprocne

Mystacea and *comata* are both very distinct species that have always been viewed as separate species. *Coronata* and *longipennis* are far closer and constitute the *longipennis* superspecies. Brooke (1969c) disputed the suggestion of Peters (1940) that the two are conspecific and this view prevails today.

15

BREEDING BEHAVIOUR

Mate selection

Most studies indicate that in the next breeding season swifts will pair up with the previous season's mate. However, as Bromhall (1980) points out, migrant species have to renew the bonds forged in the previous year. Lack (1956c) showed that Common Swift pairs remain together through the occupation of the same nesting hole. His study demonstrated that roughly a quarter of pairs arrived back at the nest on the same day, but as the other three-quarters arrived at intervals of between one and twenty-one days, this could with some justification be ascribed to chance.

In Common Swift pair-bonding apparently occurs amongst first summer birds which will occupy a nest hole and even build a nest together without laying eggs. In subsequent years the pair will defend the nest against intruders, both by screaming from the nest hole if potential intruders come close, and by physically fighting intruders. Fighting only occurs after a threat display, with one bird approaching the other with wings held slightly raised and one wing often tipped to one side, exposing the feet which will actually be used in any ensuing fight. The bill is also occasionally used but to far less effect.

Fighting is accompanied by loud screaming and often continues for several hours. Lack (1956c) recorded a fight that lasted four and a half hours and believed that these fights were sometimes fatal. Often, perhaps as an attempt to avoid confrontation, instead of entering a nest hole individuals will fly slowly around a colony brushing or banging the entrance of a nest hole to elicit a response from occupying birds. If the potential nest hole is seen to be undefended the intruder will occasionally investigate further without fear of attack. Nest sites are similarly defended from other hole-nesting species.

After pairs have been parted for the winter, initial reunion may provoke threat displays. Eventually the bond is reformed and mutual preening will occur. This may follow a supposedly submissive display where a newcomer to the nest site will approach the occupant with an exposed throat. Lack (1956c) describes how once both members of the pair are fully familiar with each other, a greeting display occurs on each occasion that a partner returns to the nest hole. This involves a short scream and usually half-raised wings. However, Lack concluded that this greeting display is closer to a mild threat display, before recognition occurs. Marin and Stiles (1993) noted a Spot-fronted Swift nestling engage in a threat display which involved raising the wings and body. They state that wing-raising is a common anti-predator tactic in other *Cypseloides* swifts, but that body raising was previously unrecorded.

A number of aerial displays have been noted. Ashy-tailed Swift has a distinctive display flight involving groups of three flying in elegant swirls, holding the wings almost immobile and arched, with occasional interspersed rapid wing beats (Sick 1993). Similar displays have been noted in a variety of species in different genera.

The display which precedes copulation in Common Swift involves the female flying in front of the male holding her wings almost vertically over her back initiating a forwards plunge with the male in hot pursuit. The female's flight will then slow as she undertakes a horizontal flight on quivering wings at which point the male, perhaps after one or two unsuccessful attempts, will mount the female with his wings held in a high, straight position whilst the female's are held horizontally. The flight path takes on a shallowly descending glide, whilst both frantically move their tails with one or other flapping its wings. Often this behaviour will occur at great height and the fall may be more rapid. Similar displays to this have been recorded in many other swift species and in addition Common Swifts will, on occasions, display in a similar way without mating. One such variation involves a lower bird flying with the slow, horizontal flight on flickering wings, whilst a higher bird rapidly stoops towards it before veering up and away again. Sick (1993) has noted attempted in-flight copulation by *Chaetura* and *Streptoprocne* swifts. Marin and Stiles (1992) observed similar displays, perhaps involving aerial mating, amongst Cypseloidinae species but remain sceptical in the absence of conclusive proof. In many species (including Common Swift) mating also occurs at the nest and swiftlets are thought to mate exclusively at the nest.

A more confusing aerial display is the 'screaming party' where both breeders and non-breeders join in without any apparent aggressive intentions. The increased frequency of these flights in the late summer led Lack (1956c) to postulate that they may be connected with the onset of migration or perhaps, at other times of year, may help unify the group.

Incubation

Both sexes take part in incubation. In Lack's study the longest period that a single bird sat, without relief, on the eggs was five hours and 45 minutes and the shortest period was only two minutes. The most frequent number of changeovers recorded in ten hours was twelve and the lowest was two. On several occasions one of the incubators left the nest before being relieved and the eggs were left unattended for up to six and a half hours. In tropical regions where the weather is warmer, such behaviour would not be remarkable. However, Lack (1956c) postulates that the embryo, and similarly the fledgling, is resistant to cooling as the adult would leave the nest in cold or windy weather, particularly in that part of the day when food was most abundant.

In the Common Swift, Lack (1956c) observed that the average incubation period (measured from laying to hatching) is 19.5 days. In periods of bad weather this average can be extended by up to 4 or 5 days. Incubation periods vary considerably from genus, as illustrated by a selection of species in the following table.

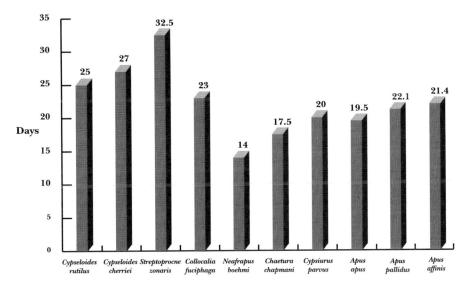

Table 2. Incubation periods for a variety of species (after Marin & Stiles 1992, Langham 1980, Brooke 1966, Collins 1968a, Fry et al. 1988, Lack 1956c, Finlayson 1979, Cramp 1985).

Langham (1980) noted in Edible-nest Swiftlet, that the first egg, was incubated for 21-29 days and the second egg for 20-25 days with an average in both laying and hatching intervals of 3 days. In Chaetura swifts, with their larger clutches, Collins (1968a) notes that the typical interval is every other day.

Moreau (1941) noted that in the case of the African Palm Swift the rather precarious position of the nest has led to an adaptation whereby the adults hold the egg against the nest by pressing during hatching. When the hatchling emerges it clings tightly to the palm with its claws.

Nestlings

From the date of hatching the nestling of the Common Swift is brooded continuously for about the first week, then for about half of each day in the second week. Brooding occurs mainly at night after this. In circumstances where food is scarce due to adverse weather conditions, both adults collect prey and during this period the young may be left for astonishingly long periods of up to several days. In these circumstances the young become torpid but can survive in this state until food is eventually brought to them.

Nestlings are fed at the nest on 'food-balls' consisting of a mass of insects bound together by saliva.

When the nestlings are newly hatched the food-ball may be broken into smaller pieces to feed to the young one by one; but when the young are old enough, each food-ball can be regurgitated directly into one of the huge gapes presented by the youngsters. The feeding process is preceded, or initiated, by sibilant screaming (less shrill than that of the adults) and waving of the open bill. The begging behaviour is initiated by the reappearance of the adult and the accompanying 'bang' as it enters the nest. Lack (1956c) noted that other sharp noises, including his own sneezing, could start the young begging.

Lack noted that after hatching with an initial average weight of 2.75 grams, the average maximum weight of 56 grams is reached in the fourth week. If the weather does not deteriorate seriously after this point weight will be lost whilst the feathers are grown. The effect of the weather upon growth is well documented. Lack notes that during a period of good weather in Switzerland one nestling reached 50 grams in just 10 days whereas an individual from the Oxford colony which Lack studied took the same time to reach 5 grams during inclement weather. Langham's (1980) study of the growth rates of nestling Edible-nest Swiftlets suggest that mass fluctuations are less likely in the tropics where changes in daily temperatures are less extreme. Similarly Marin and Stiles' (1992) studies of Cypseloidinae nestling growth curves in Costa Rica show more even growth than Lack's study. However, Langham (1980) noted in his study that second nestling chicks were more prone to fluctuations in mass, and Marin and Stiles noted an extended nestling period and considerable weight fluctuation in a Chestnut-collared Swift that suffered from botfly parasitism.

The amount of time a swift will spend in the nest averages six weeks in the Common Swift (Lack 1956c), but will extend to 56 days if the weather is poor. Lack recorded 37 days as the shortest period. In comparison to other birds of the same size swifts spend a long time in the nest, and this is thought to be a function of the need for the fledgling to be immediately self-sufficient on fledging. In particularly poor summers premature fledgings are not infrequent. Marin and Stiles' (1992) study revealed that the longest known nestling period of any swift is 65-70 days for Spot-fronted Swift. They also noted that there was no correlation between size and development rate for the four species that they studied.

Clutch size is discussed in the species accounts. It is interesting to note here that Langham (1980) did not find a statistically significant difference in clutch size for first, second or third broods of Edible-nest Swiftlet. However, Castan (1955) noted that Pallid Swifts in Tunisia averaged first clutches of 2.89 and second clutches of 1.95.

Tarburton's (1993) study into determinants of clutch size in the tropics focused on differences in clutch size between the single-egged Australian Swiftlet (*Collocalia terraereginae chillagoensis*) and the double-egged White-rumped Swiftlet (*C. spodiopygius assimilis*). The former species breeds in savanna habitat whilst the latter is found in tropical rainforest. Most savanna species have larger clutches than rainforest species and this exception to the rule was therefore an area of interesting study. A variety of discussed hypotheses failed to explain this anomaly and Tarburton considered that the reliability as well as the quantity of food determines clutch size. Greatest weight gain by the single nestlings of the Australian Swiftlet occurred after bouts of rain. Furthermore, during the two year study one year was a good season with abundant rain and insects whilst the other was a poor season. During the poor season fewer feeding visits occurred and even in the good season adults could not adequately feed young in broods that had artificially been increased to two.

It has been proved by Skutch (1935) in the case of Chimney Swift that one or even two extra adults may help at the nest. These helpers can be first summer non-breeders or even older birds.

Breeding seasons

In tropical species breeding seasons tend to coincide with wet seasons and correspondingly higher insect populations. The exact timing can vary somewhat. In Trinidad, Collins (1967) noted that Short-tailed Swifts started breeding at the very beginning of the wet season whilst Chestnut-collared Swifts did not commence until at least a month later. Marin and Stiles (1992) noted that the White-collared Swift laid eggs late in the dry season so that the early wet season abundance of insects could be used to the greatest advantage. Both Collins (1967) and Marin and Stiles (1992) note that the later nesting of *Cypseloides* swifts (i.e. 1-3 months after the start of the rainy season) relates to the appearance of appropriate nesting material and that the conditions at this time are also appropriate for the nesting materials to be firmly adhered to the substrate. Langham's (1980) study of Edible-nest Swiftlets on Penang, Malaysia showed that although there was no direct correlation between breeding season and rainfall the timing was such that the majority of eggs hatched during the dry season.

Temperate migrant species breed during summer periods when insect life is most abundant in their breeding ranges. Some tropical species such as African Palm Swift will breed throughout the year with some regional variation (Fry *et al.* 1988).

White-collared Swift

FEEDING AND ECOLOGICAL SEPARATION

Swifts prey upon what is loosely described as 'aerial plankton'. This consists of a wide variety of insect and arachnid life. Food is not randomly 'hoovered up' as might be expected and swifts will take the largest available items wherever possible, providing they are not too large to enter the gape and, in turn, the food pouch. Avoidance of stinging insects also occurs. Lacey (1910) recorded Common Swifts feeding exclusively on stingless drones around a beehive in Zaïre. However, Lack (1956c) relates an example of an Alpine Swift with several stings in its throat. It is interesting to note that the method employed to distinguish stingless insects is apparently not related to warning coloration, because as well as drones, swifts will also take insects that mimic stinging insects. The selective nature of swift feeding is reflected in the study of Cucco *et al.* (1993) which showed that insect suction traps in the study area caught a far greater percentage of flies than was present in the study species' diets.

Swifts can often be seen swerving off a line to pursue prey items, and at times they will feed with seemingly ruthless efficiency in an area of high prey density. Swifts, although primarily higher airspace feeders, (though rarely above 100m as insect numbers decline significantly above this level) are opportunist feeders and exploit swarms, hatchings or even beehives whenever possible. Bromhall (1980) describes Common Swifts feeding on large concentrations of aphids during a summer when they had reached plague proportions slicing into them in much the same way as tuna fish will attack a shoal

Figure 1. Adult Common Swift with extended throat patch due to collected food-ball.

of anchovies. A small flock of Silver-backed and Brown-backed Needletails was observed feeding on hatching mayflies by making repeated passes at breakneck speed without letting a single insect get more than three or four metres over the river (pers. obs.). However, such low-level feeding, is often forced upon swifts by cold or rainy weather, when conditions are unfavourable for the uplift of small invertebrates into the higher airspace. High-level feeding is not the optimum method for all species, however. Many tropical swift species feed most readily just over the forest canopy. Some species will actually feed by flying through the canopy, or amongst the branches of large forest trees. Swiftlets will actually feed under the canopy at dusk in much the same way as forest bat species. It may be postulated that their echolocating ability is used in these situations.

Studies into the specific breakdown of Common Swift catches show that it has a greater variety of prey items than any similarly well studied bird, with over 500 prey species recorded in Europe (Glutz and Bauer 1980). When one considers the range of insect and aphid life that could be taken by swifts in the tropics the list becomes even more daunting. Cucco *et al.* (1993) note the importance of certain insect taxa (specifically Hymenoptera: bees, wasps and ants; Diptera: true flies; Hemiptera: true bugs; and Coleoptera: beetles) which form the bulk of prey species for swifts both in the temperate and tropical regions. In Lesser Swallow-tailed Swifts and Fork-tailed Palm Swifts termites and small ants (hymenopterans) are thought to make up 90% of the diet (Sick 1993). Hymenopterans and hemipterans are particularly important for Cypseloidine swifts accounting for: 86.7% of 719 prey items of Spot-fronted Swift (Marin and Stiles 1993); 63.6% of the oesophagus and stomach contents of a White-collared Swift (Rowley and Orr 1962); 94.3% (hymenopterans only) of two White-collared Swifts oesophagi and stomachs; 91.7% (hymenopterans only) of four stomachs and two boluses of White-chinned Swift (Marin and Stiles 1992). Few treeswift stomach analyses have been undertaken. Two specimens examined by Salomonsen (1983) contained: four-winged ants (Formicidae), a large hemipteran, a click beetle (Elateridae), a small beetle and two small wasps, whilst the other had a mouth full of winged ants, small beetles and wasps. Other stomachs contained only ants.

A study (based on faecal analysis) of the diets of Common and Pallid Swifts in north-west Italy (Cucco *et al.* 1993) showed that Common Swifts ate significantly larger insects and that prey species also differed with Common eating more aphids and coleopterans, and Pallid Swift eating more dipterans and hymenopterans. This study contradicted Finlayson's (1979) findings on preferred prey size (based on food bolus analysis on Gibraltar) between the two species which showed that Pallid Swift included large insects (> 12mm), whereas Common Swift never exceeded this limit. As Cucco *et al.* note comparison of food preferences amongst species of swifts is difficult because diets vary geographically. Marin and Stiles' (1993) analysis of three closely related *Cypseloides* species showed that in the smallest species, Spot-fronted Swift, 87.3% of prey items were between 1-4mm, in the median-sized species, White-chinned Swift, 38.4% were in the 11mm category and 41.3% in the 4-5mm range, with the largest species, Black Swift, having 82.3% in the 8-11mm range (table 3).

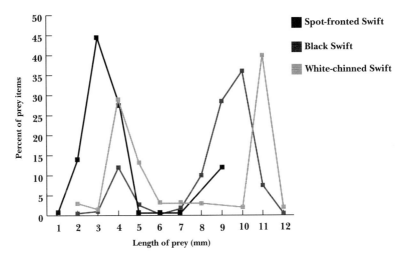

Table 3. Frequency of prey size taken by three *Cypseloides* species (after Marin & Stiles 1993, including data from Collins & Landy 1968).

When feeding young, Common Swift parents will collect 300-500 individual prey items and form them into a bolus bound together with saliva. In ideal feeding conditions up to 2000 individuals may be caught. By comparison the much larger Alpine Swift averages between 156-220 insects for each food-ball (Zehntner 1980). As would be expected the prey items of this species are larger and regularly include wasps, dragonflies and bees, as well as large numbers of smaller species. A White-collared Swift in Venezuela was recorded with 800 winged ants in its stomach (Sick 1993). Cucco *et al.* (1993) noted that feeding parent Pallid and Common Swifts brought larger prey to the nestlings as they increased in age towards fledging.

The timing of swift migration is closely related to abundance of certain invertebrate prey species. Bromhall (1980) postulated, through a study of food bolus content throughout the season, that the decline in August of the massive insect hatchings characteristic of spring and high summer leads to a lack of nutritious insects in the higher airspace (swallows and martins have no such problem as they feed primarily at lower levels). This was reflected by the relative increase in beetle (Coleoptera) numbers in a bolus collected from an adult still feeding a fledgling on the late date of 30 August. For comparison, two boluses taken in the first week of July and 30 August contained 13 beetles out of 898 prey items and 106 beetles out of 348 prey items respectively. Beetles, with the majority of their body weight taken up with a hard exoskeleton, make a poor meal in comparison with highly nutritious aphids. The figures also show that swifts will, on average, take a longer time to collect fewer prey items as the season progresses or during inclement weather. In fine conditions in high summer a parent may catch enough insects to bring a meal to the young in under an hour. Several hours may elapse in poor weather, when the adult may have to concentrate on smaller items and feed over a wider range.

The weather-affected availability of food has led to a remarkable adaptation. Lack (1956c) describes how nestling Common Swifts can survive for several days or even weeks without food. This is a result of storing large fat reserves under the skin in times of plentiful food and by drastically reducing the metabolic rate in adverse conditions. During such periods of starvation it is not unusual for nestlings to halve their weight.

Intrageneric gape size variability and thereby the ability to specialise in prey items of differing size is an important factor in range restriction. An interesting example comes from Kepler's (1972) speculations regarding swift ranges in the West Indies. White-collared Swift has a wide West Indian range but is only known as a vagrant in Puerto Rico. Only one species, Black Swift, breeds on the island and it is very rare in the winter. This occurrence coincides with the period of peak insect abundance. Furthermore, the relatively small size of the insects available and competition from Black Swift, perhaps, excludes White-collared Swift from the island.

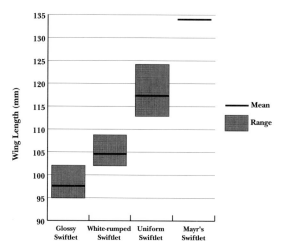

Table 4. Wing lengths of four swiftlet species from central Bismarck Archipelago (based on Salomonsen 1983).

Salomonsen's (1983) gape size analysis of the four species of swiftlet in the central Bismarck Archipelago suggests that each species has its own food niche. Tables 4 and 5 show important morphological differences. Salomonsen (1976) notes that it is a well known fact that differences in bill size are a dominant factor in determining the compatibility of related species. Interestingly this ecological separation appears to extend to free-tailed bats (Molossidae) which occur in proportionally lower numbers in those areas where swiftlets occur.

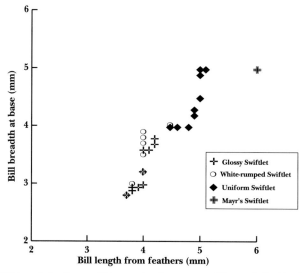

Table 5. Bill breadth:bill length for four swiftlet species from central Bismarck Archipelago (based on Salomonsen 1983).

22

Turner (1989) citing Waugh (1978) notes that, in Britain, Common Swift generally avoids competition with the three breeding hirundine species by feeding in higher airspace and that when, for whatever reason, feeding does occur within the same airspace the different sizes of prey items prevent competition. The study of Marin (1993) into vertical patterns of swift assemblages showed that the larger species in species' pairs foraged in the upper strata. In the case of Grey-rumped Swift, when feeding with the smaller Band-rumped Swift it occupied the upper strata, but when foraging with the larger Pale-rumped Swift it occupied the lower strata. Interestingly, studies into Pallid and Common Swifts suggest that the larger Pallid flies lower than Common. However, Cucco and Malacarne (1987) only noted this close to breeding colonies where Pallid tends to nest in lower cavities.

Marin (1993) notes that a number of other variables have been proposed that might be important mechanisms of ecological separation and that further study of these variables and the actual behaviour of foraging swifts may help to clarify possible competitive interactions and resultant resource partitioning mechanisms. Ecological separation allows different swift species and other aerial feeders to occur together on a worldwide scale.

Glossy Swiftlets

MORTALITY AND PREDATORS

As might be expected swifts, with their aerial habits and great speed, have few natural predators. Several falcon species, especially Eleonora's Falcon *Falco eleonorae* and the three species of hobby, will occasionally take flying birds, and Bat Hawk *Machaerhamphus alcinus*, habitually feeds upon swiftlets and bats around cave entrances. Indeed in Sarawak a bounty was put on the Bat Hawk in order to protect the valuable birds-nest industry.

Several predators such as rats or snakes, could occur at the nest, but the tameness of most swift species at the nest suggest that attacks of this sort are rare. However one particular species, African Palm Swift, seems to suffer from avian predators at their rather exposed nest sites, with adults taken by Red-winged Starling *Onychognathus morio*, eggs and nestlings taken by Fiscal Shrike *Lanius collaris* and crows *Corvus* spp., and young taken by Spotted Eagle Owl *Bubo africanus* (Fry *et al.* 1988). The protection afforded to those species that nest behind waterfalls may not always be secure as Sick (1993) recalls seeing a Great Dusky Swift washed away by a waterfall and although this individual managed to get out of the water without apparent harm, others are apparently killed like this. In the vast caverns of northern Borneo one insect species has evolved that feeds primarily on the young and eggs of swiftlets. It is a large wingless, grasshopper-like Rhaphidophore that bites through the egg in the nest or attacks the young.

Some species such as the African Palm Swift seem very vulnerable to cold with a colony of 100 birds dying between 11-13 November 1968 (Steyn and Brooke 1971). However, Donnelly (1982) suggests that when cold and wet weather persists, the death is as likely to be caused by starvation. Hard rainstorms have grounded Scarce Swifts (Scott 1979) and many Common Swifts were found dead after severe thunderstorms in Zambia. Indeed Bowen (1977) considered that storms may cause more fatalities than predators in Africa. Fifty African Palm Swifts died through electrocution after their palm tree was struck by lightning (Mortimer 1975).

Vagrant swifts, such as White-rumped Swift in Finland, Pacific Swift found on a North Sea gas terminal and Alpine Swifts in Britain have been found dead, moribund or exhausted. A vagrant Alpine Swift on the Isles of Scilly, England was grounded after colliding with a telegraph wire. These problems presumably relate to unexpected weather conditions and to feeding problems.

Many fatalities and population changes, can be both directly and indirectly attributable to man. Benson and Benson (1977) showed that many adult Common Swifts were killed for food by man in Malawi and Scarce Swift flocks often suffer from mist-netting and, due to the sheer density of some vast flocks, by children knocking them to the ground with sticks (Fry *et al.* 1988). Particularly unusual incidents include four African Palm Swifts dying after drinking from a chlorinated swimming pool (Fry *et al.* 1988) and a record of a Great Dusky Swift dying after colliding with a plane (Sick 1993). Large numbers of swifts are sometimes taken when feeding around beehives, such as on the Philippines when between January and March 1968 approximately 450 Purple Needletails were caught (Morse and Laigo 1969). All *Chaetura* species seem vulnerable to asphyxiation or death through burning and Sick (1993) describes finding destroyed nests, eggs and nestling Ashy-tailed Swifts in chimneys. In southern Thailand small children have been seen trying to hit Edible-nest Swiftlets using catapults (pers. obs.), and in Zaïre an adult Common Swift was killed using a bow and arrow (Bromhall 1980).

Many species of swift have actually benefited from the activities of man. White-rumped, Little, House and Common Swifts and various *Chaetura* swifts (which have taken advantage of chimney nesting sites) have all spread after adapting to man-made nesting sites, whilst both species of Old World palm swifts *Cypsiurus* have benefited from man expanding the ranges of certain palm species. Asian Palm Swift has also taken advantage of thatched roofs (Ali and Ripley 1970).

However, application of pesticides and habitat destruction seem likely to have had adverse effects on most species. Unfortunately due to a lack of accurate population studies over a long period the exact picture remains clouded. Round (1988) considered that Brown-backed Needletail, with its preferred nesting sites in the holes of mature forest trees, could certainly be under threat from habitat destruction as indeed all forest species must be. For example, a relict population of a predominantly West African species, Sabine's Spinetail, occurs in an isolated forest such as Kakamega in western Kenya. Formerly its East African range was presumably far more extensive. Habitat-specific species are thus clearly under threat. The over-harvesting of nests of those species of swiftlets which have edible nests in commercially viable nest sites is undoubtedly of grave consequence to the population as a whole.

Swifts have low mortality rates. Collins (1974) considered that Chestnut-collared Swift had an annual survival rate of 83%-85%, whilst Perrins (1971) considered Common Swift to have a rate of 81%-85%. Dexter's (1979) figure for Chimney Swift was 71%-81%. When it is considered that the latter two species make annual transequatorial migrations, the figures are remarkably high. All studies indicate that the greatest loss is in the first year after hatching. Beklova's (1976) study of Common Swifts in Czechoslovakia, showed that the mortality rate for first year birds was 29% and this dropped to 12% by the fourth year. Similarly, studies of a colony of Pallid Swifts at Gibraltar (Finlayson 1979) showed a first year mortality rate of 67.3% and subsequent adult mortality of 26%.

With such high survival rates it is not surprising that the oldest ringed individuals include Common Swift 21 years (Rydzewki 1978), Alpine Swift 26 years (Arn-Willi 1967), White-rumped Swift 10 years (Cramp 1985), Horus Swift 13 years (Fry *et al.* 1988), Chimney Swift 14 years (Dexter 1979) and Chestnut-collared Swift at least 10.5 years (Collins 1974).

Pacific Swift

MOULT

A thorough knowledge of moult in swifts is not essential for correct identification. However, it is vital to realise that moulting individuals can look different from fully feathered birds.

Most importantly wing shape can be affected. *Apus* swifts, when moulting, can often show a wing shape reminiscent of spinetails as secondary length can be reduced, causing an obvious step in the wing before the inner primaries broaden. Similarly, the suspension of moult in the outer primaries can lead to attenuation of the apparent difference in length of the ninth and tenth primaries, which can cause confusion (figure 2).

Tail shape can be similarly altered, especially in those species that have forked tails when the outer rectrices are not fully grown and accordingly the depth of the tail-fork and the length of the tail is diminished.

Body moult can cause differences in appearance when old, bleached and worn feathers are replaced by new, darker, glossier feathers. This can be most apparent in paler species, such as Asian Palm Swift, where distinct spotting can be apparent on the plumage during moult.

The following is a brief summary of present knowledge of swift moult. Cramp (1985), de Roo (1966), Langham (1980), Marin and Stiles (1992), Medway (1962), Somadikarta (1968) and Stresemann and Stresemann (1966) give more detailed accounts.

Figure 2. Wings of worn (upper) first summer and fresh (lower) adult Common Swift showing difference in pattern. Note worn (and consequently shortened) retained longest primary in otherwise fresh wing.

It is apparent that most migratory species time their moult cycles to coincide with arrival in the winter quarters. This is achieved by starting primary moult on the breeding grounds and then suspending this until arrival on the wintering grounds, or by not beginning moult until the wintering grounds are reached. Tropical species, such as the swiftlets, can have more prolonged moult cycles.

Cypseloidinae

Few details have been published about moult sequence in this subfamily. Marin and Stiles (1992) noted the resident *Streptoprocne* and *Cypseloides* swifts that they studied showed considerable overlap between their breeding and moulting seasons, whereas the migratory Black Swift showed no such overlap.

Collocalia

Studies of Edible-nest Swiftlet in peninsular Malaysia (Langham 1980) and Black-nest Swiftlet on Borneo (Medway 1962) indicate that swiftlet moult is a protracted affair. Primaries are moulted descendantly (innermost outwards), secondaries and tertials ascendantly and rectrices centripetally, starting with the outer pair and working inwards. Primaries are moulted before the secondaries and rectrices, although Langham (1980) noted this was often complicated by a subsequent wave of primary moult. In the Edible-nest Swiftlets studied, moult was noted in a large proportion of the birds throughout the seven month study period with no apparent seasonal pattern to the primary moult, whereas in the Bornean Black-nest Swiftlets primary replacement was most apparent in April and May, at the end of breeding.

Somadikarta's (1967) study of moult in Giant Swiftlet showed that the moult cycle avoids the reproductive period as only one of the 58 skins studied was in moult during this period. Wing moult in males starts late in December whilst females start in late January, with both suspending moult from late August to mid September. Only two primaries are growing at the same time and the moult is descendant. Secondary moult is initiated when the fifth primary is in pin or at 'brush' stage, starting with the first secondary, then the seventh, with the others moulting centripetally. Tail moult starts with the fifth (simultaneously with the growing out of the seventh or eighth primary) and is centripetal. He considered that the remiges and rectrices were not moulted in the first year of life and that juvenile body moult occurs when the fifth to tenth primaries are in moult.

Salomonsen (1983) notes that in all swiftlets the tenth primary does not moult until the ninth primary is fully grown and, after moulting, grows back very slowly. Tarburton (1993) notes that, unlike other swifts, some swiftlets (at least Australian, White-rumped, Black-nest and, as shown above, Edible-nest Swiftlets) synchronise breeding and moult.

Hirundapus

White-throated Needletail normally has a complete moult in the winter quarters. The post-juvenile moult begins in the winter quarters and the remiges are replaced either in the following summer or, in the case of those that suspend moult for the autumn migration, in the second winter at which stage the adult moult cycle is entered. Some adults are known to start primary moult on the breeding grounds which, after suspension for migration, is completed in the winter. Silver-backed Needletail probably has a similar moult cycle and it is interesting to note that this can serve to draw attention to the species during the South-east Asian spring as Brown-backed Needletail is in wing moult during this period and its typical *Hirundapus* wing-shape is distorted (figure 3).

Figure 3. Brown-backed Needletail with moulting secondaries. Note step effect on trailing edge of inner wing.

Chaetura

Studies of Chapman's Swift (Collins 1968a) show that primary moult is descendant and that secondary (ascendant) and tail (centripetal) moult occurs after initiation of primary moult and ends before its completion. The tertials, however, are moulted as a group early on in the secondary moult. Post-juvenile moult (late September and October) involves only the body feathers.

Tachymarptis

Alpine Swift (in the Western Palearctic) has a complete moult to winter plumage starting with the first primary (primaries moult descendantly) in June-July and ending on the wintering grounds, probably in December-February. Primary moult suspension occurs in some whilst others continue to moult on migration. The post-juvenile moult is complete with the exception of some wing-coverts and the primaries which are moulted in the following summer.

Apus

The five plain Western Palearctic species (and probably White-rumped Swift) have very similar moult cycles, with a complete moult in the winter quarters (Pallid and White-rumped can start inner primary replacement whilst raising a second brood and then suspend moult until the completion of migration), and partial post-juvenile moult to first winter plumage and subsequent complete moult on the wintering grounds in the second winter.

In Common Swift the post-juvenile moult involves body feathers, rectrices and smaller wing-coverts. Remiges and larger wing-coverts are not replaced until the second winter. This is important for identification as this retention leads to increased contrast in the wing and a more Pal-

Figure 4. Underwing of moulting Pallid Swift showing worn (and shortened) longest primary, growing ninth primary and fresh first and second primaries.

lid-like pattern. It seems likely that the other plain *Apus* species have similar post-juvenile moult strategies.

Little Swift in the Western Palearctic (and probably House Swift and other Little Swift populations) moults during the breeding season, starting with the innermost primary in late spring and ending with the outermost in mid winter. Body moult occurs during the same period. Post-juvenile moult starts soon after fledging with the replacement of the body feathers. Primary moult starts in mid winter, after the completion of the tail moult, and continues until spring. During this period, unlike the adults, body moult does not occur.

More sedentary African species and populations may employ different strategies. For example Ash's (1981) studies of Forbes-Watson's Swift detected six moult patterns and revealed that primary moult occurred from two centres descendantly.

Moult has not been thoroughly studied in the following genera: *Schoutedenapus, Mearnsia, Zoonavena, Telacanthura, Rhaphidura, Neafrapus, Aeronautes, Tachornis, Panyptila, Cypsiurus* and *Hemiprocne*.

Alpine Swift

FLIGHT

The structure of swifts is highly specialised for high speed flight. Compared with most other aerial feeders such as hirundines, they have low manoeuvrability and are unable to fly at lower speeds.

The carpus bones are much the longest wing bones, as can be seen in figure 5, and they account for the great length of the primaries compared with other bird families. These long primaries are used to produce a predominantly downwards and forwards force whilst the short secondaries and inner primaries provide a far less powerful lift but with a still considerable forward dimension. The primaries are tightly gripped together with the outer primaries bent slightly backwards compared to the inner primaries which are bent slightly outwards by the tension.

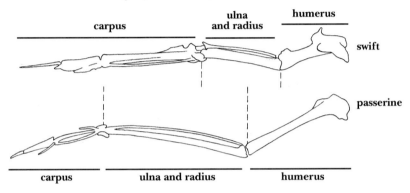

Figure 5. Wing bones of a swift (upper) and passerine (lower) (Based on Lack 1956c).

Descriptions of flight action in the species accounts should be treated with some caution as wind conditions and prey types can affect flight, making identification based solely on flight technique alone unsafe (Chantler 1993). In a recent review of this paper Brooke (1993) notes that the appearance of the flight of a swift is a function of its body mass and wing length, not its taxonomic position, and secondarily of what it is doing. This needs to be clarified; the phenotype of each swift taxon is individual and its unique body mass, wing length and shape, and tail length and shape will affect its flight action. For example, Pallid Swift with its greater mass, broader wings and shorter, less forked tail will be seen to fly differently from the lighter, thinner-winged, longer, more forked-tailed Common Swift when the two are observed together under the same conditions. In summary, wing length and body mass are functions of the phenotype and therefore it can be argued that the appearance of a swift's flight can be ascribed to its taxonomic position.

A good example of flight action being used to support a case for taxonomic review is given in King (1987) who observed that the flight of Giant Swiftlet *Collocalia gigas* (which he calls Waterfall Swift *Hydrochous gigas*) 'differs dramatically from that of the *Collocalia* and *Aerodramus* swiftlets by its smooth directness, without any of the jerkiness associated with swiftlets' side to side movements; *Hydrochous* glides with its wings more near the horizontal (not so down-turned)'.

CONSERVATION

Currently 14 swift taxa are included in the World List of Threatened Birds (Collar *et al.* 1994). Eight of these are considered to be Vulnerable: White-chested Swift *Cypseloides lemosi*, Seychelles Swiftlet *Collocalia elaphra*, Volcano Swiftlet *C. (brevirostris) vulcanorum*, Whitehead's Swiftlet *C. whiteheadi*, Sawtell's Swiftlet *C. sawtelli*, Tahiti Swiftlet *C. leucophaeus* (excluding *C. (l.) ocista*), Schouteden's Swift *Schoutedenapus schoutedeni* and Dark-rumped Swift *Apus acuticauda*. Three taxa are considered to be Data Deficient Species, defined as 'neither a threatened nor a non-threatened category; it acknowledges that the species it contains are potentially either': White-fronted Swift *Cypseloides storeri*, Mayr's Swiftlet *Collocalia orientalis* and Fernando Po Swift *Apus (barbatus) sladeniae*. Three species are listed under the Near-threatened category: Rothschild's Swift *Cypseloides rothschildi*, Waterfall Swift *Hydrochous gigas* (Giant Swiftlet *Collocalia gigas*) and Mascarene Swiftlet *Collocalia francica*. The contentious taxonomic nature of four of these forms does not detract in any way from their importance and the need to discover their ecological requirements.

It has become increasingly apparent in recent years that the trade in swiftlet nests for the bird's-nest soup industry is becoming a conservation problem. This lucrative trade poses considerable problems for the future of those species that produce edible nests. Hong Kong, which is now the world's major recipient of nests, imports an estimated 100 tons worth $25million from South-east Asia. Thailand is now seen as the country with the most sought-after nests and Thai villagers can make up to $1200 for a kilogram of nests. Nests are harvested three times in the four month breeding season and swiftlet numbers are now in decline. At the Convention on International Trade in Endangered Species (CITES) in November 1994 this problem was highlighted and the World Wide Fund for Nature has called for the restriction of harvesting to the period before laying followed by a four month close season so that fledging can occur and the populations sustained.

The future of the order as whole is closely linked to the future of all bird species in a rapidly changing world. The high mobility of swifts and their great longevity may help to insulate them from the kind of short-term population crashes that other species have had to suffer, providing of course that feeding habitats and nesting sites remain unscathed.

Böhm's Spinetail

UNDESCRIBED SPECIES?

In recent years there have been several claims of previously undescribed swifts.

One of the most intriguing series of records concerns the so-called 'Beidaihe Swift'. On 26 April 1985, in a northward migration of swifts and hirundines, a small, dark swift, which in many ways resembled Himalayan Swiftlet, was observed at Beidaihe, Hebei Province in northern China (Williams *et al.* 1986). It differed from the latter species in appearing slightly smaller with shorter tail (fork not always discernible), broader shorter wings, darker below with less contrast between the body and wing, little contrast between the coverts and remiges on the underwing, and a darker rump. The same bird, or another, was seen on 29 April. On the 20 March 1992 two similar individuals were seen at Mai Po marshes, Hong Kong, with one remaining to 22 March (P. Leader *in litt.*). It is hard to say whether or not these records concern an undescribed species or an unrecognised swiftlet, as there have been several recent records of unidentified swiftlets from Hong Kong (Kennerley 1991). Furthermore, in 1983 it was found that Edible-nest Swiftlet breeds north to Hainan Island (Xian and Zhong 1983).

Figure 6. Little Swift carrying a feather, giving the appearance of having a pale whisker.

Swifts observed in South Africa by Jeff Blincow (1992) were also considered possibly to represent a previously undescribed species. These were swifts with prominent white whiskers which otherwise resembled Little Swifts. Although they may prove to be a new species there is a possible explanation. In 1993 whilst watching a Little Swift colony at Malindi, Kenya (pers. obs.) birds carrying pale feathers for nesting material (often for up to 15 minutes) showed this feature, the speed of the flight positioning the feathers so that they exactly mimicked a whisker (figure 6). This was so realistic that it was easy to imagine that it could be mistaken for a plumage feature. Ian Sinclair (pers. comm.) has noted this phenomenon on both Little and Horus Swifts and, in addition to the explanation of feather carrying, suggests that on occasions when swifts have bulging throat pouches the displacement of the feather tracts from around the throat could lead to pale feather bases becoming exposed and appearing as short pale lines. Dave Buckingham (pers. comm.) noted these features on a House Swift in Malaysia and considered that feather carrying was a plausible explanation. In summary, observers faced with a swift that appears to show pale whiskers should check that they show on each side of the gape (and if so, that they are in symmetry). The shape and length of the whiskers should also be carefully noted as well as other plumage features.

During spring 1993 Inskipp and Inskipp (1994) observed several swiftlets in Bhutan that may prove to be a new species. These birds differed from Himalayan Swiftlet in their slightly larger size, longer-winged appearance, and blacker plumage with only a slight contrast on the rump. The flight was also stronger than Himalayan Swiftlet.

Finally, Williams (1980) remarks upon a very large all black swift seen on Marsabit Mountain in the Northern Frontier Province of Kenya, but no further information has been forthcoming. More recently, a very large black spinetail has been reported by Brian Finch from Irangi in eastern Zaïre.

Due to their subtle plumages and unobliging habits swifts will often remain unidentified and field claims of new species, or rare species outside their normal ranges, must be backed up with very detailed descriptions.

WATCHING SWIFTS

Contrary to popular belief and in spite of their high mobility swifts can be viewed over long periods and in great detail, with just a little effort. It is usually a matter of finding suitable habitat and using a basic knowledge of their habits. In addition, certain weather conditions can help attempts to obtain good views. Watching over waterways or lakes can be particularly rewarding, especially in the evening or when dull weather brings swifts low over the water to feed. Hatching insects, such as flying ants, will often attract the arrival of large numbers of swifts, oblivious of any human observers and often at close range. Grass fires will often attract large numbers of swifts, as well as other insectivorous species. If feeding conditions are suitable swifts will remain loyal to certain sites for long periods, behaving in a very predictable manner and facilitating close approach.

Cliff tops, or rocky bluffs overlooking gorges, can be used to great advantage during migration or at colonies. By gaining height the birds can be seen from a variety of angles, not just from below, facilitating easier identification. Swifts tend to move along gorges and cliff tops at close range, feeding in the lee of the cliffs to avoid strong winds.

Some species are very loyal to certain habitats, such as damp gorges and waterfalls in the case of the *Cypseloides* swifts, baobabs for Mottled Spinetails, or palms in the case of the *Tachornis* and *Cypsiurus* swifts. Breeding swifts can often be watched more easily; for example, in many Old World cities the untidy nests of Little and House Swifts are regularly visited by their occupants allowing easy observation. In Africa and Asia bridges provide ideal situations to observe swifts, providing both nesting sites and feeding areas. Swiftlet caves in South-east Asia can be very easy to locate and often provide the only safe means of identification since the birds can be observed on their, mainly, diagnostic nests. Similarly, several species have regular and well known roosting sites. This is particularly important in the case of the New World *Chaetura*, *Cypseloides* and *Streptoprocne* swifts where prolonged views of large numbers and often several species can be obtained at such roosts.

Forest species should be looked for in clearings, rides cut through the forest and particularly over waterways. These breaks in the forest are often very rich in insect life and species that are traditionally only seen high over the canopy will often fly low to take advantage of the abundant food supply.

However, despite this optimism most views of swifts can be very frustrating and observers should not be tempted to stretch the limits of identification with fleeting views of fast moving swifts. Swift identification can often be like completing a jigsaw, with a number of different features becoming apparent over a period of time leading towards a positive identification. The following pitfalls (mainly taken from Chantler 1993) should be borne in mind whilst watching swifts.

Figure 7. Partial albinism in Common Swift (left) and Grey-rumped Treeswift (perched and right).

PARTIAL ALBINISM

Partial albinism has been recorded in many species (figure 7). Such birds tend to show white on areas of the body that are important to identification, i.e. the belly, throat and rump. Any suspected partially albinistic swift should be carefully inspected for signs of asymmetry in the white plumage, and for additional plumage and structural features to confirm the identification.

Another note of caution is that often in the breeding season swifts can be seen carrying feathers to the nest. When a large white feather is seen hanging from the bill of a dark swift, covering the throat patch, the effect can be very striking.

EFFECTS OF LIGHT ON PLUMAGE

The effect of light on the plumage of any species and the resulting changes in appearance is a problem all birders should be aware of. The problem is compounded when birds are viewed in flight as changing body-wing angles must also be taken into account. As swifts are seen almost exclusively in flight, they are therefore more prone to misidentification in the field. The following is a list of the most important light-affected characters. We have endeavoured to describe and illustrate the effects of light on these characters in the individual species accounts, with particular reference to field identification, but the following general comments should be borne in mind.

Body-wing contrast

In the discussions and descriptions mention is frequently made of the contrast between the body and wings; this is hard to assess in the field and is influenced by the effects of light falling on the various feather tracts and the angle of the wings relative to the body. This is particularly true of the underwing and it is quite possible that in a single individual, within the space of a few minutes, the darkest underwing-coverts could appear darker, paler or equal in tone to the underparts!

Translucency

Although not specifically mentioned in all species accounts, any species, when seen from below against bright sunshine, can show translucency on the remiges and rectrices. This is usually most visible on the inner primaries and secondaries (especially the trailing edge to the secondaries). Strong light can also make the underwing-coverts appear very pale.

Secondaries

In the hand or in museum specimens, the secondaries of many species can appear darker than the greater wing-coverts (especially on fresh individuals). This is, however, rarely visible in the field, presumably because of the translucency of the secondaries.

Trailing edges

Another light-affected phenomenon is the appearance in any species of dark trailing edges to the wing feathers (particularly on the remiges and greater coverts of the underwing). Good examples are the *Chaetura* and *Hirundapus* swifts which actually have very plain greater wing-coverts but often show a dark terminal bar to these coverts in the field. This is caused by the area of overlap between the feathers and possibly by the slight change in angle at this area of overlap. For example, all of the *Apus* swifts show some degree of dark subterminal crescents on the wing feathers but perhaps these are overemphasised as a result.

Outer primaries

In all species the outer primaries can appear darker than the inner primaries and secondaries (on both the upper and underwing). This is due to the darker outerwebs being more densely clustered on the outer primaries whereas on the inner primaries the pale innerwebs can be clearly seen when the wing is spread.

SWIFT TOPOGRAPHY

A thorough knowledge of swift topography is essential to aid identification. Two non-standard terms are employed in this book: eye patch for the coarse, invariably black bristle feathers in front of and just above the eye (supraloral patch is used when a paler area can be recognised within this dark patch), and leading edge-coverts for the small tract of feathers on the leading edge of the wing (which are more typically called marginal coverts).

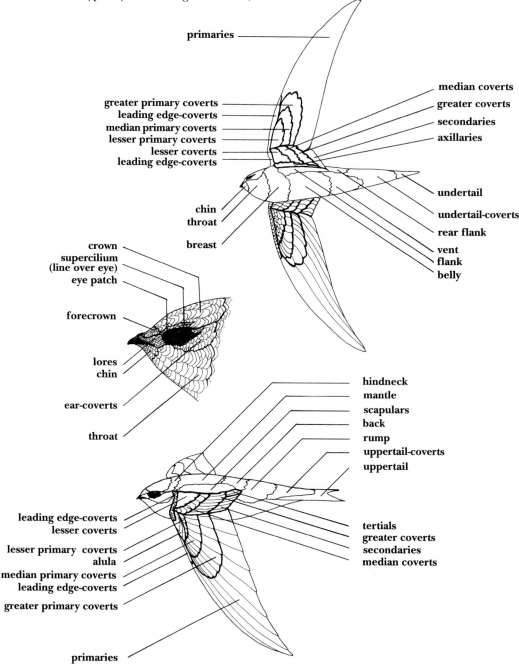

Figure 8. Topographical charts of swift plumage tracts (Gerald Driessens).
Reproduced from *Dutch Birding* 15 : 97-135 (1993).

PLATES
1-24

2   **Tepui Swift** *Cypseloides phelpsi*   **Text and map page 92**

Restricted South American range in S Venezuela and bordering areas of Brazil and Guyana. See also Plate 2.

  2a   **Adult male** Very black plumage with bright orange collar, extending across ear-coverts and onto chin. Note distinctly forked tail.
  2b   **Adult female** Collar slightly paler.
  2c   **Juvenile** Note prominent paler fringing from throat to tail-coverts (at close range). Very nondescript at a distance but forked tail still apparent when tail closed.

1   **Chestnut-collared Swift** *Cypseloides rutilus*   **Text and map page 91**

Disjunct Central (north to C Mexico) and South American (south to Bolivia) range. Also on Trinidad. See also Plate 2.

  1a   **Adult male** (nominate; Venezuela to the Guianas and on Trinidad) More reddish collar than Tepui Swift, with dull brown markings on mid-throat and ear-coverts. Plumage less black and tail less forked. In the distance note dark plumage and prominence of collar.
  1b   **Adult female with partial collar** (nominate) Note reddish collar admixed with brown.
  1c   **Adult female lacking collar** (nominate) Note paler plumage than male, or females with rufous. Tail appears square (which can be caused by wear). Females also tend to have shallower tail-forks than adult males.
  1d   **Juvenile** (nominate) Note generally paler plumage and shallower tail-fork. In the distance, plumage very uniform and paler than most Spot-fronted and White-chinned Swifts.

3   **Black Swift** *Cypseloides niger*   **Text and map page 93**

For comparison. See also Plate 2.

  Adult (*costaricensis*; Mexico to Costa Rica) Note large size and forked tail (usually). This form is darker and smaller than *C. n. borealis*. Paler headed than other species on plate.

7   **Spot-fronted Swift** *Cypseloides cherriei*   **Text and map page 97**

Central America (from Costa Rica) to N South America (Venezuela, Colombia and Ecuador). Local. See also Plate 2.

  7a   **Adult** Blackish plumage with striking facial pattern. Short square tail (cf. Chestnut-collared Swift). Darker than collarless Chestnut-collared Swifts.
  7b   **Juvenile** White fringes on underparts.

8   **White-chinned Swift** *Cypseloides cryptus*   **Text and map page 99**

Central America (north to Costa Rica) and N South America south to Peru and east to Guyana. See also Plate 2.

  8a   **Adult** Blackish plumage. Pale chin hard to see in the field. Larger than Spot-fronted Swift and with a shorter tail. Darker than collarless Chestnut-collared Swifts. Tail always square unless fully spread.
  8b   **Juvenile** White fringes on underparts.

9   **White-fronted Swift** *Cypseloides storeri*   **Text and map page 100**

Very localised. Only known from SW Mexico. See also Plate 2.

  Adult Very similar structure and plumage to White-chinned Swift. Differs only in extensive pale forehead.

3   Black Swift *Cypseloides niger*   **Text and map page 93**

Extensive range. W North America north to Alaska and south to California. In Central America from Mexico south to Honduras. Also on Greater and Lesser Antilles. Northern populations are migratory, wintering mainly in Central America. See also Plate 1.

  3a   **Adult** (*borealis;* Alaska south to California) Strikingly black plumage with paler, greyer head (especially forehead). Note tail-fork. In the distance, note long-winged appearance.

  3b   **Juvenile** (*borealis*) Broad white fringes on underparts, giving spotted appearance.

4   White-chested Swift *Cypseloides lemosi*   **Text and map page 95**

SW Colombia; local.

  4a   **Adult male** Distinctive blackish plumage with white chest patch. Very dark above, with slightly paler forehead.

  4b   **Adult female** Reduced chest patch and noticeable underpart fringing. Note slightly less forked tail.

  4c   **Juvenile** As with female, chest patch can be missing and the tail-fork is reduced.

6   Sooty Swift *Cypseloides fumigatus*   **Text and map page 96**

C South America, in E Bolivia, SE Brazil and Paraguay.

  Adult Very nondescript plumage above and below. Note square tail.

5   Rothschild's Swift *Cypseloides rothschildi*   **Text and map page 96**

SC South America, in NW Argentina and S Bolivia.

  Adult Similarly unstriking as last species but distinctly paler brown.

CYPSELOIDES HEAD PATTERNS

3   Black Swift *Cypseloides niger*
   Adult (*borealis*) Note pale plumage, especially upper head.

4   White-chested Swift *Cypseloides lemosi*
   Adult Darker head than Black Swift, due to less extensive pale fringing.

2   Tepui Swift *Cypseloides phelpsi*
   Juvenile Lacks collar of adults. Note fringing on throat (lacking in juveniles of Chestnut-collared Swift).

1   Chestnut-collared Swift *Cypseloides rutilus*
   Juvenile (nominate) Lacks collar of adults. Very uniform head.

6   Sooty Swift *Cypseloides fumigatus*
   Adult Very uniform, as in Rothschild's Swift.

5   Rothschild's Swift *Cypseloides rothschildi*
   Adult Paler than Sooty Swift.

7   Spot-fronted Swift *Cypseloides cherriei*
   Adult Note prominent head markings.

8   White-chinned Swift *Cypseloides cryptus*
   Adult Pale chin. Note slightly paler lores and line over eyes.

9   White-fronted Swift *Cypseloides storeri*
   Adult As White-chinned Swift but extensive pale fringing on the forehead.

Gerald Driessens
'94

10 Great Dusky Swift *Cypseloides senex*   **Text and map page 101**

SE South America: E Paraguay, NE Argentina and C and S Brazil.

> **10a**   **Adult** Square or slightly rounded tail. On occasions the forehead can appear quite whitish. Rump slightly paler than mantle.
>
> **10b**   **Juvenile** Atypically large *Cypseloides*. Brown plumage with paler throat and forehead. Age indicated by comparatively pale median primary coverts.

11 White-collared Swift *Streptoprocne zonaris*   **Text and map page 102**

Extensive Neotropical range: Greater Antilles and Trinidad; Central America south from S Mexico, continuously through South America to WC Argentina.

> **11a**   **Adult male** (*subtropicalis*; mid-elevations from Santa Marta, Colombia and Merida, Venezuela, south to Peru) Diagnostic black plumage with strong white collar. Note difficulty in seeing true extent of collar in side views. Head can appear slightly paler. When worn or moulting the tail shape can appear less forked and more square. In these instances concentrate on ascertaining the extent of the collar, as most worn birds will have undergone post-juvenile body.moult.
>
> **11b**   **Adult female** (*subtropicalis*) Rather mottled collar.
>
> **11c**   **Juvenile** (*subtropicalis*) Shows darker fringes to white collar feathers.
>
> **11d**   **Adult** (nominate; S Brazil, Bolivia and W Argentina) Larger and less black than *subtropicalis*.
>
> **11e**   **Adult** (*albicincta;* Venezuela and Guyana) Note size relative to other races.
>
> **11f**   **Adult** (*mexicana*; S Mexico to Belize) Plumage duller, less glossed than *subtropicalis*.

12 Biscutate Swift *Streptoprocne biscutata*   **Text and map page 105**

C South America: E Brazil.

> **Adult** Dark plumage with distinct white nape and upper-breast patch. Square tail. From the side the broken collar can be observed. In distant views harder to ascertain exact pattern of white patches.

13 White-naped Swift *Streptoprocne semicollaris*   **Text and map page 106**

Endemic to Mexico.

> **Adult** Very heavy appearance. Square tail. White restricted to nape. From below clearly lacks any white on lower throat/upper breast region. Hard to ascertain extent of white markings from the side and behind. Note tail shape can vary with angle and wear and can appear quite rounded on occasions.

PLATE 4: WHITE-BELLIED SWIFTLETS AND GIANT SWIFTLET

Underpart pattern (whitish abdomen, dark breast and undertail-coverts) of 15, 16 and 17 excludes all other swiftlets.

16 Linchi Swiftlet *Collocalia linchi* **Text and map page 112**

Sundaic endemic: Java (and satellite islands), upland Sumatra and Mount Kinabalu, Sabah.

> **Adult** (nominate; Java, Madura, Bawean and Nusa Penida) Very similar to Glossy Swiftlet; differs in green gloss on remiges, rectrices and dark feathers of upperparts. Note underpart pattern and square tail.

15 Glossy Swiftlet *Collocalia esculenta* **Text and map page 109**

Extensive range. SE Asia, from peninsular Thailand and Malaysia, Andaman and Nicobar Islands, Greater and Lesser Sundas, Philippines, to E Indonesia and New Guinea.

> **15a** **Adult** (*cyanoptila*; Malay Peninsula, E Sumatra and the lowlands of Borneo) Blue-glossed upperparts. Underparts extensively white on belly with dark chest and undertail-coverts. Square tail.
>
> **15b** **Adult** (*spilura*; N Moluccas) Like adult *manadensis* (N Sulawesi), this race has extensively dark underparts.
>
> **15c** **Adult** (*marginata*; Philippines, from C Luzon south to Bohol and Leyte) Note mottled, pale grey rump. This is individually variable with some birds appearing whiter-rumped.
>
> **15d** **Adult** (*nitens*; lowland New Guinea, Yapen, Karkar and Long Islands, Misool and Batanta) Brilliant blue-glossed plumage. Rectrix spots individually variable (very hard to see in the field).

17 Pygmy Swiftlet *Collocalia troglodytes* **Text and map page 114**

Restricted to the Philippines where it is widespread.

> **Adult** Plumage less glossed than Glossy Swiftlet. Underparts typically a mixture of brown and white, lacking the well-defined white belly of many Glossy Swiftlets. Narrow, highly contrasting white rump.

14 Giant Swiftlet *Collocalia gigas* **Text and map page 108**

Rare and local. Breeds on Java (has bred peninsular Malaysia). Also recorded Sumatra and Borneo.

> **Adult** Large size and uniform plumage. In flight, proportions more *Apus*-like than other swiftlets. Note deeply forked tail.

16

16

15a

15a

15c

15b

15d

17

17

17

14

14

14

Gerald Driessens
'94

PLATE 5:   SWIFTLETS — INDIAN OCEAN AND OCEANIA

18   Seychelles Swiftlet *Collocalia elaphra*   **Text and map page 115**

Seychelles, Indian Ocean.

> **Adult** Underparts browner than Mascarene Swiftlet. Rump barely paler than upperparts. Tail forked.

19   Mascarene Swiftlet *Collocalia francica*   **Text and map page 115**

W Mascarenes, Indian Ocean.

> **Adult** Underparts paler than Seychelles Swiftlet. Rump clearly paler than rest of upperparts. Tail forked.

32   Palau Swiftlet *Collocalia pelewensis*   **Text and map page 129**

Palau Island, W Pacific.

> **Adult** Underparts slightly darker below throat. Rump paler than upperparts. Tail forked.

33   Guam Swiftlet *Collocalia bartschi*   **Text and map page 130**

S Mariana Islands, W Pacific. Also introduced to Oahu, Hawaiian archipelago.

> **Adult** Silvery-grey throat and upper breast, remainder of underparts darker. Rump uniform with upperparts. Shallow tail-fork.

34   Caroline Swiftlet *Collocalia inquieta*   **Text and map page 130**

Caroline Islands, W Pacific.

> **Adult** (*ponapenis*; Ponape Island) Dark underparts; little or no contrast with throat. Upperparts very dark. Can show indistinctly paler rump.

35   Sawtell's Swiftlet *Collocalia sawtelli*   **Text and map page 131**

Atiu, Cook archipelago, SW Pacific.

> **Adult** Pale grey-brown underparts, contrasting strongly with dark upperparts. Pale grey rump. Tail forked.

36   Polynesian Swiftlet *Collocalia leucophaeus*   **Text and map page 132**

Tahiti and Moorea, E Society Islands and the Marquesas archipelago, Pacific Ocean.

> **Adult** (nominate; Moorea and Tahiti) Darker underparts than Sawtell's Swiftlet, creating less contrast with upperparts. Upperparts less blackish with less contrasting rump.

Gerald Driessens '84

20 Indian Swiftlet *Collocalia unicolor*   **Text and map page 116**

Peninsular India and Sri Lanka.

> **Adult** Supraloral patch extensively white. Throat slightly paler than Himalayan Swiftlet. Underparts appear very uniform. Rump uniform or very indistinctly paler. Shallower tail-fork than Himalayan Swiftlet.

26 Himalayan Swiftlet *Collocalia brevirostris*   **Text and map page 121**

Extensive range. Himalayas, east to S China and throughout SE Asia (north of peninsular Thailand). Also an isolated population on Java. Partially migratory, wintering south to the Malay Peninsula (possibly to Sumatra). See also Plate 7.

> **26a**   **Adult** (nominate; Himalayas from Himachal Pradesh east to Burma and Thailand) Extensively white supraloral patch. Throat slightly darker than Indian Swiftlet. Grey rump band clearly paler than upperparts. Underparts very uniform at long range. Note depth of tail-fork.
>
> **26b**   **Adult** (*innominata*; C China, migrating south to Malay Peninsula) Rump greyer than nominate race.

37 Black-nest Swiftlet *Collocalia maxima*   **Text and map page 133**

Extensive range. Southern SE Asia and Greater Sundas. See also Plate 7.

> **Adult** (nominate; S Burma, peninsular Thailand, Malay Peninsula and W Java) Underparts average darker than Himalayan Swiftlet or *C. fuciphaga amechana*, and clearly darker than *C. f. germani*. Indistinctly pale grey rump (sometimes paler than illustration). Tail normally less deeply forked than Himalayan Swiftlet and can appear slightly rounded when fully spread.

38 Edible-nest Swiftlet *Collocalia fuciphaga*   **Text and map page 135**

Extensive range. Coastal SE Asia north to Hainan Island, Malay Peninsula, Andaman and Nicobar Islands, Greater and Lesser Sundas, S Philippines. See also Plate 7.

> **38a**   **Adult** (*germani*; Hainan south to coastal Indochina, Mergui Archipelago (Burma), Malay Peninsula and S Philippines) Whitish rump band and pale grey-white underparts. Tail more deeply forked on average than Black-nest Swiftlet. Greater contrast between pale and dark areas of plumage than other species in region.
>
> **38b**   **Adult** (*amechana*; Malay Peninsula, south of *germani*) Greyer rump than *C. f. germani*. Averages paler below than Black-nest Swiftlet.

Gerald Driessens
'84

PLATE 7:   SWIFTLETS — GREATER SUNDAS

38   Edible-nest Swiftlet *Collocalia fuciphaga*   **Text and map page 135**

Extensive range. Coastal SE Asia north to Hainan Island, Malay Peninsula, Andaman and Nicobar Islands, Greater and Lesser Sundas, S Philippines. See also Plate 6.

38a   **Adult** (*vestita*; Sumatra and Borneo) Note white nest. Similar to next two species but deeper tail-fork and slightly paler rump often visible.

38b   **Adult** (*amechana*; Malay Peninsula south of *germani*; either this subspecies or *germani* also occurs in coastal N Borneo) Pale grey-white rump.

30   Mossy-nest Swiftlet *Collocalia salangana*   **Text and map page 125**

Greater Sundas.

Adult (nominate; Java) Note nest made mainly of vegetable matter. Underparts very finely shaft-streaked, but absent in some. Uniform upperparts and only shallowly forked tail.

37   Black-nest Swiftlet *Collocalia maxima*   **Text and map page 133**

Extensive range. Southern SE Asia and Greater Sundas. See also Plate 6.

Adult (*lowi*; Sumatra, Nias, N and W Borneo, Labuan Island, W Java, Palawan) Note white nest with many impurities, mainly feathers. Dark rump. Tail-fork deeper than Mossy-nest Swiftlet, but usually shallower than Edible-nest Swiftlet.

26   Himalayan Swiftlet *Collocalia brevirostris*   **Text and map page 121**

Extensive range. See also Plate 6.

Adult (*vulcanorum*; volcanic peaks on Java) Dark underparts, especially throat, and pale grey rump.

PLATE 8:   SWIFTLETS — PHILIPPINES AND MOLUCCAS

21   Philippine Grey Swiftlet *Collocalia mearnsi*   Text and map page 117

Philippines, mainly submontane.

Adult Small grey-brown swiftlet with uniform rump. Grey-white supraloral patch. Tail-fork shallower than *amelis* race of Uniform Swiftlet. Rump uniform with upperparts.

31   Uniform Swiftlet *Collocalia vanikorensis*   Text and map page 126

Philippines, E Indonesia, New Guinea and Melanesia. See also Plate 9.

31a   Adult *(amelis;* Philippines, including Luzon, Mindoro, Cebu, Bohol and Mindanao) Very similar to Philippine Grey Swiftlet. Tail-fork slightly deeper and body length may average longer. Rump uniform with upperparts.

31b   Adult *(palawanensis;* Palawan) Plumage similar to *C. v. amelis*, but averages larger.

31c   Adult *(waigeuensis;* Waigeo, Morotai, Halmahera, Misool, Batanta and probably Salawati) Dark rump (may be slightly paler in some birds); throat and upper breast pale, contrasting with darker brown underparts.

27   Whitehead's Swiftlet *Collocalia whiteheadi*   Text and map page 123

Rare and local. Known only from Mount Data, Luzon and Mount Apo, Mindanao, Philippines.

Adult (nominate; Mount Data, Luzon) Note very large-headed appearance. Extensive pale frosting on lores and in front of eye. Pale bases to supraloral patch not visible. Rump uniform with upperparts. Lacks any distinctive markings below. Note deeply forked tail.

22   Moluccan Swiftlet *Collocalia infuscata*   Text and map page 118

Wallacea, Indonesia.

22a   Adult *(ceramensis;* Buru, Ambon and Seram) Pale underparts and supraloral patch. Broad whitish rump patch. Note rather uniform underparts and forked tail.

22b   Adult *(infuscata;* Halmahera, Morotai and Ternate) Rump less distinct than other two races and occasionally uniform with upperparts. Underparts paler and more uniform than *waigeuensis* race of Uniform Swiftlet.

21

21

21

31a

31a

31b

27

27

27

22a

22a

22b

31c

Gerald Driessens
'94

Note that the four smaller species in this region (22, 24, 23 and 30) have proportionally longer and more deeply forked tails than the three larger species (27, 28 and 39).

23   Mountain Swiftlet *Collocalia hirundinacea*   Text and map page 118

New Guinea and some satellite islands.

> **Adult** (nominate; New Guinea (except range of *excelsa*), Goodenough and Karkar Islands) Upperparts uniform (slightly paler rump in many individuals). Underparts average paler and are more uniform than in *granti* race of Uniform Swiftlet.

25   Australian Swiftlet *Collocalia terraereginae*   Text and map page 121

Queensland, NE Australia.

> **Adult** (nominate; N Queensland) Grey underparts and rump. Forked tail.

24   White-rumped Swiftlet *Collocalia spodiopygius.*   Text and map page 119

Islands in Papuasia, Melanesia and Polynesia.

> **24a   Adult** *(delichon*; Manus) Deep black upperparts with striking broad white rump band. Forked tail.

> **24b   Adult** *(ingens*; S New Hebrides) Whitish underparts with dark throat. Note tail more obviously forked when less spread.

> **24c   Adult** *(townsendi*; Tonga) Sooty below with grey rump band.

28   Bare-legged Swiftlet *Collocalia nuditarsus.*   Text and map page 124

New Guinea south of the watershed, except in the W highlands of the Schrader range.

> **Adult** Large sooty-grey swiftlet. Extensive pale fringing on lores as with the larger species in this region. Rump uniform with upperparts. Shallowly forked tail. Uniform underparts.

29   Mayr's Swiftlet *Collocalia orientalis*   Text and map page 125

Rare and local. Lelet Plateau, C New Ireland and Guadalcanal, Solomons.

> **Adult** (nominate; Guadalcanal) Large species. Uniform underparts unlike Papuan Swiftlet. Noticeably pale grey rump band. Shallowly forked tail can appear square when open. Note dark throat and extensive pale frosting on lores and over eye.

31   Uniform Swiftlet *Collocalia vanikorensis*   Text and map page 126

Extensive range. See also Plate 8.

> **31a   Adult** (*granti;* lowland New Guinea, Yapen Island, D'Entrecasteaux and Aru Islands) Underparts brown but throat and upper breast contrastingly paler.

> **31b   Adult** (*pallens;* New Britain, New Ireland, Djaul and New Hanover) Similar to *granti* but contrast on underparts even greater.

39   Papuan Swiftlet *Collocalia papuensis*   Text and map page 136

North of the watershed in New Guinea. Also south of the watershed around Port Moresby. Local.

> **Adult** Large swiftlet closely related to Bare-legged and Mayr's Swiftlets but silvery-grey throat contrasts with dark underparts. Upperparts warmer and more glossed than Bare-legged Swiftlet. Rump uniform with upperparts. Tail shallowly forked.

48 Black Spinetail *Telacanthura melanopygia*   **Text and map page 147**

W and C sub-Saharan Africa; local.

> **Adult** Distinctive spinetail wing shape and *Telacanthura* tail shape (relatively long body beyond wing, for a spinetail, and prominent spiky tail). Predominantly dark plumage. May appear paler on the throat but mottling on throat only viewable at close range. Dark and highly streamlined at distance.

47 Mottled Spinetail *Telacanthura ussheri*   **Text and map page 145**

Widespread in sub-Saharan Africa.

> **47a** **Adult** (nominate; W Africa from Senegambia to Nigeria) Typical *Telacanthura* structure. Note pale mottled throat, white vent area and white rump. Compared to Little Swift *Apus affinis* (88) note Mottled Spinetail's blacker upperparts(especially very uniformly dark wings and forehead), darker throat and characteristic wing and tail shape.

> **47b** **Adult** (*stictilaema;* E Africa) Throat patch browner than nominate race. Also tail slightly longer and wings shorter. Slight cleft in tail when held tightly closed is typical of this species.

88 Little Swift *Apus affinis*   **Text and map page 209**

For comparison. See also Plate 23.

> **Adult** (nominate; E Africa and India) Upperparts less black, paler throat and different wing and tail shape.

51 Cassin's Spinetail *Neafrapus cassini*   **Text and map page 150**

W and C sub-Saharan Africa; local.

> **Adult** Striking underpart pattern. Note more extensive black on rear flanks than Böhm's Spinetail and larger size. Narrow rump band.

52 Böhm's Spinetail *Neafrapus boehmi*   **Text and map page 151**

E and S Africa.

> **Adult** (nominate; W and E Angola, Zaïre, W Tanzania and N Zambia) Note general similarity in structure to latter species but smaller size. Underparts differ in less clear-cut division between the white underparts and the dark breast, and whiter rear flanks. Broader rump than Cassin's Spinetail (extending to the proximal uppertail-coverts).

50 Sabine's Spinetail *Rhaphidura sabini*   **Text and map page 149**

W and C Africa.

> **Adult** Extensively white underparts, sharply divided from black upper breast. Elegant appearance. Note very black upperparts and wings, contrasting strongly with white rump and long uppertail-coverts.

49 Silver-rumped Spinetail *Rhaphidura leucopygialis*   **Text and map page 148**

For comparison. See also Plate 12.

> **Adult** Note similarity in general appearance between this species and the allopatric Sabine's Spinetail, but underpart pattern is distinct.

48

51

52

47a

88

47b

50

48

51

51

50

49

52

50

47

52

50

47

52

49

88

Gerald Driessens
'93

PLATE 11: AFRICAN ISLAND SPINETAILS AND BATES'S SWIFT

45 São Tomé Spinetail *Zoonavena thomensis* Text and map page 143

São Tomé and Príncipe, Gulf of Guinea.

> **Adult** Typical spinetail jizz. Dark areas of plumage glossy. Broad pale rump band contiguous with pale underparts. Indistinct cut-off between brown breast and brownish-white, heavily streaked underparts. Note tail-coverts do not cloak tail.

44 Madagascar Spinetail *Zoonavena grandidieri* Text and map page 142

Madagascar and Comoros.

> **Adult** (nominate; Madagascar) Dark upperparts with narrow, heavily streaked whitish rump. Note slightly capped effect. Tail slightly cleft in certain postures. Underparts rather uniform for a spinetail.

92 Bates's Swift *Apus batesi* Text and map page 217

Sub-Saharan W Africa.

> **92a** **Adult** Diminutive highly glossed black *Apus* swift. Note deeply forked tail.
> **92b** **Juvenile** Shorter tail and fringed underparts. Note indistinct throat patch and dark underparts typical of this species.

58

Gerald Driessens
'93

46   White-rumped Spinetail *Zoonavena sylvatica*   **Text and map page 143**

Indian subcontinent.

> **Adult** White rump contiguous with white underparts. Division between dark and pale underparts indistinct. Note square tail when closed. White undertail-coverts do not cloak tail as in some spinetails.

---

49   Silver-rumped Spinetail *Rhaphidura leucopygialis*   **Text and map page 148**

SE Asia and Greater Sundas.

> **Adult** Very dark upperparts cause rump and uppertail-coverts to look very white in comparison. Grey tinge and dark shaft streaks only visible at close range. Very dark below (although white rump can be seen from the side). Tail appears quite full when spread. Black tail often viewable along trailing edge of long tail-coverts. White tips of uppertail-coverts can sometimes be seen beyond tail from below.

---

89   House Swift *Apus nipalensis*   **Text and map page 212**

For comparison. See also Plate 23.

> **Adult** (*subfurcatus*; Malay Peninsula, Borneo, Sumatra) In comparison with preceding, sympatric spinetails has different wing and tail shape, dark underparts (cf. White-rumped Spinetail), dark uppertail-coverts (cf. Silver-rumped Spinetail) and white throat.

---

42   Philippine Spinetail *Mearnsia picina*   **Text and map page 140**

Philippines.

> **Adult** Highly glossed, black plumage and short-tailed appearance. Striking, and unique, white underwing patches.

---

43   Papuan Spinetail *Mearnsia novaeguineae*   **Text and map page 141**

New Guinea.

> **43a   Adult** (nominate; S and SE New Guinea) Highly glossed black plumage. Striking underpart pattern: grey-brown throat and white underparts with black undertail-coverts. Distinctive body and wing shape viewable from long distance.

> **43b   Adult** (*buergersi*; N New Guinea) Paler race with whiter underparts than nominate race. Note in some lights a flash in the rectrices can be apparent (due to the translucency of these feathers).

Gerald Driessens
'93

PLATE 13: NEEDLETAILS

Note distinctive structure and horseshoe mark on the underparts of all species.

53   White-throated Needletail *Hirundapus caudacutus*   Text and map page 153

Summer visitor from C Siberia to Japan, wintering mainly in Australia. Resident population in the Himalayas.

53a   **Adult** (nominate; Siberia, N China and SW Mongolia; winters south to Australia) Highly contrasting white throat patch and prominent white on forehead and lores. Note silvery-white saddle and white on inner web of tertials. Appears 'sharp-ended' when tail is closed.

53b   **Juvenile** (nominate) Pale saddle less striking, white on tertials more restricted and paler lores. Dark fringing on white areas of the underparts indicate that this is a juvenile.

53b   **Adult** (*nudipes;* Himalayas) Lacks white on lores and forehead.

54   Silver-backed Needletail *Hirundapus cochinchinensis*   Text and map page 155

E Himalayas and SE Asia. Partially migratory. Winters south to Java.

54a   **Adult** (nominate; E Himalayas and SE Asia) Indistinct throat patch, merging with darker brown upper breast. Impression of darker ear-coverts can cause throat patch to look more distinct. Slightly paler underparts and paler throat patch than Brown-backed Needletail. Saddle similar to White-throated Needletail. Tertial markings greyer and less extensive, and lores and forehead dark. Tail similar to White-throated Needletail.

54b   **Adult** (*rupchandi*; C Nepal) Darker throat.

55   Brown-backed Needletail *Hirundapus giganteus*   Text and map page 157

Indian subcontinent, SE Asia, Philippines (Palawan and Culion) and Greater Sundas.

55a   **Adult** (nominate; Malay Peninsula, from S Thailand, to the Greater Sundas and Palawan) Massive appearance. Indistinct throat patch. Dark lores indicate nominate race. Note prominence of rectrix spines and pointed (sharply rounded) shape of closed tail. Significantly paler underparts than Purple Needletail. Plumage less glossy and saddle less distinct than Silver-backed or White-throated Needletails.

55b   **Adult** (*indicus*; N and W of nominate) Shows prominent white supraloral spot.

56   Purple Needletail *Hirundapus celebensis*   Text and map page 158

NE Sulawesi and Philippines.

56a   **Adult** Dark, highly glossed plumage. Lacks pale throat patch or pale saddle. Prominent white supraloral spot indicates an adult. Tail shape similar to Brown-backed Needletail.

56b   **Juvenile** Less well-defined supraloral spot indicates a juvenile.

Note distinctive wing shape in all three species and the uniformity of the rump and uppertail-coverts.

61 Chimney Swift *Chaetura pelagica*   Text and map page 166

Summer visitor to E North America. Migrates through Central America and across the Gulf of Mexico. Winters mainly in W South America.

Adult Greyish throat patch. Underparts dark from mid-breast. Less strikingly capped than Vaux's Swift. Generally the most uniform *Chaetura* above with little contrast between the saddle and rump. Less black above than Chapman's Swift. Note tail shapes: square when closed, more rounded when open.

62 Vaux's Swift *Chaetura vauxi*   Text and map page 168

Summer visitor to W North America (north to Alaska). Resident in Central America and N Venezuela.

62a   **Adult** (nominate; W North America; migrates to Mexico and Guatemala) Extensively pale underparts, only dark on the undertail-coverts. Pale throat very apparent in head-on view. Appears more capped than Chimney Swift. Olive-brown upperparts with pale saddle (more striking than in Chimney Swift).

62b   **Adult** *(tamaulipensis*; E Mexico) Darker belly and undertail-coverts than nominate race.

62c   **Adult** *(richmondi*; S Mexico to Costa Rica and W Chiriqui, Panama) Blacker upperparts than nominate race and darker grey rump. Underparts also darker but white throat patch distinct.

62d   **Adult** *(gaumeri*; Yucatan and Cozumel Island, Mexico) Similar to *richmondi*, but slightly paler on rump and a little smaller.

63 Chapman's Swift *Chaetura chapmani*   Text and map page 170

Wide ranging from Panama across N South America and on Trinidad. Also C Brazil. Local.

Adult (nominate; N South America and Trinidad) Dark, rather uniform underparts. Less capped appearance than Vaux's or Chimney Swifts. Glossy black upperparts with contrastingly pale grey rump and uppertail-coverts. Note contrast between tail and uppertail-coverts obvious only when tail is fully spread. In distant views uniformity of plumage very apparent.

Gerald Driessens
'93

PLATE 15: *CHAETURA* SWIFTS 2

58 **Lesser Antillean Swift** *Chaetura martinica* Text and map page 162

Endemic to the Lesser Antilles.

> **Adult** Dark underparts with indistinctly paler throat. Comparatively dark throat reduces the capped effect apparent in other *Chaetura* species. Narrow but distinct rump band contrasts with dark mantle and uppertail-coverts. At distance appears very dark.

65 **Ashy-tailed Swift** *Chaetura andrei* Text and map page 172

Resident in N South America. Summer visitor to C South America, wintering north to Panama.

> **Adult** (*meridionalis*; breeds E and SE Bolivia, E Brazil, N Paraguay and NW Argentina. Winters north to Panama) Large, dark *Chaetura*. Appears rather short-tailed but less so than Short-tailed Swift. Olive-brown above with moderate contrast on paler, greyer rump and uppertail-coverts. Paler rump is still noticeable at long distance. Pale throat emphasised by dark underparts. Note paler undertail-coverts (this feature is only shared, in the *Chaetura*, by the very different Short-tailed Swift).

64 **Short-tailed Swift** *Chaetura brachyura* Text and map page 171

Wide range from Panama and the S Lesser Antilles in the north and throughout South America, mainly east of the Andes, south to C Brazil.

> **64a** **Adult** (nominate; N South America, from Panama south to Bolivia and C Brazil) Very short-tailed, with exaggerated wing shape: short secondaries, broad inner primaries and hooked outer primaries. Long-winged appearance emphasised by short tail. Dark plumage (especially throat) with pale undertail-coverts. Pale rump and uppertail-coverts contrast strongly with dark upperparts and wings.

> **64b** **Adult** (*ocypetes*; SW Ecuador and NW Peru) Paler throat.

58

58

58

65

65

64a

64a

64b

Gerald Driessens
'93

PLATE 16: *CHAETURA* SWIFTS 3

57   Band-rumped Swift *Chaetura spinicauda*   **Text and map page 160**

S Central America and N South America.

57a   Adult (nominate; E Venezuela, the Guianas and N Brazil) Highly contrasting, narrow white rump band, not extending onto uppertail-coverts. Plumage browner and less glossed than Grey-rumped Swift. Rather uniform underparts, throat slightly paler.

57b   Adult (*aethalea*; C Brazil) Dark below with pronounced pale throat patch. Rump more extensive and a little darker than nominate. Plumage generally blacker than nominate.

57c   Adult (*fumosa;* W Costa Rica, W Panama and N Colombia) Similar below to *aethalea*, but above rump more extensive and slightly darker. Plumage generally blacker than nominate.

59   Grey-rumped Swift *Chaetura cinereiventris*.   **Text and map page 163**

Wide range in Central and South America. Also S Caribbean.

59a   Adult (*guianensis*; E Venezuela and the Guianas) Bluish gloss to blackish plumage and cold, pale grey on rump and proximal uppertail-coverts. Note that uppertail-coverts do not entirely cloak tail, so that even if tail is not fully spread there is considerable contrast between the pale uppertail-coverts and the dark tail. Underparts typically grey, paler on throat with black on undertail-coverts.

59b   Adult (*sclateri;* N Brazil, S Venezuela and S Colombia) Very dark underparts, only throat is clearly paler. Grey rump band.

59c   Adult (*phaeopygos*; E Nicaragua to Panama) Rump colour intermediate between darkest and palest races.

59d   Adult (nominate; E Brazil) Pale underparts and comparatively narrow grey-white rump band.

60   Pale-rumped Swift *Chaetura egregia*   **Text and map page 165**

WC South America, east of the Andes; local.

Adult Black plumage glossed bronze rather than bluish as in Grey-rumped Swift. Whitish rump band, extending onto uppertail-coverts. Pale throat, but underparts more extensively dark than most races of Grey-rumped Swift. Appears longer-winged than Grey-rumped Swift.

57a

57a

57a

57b

57c

59d

59b

59a

59c

59a

60

60

60

Gerald Driessens
'93

PLATE 17: *AERONAUTES* SWIFTS

Note the typical *Aeronautes* wing shape: broad inner secondaries and sharply pointed outerwing.

66 White-throated Swift *Aeronautes saxatalis*   Text and map page 174

Extensive North and Central American range.

66a **Adult male** (nominate; North America, partly migratory) Upperparts jet black with paler head. At distance head can appear very pale. Note prominence of flank patches and belly stripe. Broad white trailing edge on secondaries and inner primaries.

66b **Adult female** (nominate) White trailing edge typically less prominent than on male.

66c **Adult** (*nigrior*, S Mexico, Guatemala, El Salvador and Honduras) Tail shape rather broad and spiky when fully opened. Dark-headed appearance typical of southern race.

67 White-tipped Swift *Aeronautes montivagus*   Text and map page 175

N South America (Andes and Tepuis) south to N Bolivia.

67a **Adult male** (nominate; Andes and N Cordilleras of Venezuela) Dark-capped, broad white throat patch, white ventral and flank patches and white trailing edge on secondaries. Extensive white tail tips only on male. From side-on view extensive white flank-tufts can give impression of pale rump.

67b **Adult female** (nominate) Lacks white tail tips. Tail shows shallow fork when fully spread.

68 Andean Swift *Aeronautes andecolus*   Text and map page 177

Restricted to the sub-equatorial Andes.

68a **Adult** (*parvulus*; W Peruvian Andes to N Chile) Deeply forked tail. Underparts extensively whitish; less black on face than White-tipped Swift. Prominent white collar and rump. Highly contrasting plumage is typical of this race.

68b **Adult** (nominate; C Bolivia to W Argentina) Buff tinge to underparts.

66a

66b

66a

66c

66a

66b

67a

67b

67a

67b

67b

67b

68a

68b

68a

68a

68b

68a

68b

Gerald Driessens
'93

PLATE 18: *TACHORNIS* AND *PANYPTILA* SWIFTS

69   Antillean Palm Swift *Tachornis phoenicobia*   **Text and map page 178**

Greater Antilles.

> **Adult** (nominate; Hispaniola and Jamaica) Extensive white throat and belly patch. Broad white rump band, separated from belly patch by dark rear flanks. Comparatively short and shallowly forked tail. At distance note underpart pattern and capped appearance.

70   Pygmy Swift *Tachornis furcata*   **Text and map page 179**

Rare and local. NE Colombia and W Venezuela.

> **Adult** (nominate; NE Colombia and adjacent NW Venezuela, south of Lake Maracaibo) Dark upperparts. Note very thin shape of tail when closed. When tail spread, deep fork is visible. Underparts similar to Fork-tailed Palm Swift, but better defined breast band separates pale throat from pale belly. Smaller and slighter than Fork-tailed Palm Swift. White flashes at the base of the tail visible when banking.

71   Fork-tailed Palm Swift *Tachornis squamata*   **Text and map page 179**

South America east of the Andes. Also on Trinidad.

> **Adult** (nominate; E Peru to Venezuela) Pale underparts appear rather dirty or mottled. Breast band typically incomplete at centre. Lightly scaled upperparts hard to see except at very close range, appearing dark. Broader-winged and broader-tailed than tiny allopatric Pygmy Swift.

72   Great Swallow-tailed Swift *Panyptila sanctihieronymi*   **Text and map page 181**

Restricted Central American range: S Mexico to S Honduras.

> **Adult** Piebald plumage diagnostic of *Panyptila*. Broader-winged and broader-tailed than Lesser Swallow-tailed Swift and collar typically more prominent. Tail only appears deeply forked when spread; usually appears spike-like, but broader than Lesser Swallow-tailed Swift.

73   Lesser Swallow-tailed Swift *Panyptila cayennensis*   **Text and map page 182**

Extensive Neotropical range, from S Mexico south throughout Central America into N South America. Also on Trinidad.

> **Adult** Plumage similar to Great Swallow-tailed Swift. Despite slender proportions appears quite 'bull-headed'. Note comparatively narrow collar on nape.

Gerald Driessens
93

In both *Schoutedenapus* species note the thin wings tapering sharply from the secondaries, the needle-thin tail (usually held tightly closed) and slim bodies with rather bulbous heads. Note the pale plumage and very long wings and tail of the *Cypsiurus* species.

40 Scarce Swift *Schoutedenapus myoptilus*   Text and map page 137

Sub-Saharan Africa, mainly E Africa.

40a   **Adult** (nominate; E Africa, from Ethiopia south to Zimbabwe) Nondescript plumage with paler throat. Pale throat visible at considerable distance when caught in light; overall appearance can be silvery-grey in these conditions. Very uniform upperparts. Slightly capped appearance. Deeply forked tail can be seen when tail is spread. Fluttering flight is a feature of this species when seen in high flying flocks.

40b   **Adult** (nominate) Moulting bird showing short tail.

40c   **Adult** (*chapini;* Zaïre, Rwanda, SW Uganda) Plumage darker than nominate race.

41 Schouteden's Swift *Schoutedenapus schoutedeni*   Text and map page 139

Extremely rare and local. E Zaïre.

Adult Throat almost uniform with underparts. Lacks slightly capped appearance of Scarce Swift. Appears very dark at distance. Flight pattern assumed to be closely similar to Scarce Swift.

83 African Swift *Apus barbatus*   Text and map page 202

For comparison. See also Plate 23.

Adult (*roehli*; E Africa) Note heavier appearance, proportionally shorter tail, broader wings and body typical of larger *Apus*. Well defined throat patch.

91 White-rumped Swift *Apus caffer*   Text and map page 215

For comparison. See also Plate 22.

Adult Note similar appearance of this species in tail shape but head profile more elegant and wings more evenly 'full'. Throat patch very prominent.

74 African Palm Swift *Cypsiurus parvus*   Text and map page 184

Widespread in sub-Saharan Africa and Madagascar. Also SW Arabia.

Adult (nominate; Senegambia to Ethiopia and S Sudan. Also in SW Arabia) Very pale underparts with uniform throat and underwing-coverts appearing darker than body. Note light streaking on throat. Tail typically held tightly closed but deeply forked when spread. Despite very pale plumage the effect of light can produce a very different impression.

75 Asian Palm Swift *Cypsiurus balasiensis*   Text and map page 186

Extensive S and SE Asian range, extending throughout the Greater Sundas and the Philippines.

Adult (nominate; Indian subcontinent, including Sri Lanka) Tail shorter than African Palm Swift. Throat not streaked. Upperparts appear darker and greyer than African Palm Swift.

Note heavy appearance of the two *Tachymarptis* species, especially the broad wings, thickset bodies and, in Alpine, a comparatively short, shallowly forked tail.

76 Alpine Swift *Tachymarptis melba*   Text and map page 188

Extensive Old World range. Summer visitor to W Palearctic and S Africa. Resident in Indian subcontinent and Sri Lanka, sub-Saharan Africa and Madagascar.

76a   **Adult** (nominate; W Palearctic east to W Himalayas, wintering mainly in sub-Saharan Africa) Pale upperparts, but lacking prominent saddle of Pallid Swift. Highly distinctive white throat and belly patch.

76b   **Adult** (*nubifuga*; Himalayas, wintering to C India) Darker than nominate race with a broader breast band and less extensive throat patch. Note that in all races of Alpine Swift the throat patch can be very hard to discern.

76c   **Adult** (*bakeri*; Sri Lanka) Darker than *nubifuga* but with a narrower breast band.

77 Mottled Swift *Tachymarptis aequatorialis*   Text and map page 190

Sub-Saharan Africa. Mainly E Africa with a disjunct W African range.

77a   **Adult** (nominate; Ethiopia to Angola, and Zimbabwe east of *gelidus*) Heavily barred underparts. Indistinct pale throat patch. Darker mantle contrasts slightly with head and markedly with rump. Innerwing paler than outerwing.

77b   **Adult** (*furensis*; Darfur, W Sudan) Very pale whitish belly but darker crescents still apparent.

77c   **Adult** (*lowei*; W Africa, from Sierra Leone to Nigeria) Pale belly somewhat intermediate between *furensis* and nominate race.

86 Pacific Swift *Apus pacificus*   Text and map page 206

E Palearctic range. Summer visitor to N Asia, wintering south to Australia. Resident populations in the Himalayas and SE Asia.

Adult (nominate; breeds Siberia, N China and S Japan, winters south to Australia) Prominent white rump band. Upperparts blackish with slightly paler head. Underparts appear greyish when viewed in sunlight at distance and scaly plumage only visible at close quarters. Throat patch not as distinct as in some species. When tightly closed tail appears spike-like, but broader than similar tail posture of White-rumped Swift. Tail broad and heavily forked when spread.

87 Dark-rumped Swift *Apus acuticauda*   Text and map page 208

Rare and local. Breeds with certainty only in the Khasi Hills, Meghalaya, NE India, and presumably also in Mizoram. Also recorded in Nepal (once), and possibly Myitkyina, Burma. Rare winter visitor to NW Thailand.

Adult Deeply black upperparts. Outer tail less heavy than Pacific Swift due to the strongly emarginated outer rectrices. Underparts similar to Pacific Swift, but throat more heavily streaked, appearing darker.

PLATE 21: WESTERN PALEARCTIC *APUS* SWIFTS

79 Common Swift *Apus apus*   **Text and map page 194**

Summer visitor with an extensive Palearctic range. Winters mainly in sub-Saharan Africa.

79a   **Adult** (nominate; west of species range east to Lake Baikal and south towards Iran) Black-brown underparts and indistinct throat patch. Comparatively uniform upperparts. Note saddle effect still visible but most contrast between lower back and rump.

79b   **First summer (nominate)** Worn bird showing greater contrast between inner and outer wing.

79c   **Juvenile** (nominate) Broad white fringes to underwing-coverts and large, contrastingly white throat. Note white forehead in head-on view and blacker plumage than worn adult.

79d   **Adult** (*pekinensis*; Iran east to N China) Averages paler on forehead and has more inner/outer wing contrast than nominate form.

82 Pallid Swift *Apus pallidus*   **Text and map page 199**

Predominantly W Palearctic range, extending to W Asia and the Middle East. Mainly migratory, wintering in Africa.

82a   **Adult** (*brehmorum*; Europe, except range of *illyricus,* coastal N Africa, from Morocco to W Egypt, Canaries and Madeira) Bulkier appearance of Pallid Swift compared to Common Swift and note differences in tail and wing shape. Broad, poorly defined whitish throat and barred underparts (paler than Common Swift). Greater contrast on underparts and smaller area of darkest underwing-coverts than Common Swift. Note tendency of the darkest underwing-coverts to appear darker than underbody. Pattern of upperparts appears more broken than Common Swift: small saddle contrasting with wing-coverts, lower back and head, and significant contrast between outer primaries and greater coverts. Appears very pale-headed when viewed head-on.

82b   **Adult** (nominate; Banc d'Arguin east through Saharan highlands in Egypt, through Middle East to Pakistan) Palest race. Averages a larger throat patch than *brehmorum.*

82c   **Adult** (*illyricus*; former Yugoslavian coast and possibly E Italian coast) Darkest race. Note darker underparts enhance contrast of throat patch.

80 Plain Swift *Apus unicolor*   **Text and map page 197**

Madeira and Canary Islands. Winter records from Morocco and Mauritania.

Adult Dark plumage below with indistinct, mottled throat patch. Note barred appearance of underparts (cf. Common Swift). Similar to Common Swift above, lacking pale-headed appearance (especially forehead) or the strongly contrasting saddle of Pallid Swift. Note rakish appearance, with very streamlined body (lacking the 'bulk' of the previous, larger species), long deeply forked tail and thin wings.

78 Alexander's Swift *Apus alexandri*   **Text and map page 193**

Cape Verde Islands.

Adult Broad, but ill-defined throat patch. Paler below than other Western Palearctic *Apus* species. More prominent saddle than Plain Swift. Note structure of this small species: comparatively short-winged, with a shallow tail-fork.

PLATE 22: AFRICAN *APUS* SWIFTS

83 African Swift *Apus barbatus* Text and map page 202

Widely distributed in E and S sub-Saharan Africa.

83a **Adult** (nominate; S Africa, migratory) Bulky structure, similar to Pallid Swift. Deeply black underparts and small whitish throat patch. Prominent black saddle and pale innerwing panel.

83b **Adult** (*balstoni*; Madagascar) Very dark race.

83c **Adult** (*roehli*; widespread in E Africa) Blacker than nominate form and Nyanza Swift (which has a less striking throat patch).

83d **Adult** (*hollidayi*; Victoria Falls area) Saddle uniform with lesser coverts.

81 Nyanza Swift *Apus niansae* Text and map page 198

Sub-Saharan E Africa.

81a **Adult** (nominate; W Ethiopia, E Uganda, W Kenya and N Tanzania) Similar in structure to Common Swift (small rounded head and scythe-shaped wings) except for shorter, more shallowly forked tail and less sharply pointed rectrices. Underparts brown with ill-defined pale throat patch. Very strong contrast between inner and outer wing. Saddle indistinct compared with Pallid or African Swifts.

81b **Adult** (*somalicus*; N Somalia and adjacent areas of Ethiopia; some winter in Kenya) Paler than nominate form. Throat patch better defined.

85 Bradfield's Swift *Apus bradfieldi* Text and map page 205

SW Africa.

85a **Adult** (nominate; Angola and Namibia) Structure similar to Pallid or African Swifts. Saddle least distinct of the plainer *Apus* species. Upperbody appears paler than outer wing and tail. Very pale underparts with well-defined barring. Pale throat patch indistinct.

85b **Adult** (*deserticola*; S areas of range) Darker body than nominate form resulting in more prominent throat patch.

84 Forbes-Watson's Swift *Apus berliozi* Text and map page 204

Breeds Somalia and Socotra, some wintering on the Kenyan coast.

84a **Adult** (nominate; Socotra Island) Very similar to Pallid Swift but saddle contrasts less with head and wings more uniform (especially outer greater primary coverts). Lacks the strikingly paler innerwing of Nyanza or Black Swifts. Lores and forehead darker than Pallid Swift, but paler than most Common Swifts. Heavily marked underparts, darker than Nyanza or Pallid Swifts, but paler than African Swift. Throat patch well-defined. Pale-headed appearance head-on, but note that forehead and lores are clearly darker than throat.

84b **Adult** (nominate) Worn individual showing browner underparts and less well-defined throat patch.

Species included for comparison.

82 **Pallid Swift *Apus pallidus*** (See also Plate 21)
Adult (*brehmorum*) Underparts similar to Bradfield's Swift, but pale throat more extensive. Note similarity to Forbes-Watson's Swift but paler forehead and lores on Pallid Swift.

40 **Scarce Swift *Schoutedenapus myoptilus*** (See also Plate 19)
Adult (nominate) Very different structure from the uniform *Apus* species. Pale throat patch very poorly defined.

79 **Common Swift *Apus apus*** (See also Plate 21)
Adult (nominate) Note more uniform underwing and darker body than Nyanza or Bradfield's Swifts. Tail more deeply forked than other uniform African species.

77 **Mottled Swift *Tachymarptis aequatorialis*** (See also Plate 20)
Adult (nominate) Note difference in structure, especially body and wing shape, coupled with the relatively shorter, larger, shallowly forked tail.

82

40

83c

83a

83a

83d

83b

81a

81a

79

81b

77

85a

85b

82

85a

85a

84b

85a

84a

84a

84a

82

Gerald Driessens
'94

PLATE 23: WHITE-RUMPED *APUS* SWIFTS

88 Little Swift *Apus affinis*   Text and map page 209

Extensive sub-Saharan African and S Asian range. Scattered populations in the S Western Palearctic. Partially migratory in the north and south of range.

88a **Adult** (*galilejensis;* N Africa, through Middle East to Pakistan and south of Sahara from E Sudan to NW Somalia) Structure diagnostic: square tail, chunky body with rather blunt wing tips. Broad white throat patch and pale undertail-coverts. Note translucency of tail when spread. Tail appears slightly rounded when fully spread. Pale head, black saddle and wing-coverts, and broad white rump patch (broader than House Swift). Even when overhead extensive white is visible around rear flanks. Pale forehead apparent in head-on view.

88b **Adult** (*galilejensis*) Wing shape distorted by moult.

88c **Adult** (nominate; India and E Africa) Darker plumage than *galilejensis*, especially tail-coverts.

89 House Swift *Apus nipalensis*   Text and map page 212

SE Asia, Greater Sundas and N Philippines.

Adult (*subfurcatus;* Malay Peninsula, Borneo, Sumatra) Structure as Little Swift, but longer tail, shallowly forked. Very dark plumage, especially tail and tail-coverts. Narrower rump band than Little Swift and darker forehead. Note tail-fork is hard to discern when tail widely spread.

90 Horus Swift *Apus horus*   Text and map page 213

Sub-Saharan Africa, mainly E and S Africa. Summer visitor to most southern parts of range.

Adult (nominate; throughout range except SW Angola) Resembles a large, forked-tailed Little Swift. Plumage closely similar to African races of Little Swift. Note paler underwing-coverts. Tail broader and relatively shorter than White-rumped Swift. Note prominence of white rump which extends onto rear flanks.

91 White-rumped Swift *Apus caffer*   Text and map page 215

Widely scattered range in sub-Saharan Africa. Also small, isolated, non-resident populations in the W Palearctic in S Spain and Morocco.

Adult Very elegant in comparison with previous species: slender body with long wings and long, deeply forked tail. Very black plumage (darker than Little Swift), with narrow but highly contrasting white rump. Note dark forehead. Narrow white trailing edge to remiges visible from below. Deeply forked tail when spread, but tail often held tightly closed, appearing spike-like, for long periods.

95 **Moustached Treeswift** *Hemiprocne mystacea*   **Text and map page 222**

E Indonesia and New Guinea.

95a **Adult male** (nominate; New Guinea and W Papuan islands) Highly distinctive profile. Large, with long wings and long, very deeply forked tail. Note typical posture of bird and position of tail and wings when perched. Prominent tertial patches and facial 'streamers'. Note chestnut mark on rear ear-coverts.

95b **Adult female** (nominate) Lacks chestnut on ear-coverts. Translucent panel in remiges often visible on underwing.

95c **First winter** (nominate) Brown crescents partly replaced by grey plumage.

95d **Juvenile** (nominate) Note highly cryptic plumage and structural underdevelopment.

95e **Adult male** (*aeroplanes*; Bismarck Archipelago) Note paler upperparts.

94 **Grey-rumped Treeswift** *Hemiprocne longipennis*   **Text and map page 220**

SE Asia (mainly south of the Isthmus of Kra) and Greater Sundas.

94a **Adult male** (*harterti*; S Burma, peninsular Thailand, W Malaysia, Sumatra and Borneo) Note distinct contrast between the dark grey chest and the whitish belly. Darker chested than Crested Treeswift. Note also contrastingly dark underwing-coverts. Ear-coverts dull brick-orange (not extending to chin). Note greenish hue on the upperparts (lacking in Crested Treeswift). Wing tips extend beyond tail when perched.

94b **Adult female** (*harterti*) Dark ear-coverts (lacking orange). This race is normally darker than the nominate.

94c **Adult female** (nominate; Java and Bali) Generally lightest upperparts.

94d **Adult female** (*wallacii*; Sula Islands and Sulawesi) Greatest contrast between dark and pale areas of plumage.

94e **Adult female** (*perlonga*; Simeulue, Sumatra) Least contrasting rump patch.

94f **First winter** (*harterti*) Some brown crescentic markings replaced by bluish feathers.

94g **Juvenile** (*harterti*) Note cryptic plumage.

93 **Crested Treeswift** *Hemiprocne coronata*   **Text and map page 219**

Extensive range. Indian subcontinent and SE Asia.

93a **Adult male** Note difference in underpart and underwing pattern between this species and the latter: less distinction between dark and pale areas of the body and less contrasting underwing-coverts. Lacks the more contrastingly pale rump of Grey-rumped Treeswift. Note that the tail extends well beyond the wing tips in this species. Dull orange ear-coverts extend to chin. Crest is darker than head and is often raised when perched (but laid flat in flight).

93b **Adult female** Dark ear-coverts and faint whitish moustache, making the ear patch more contrasting than Grey-rumped Treeswift.

93c **Juvenile** Note cryptic plumage.

96 **Whiskered Treeswift** *Hemiprocne comata*   **Text and map page 223**

SE Asia (peninsular Thailand and W Malaysia), Philippines, Sumatra and Borneo.

96a **Adult male (nominate**; peninsular Thailand and W Malaysia, Sumatra and Borneo) Highly distinctive plumage and less powerful structure than preceding species. Note rather bronzy plumage, wispy facial 'streamers' and extensive white tertial patches. Tail deeply forked when spread. Chestnut ear-coverts on males only. Tends to fly less than other species, preferring to make short fly-catching sorties.

96b **Adult male** (nominate) Darker underparts created by back-lighting.

96c **Adult female** (nominate) Lacks chestnut ear-coverts.

96d **Adult female** (*stresemanni;* Pagai Island, Sumatra) Darker, more olive-green upperparts.

96e **Adult female** (*nakamurai;* Mindanao and Basilan, Philippines) Larger than nominate race, with greener upperparts.

SYSTEMATIC SECTION

CYPSELOIDES

This genus of ten medium to large new world species is one of the least well known. Two species, Tepui and Chestnut-collared Swifts, are unique amongst the Apodidae in having bright orange collars in adult plumage (females of both species take longer to acquire the full collar), whilst all other species are mid-brown to blackish-brown with a tendency to become paler around the head, especially the forehead and chin.

Separation from sympatric genera

Despite the lack of distinctive plumage features, *Cypseloides* species possess a distinctive structure and jizz. The slender, streamlined *Panyptila* and *Tachornis* swifts have long deeply-forked tails and slender, pointed wings, quite unlike the relatively short and square, or at best shallowly-forked, tails of this genus. In the Caribbean, Antillean Palm Swift is shorter-tailed but its tiny size, attenuated proportions and dashing flight are quite unlike the sole *Cypseloides* species found there. *Chaetura* swifts are closer in tail shape to the square-tailed members of this genus. However, with the exception of White-chinned and White-fronted Swifts the tails of *Chaetura* swifts are considerably shorter. The two genera differ most markedly in wing shape. *Cypseloides* species lack the typical butter-knife wing shape of *Chaetura* swifts, having instead broader, rather straight wings that appear blunter-tipped and rather long. This genus has a relatively small-headed appearance in comparison with the protruding head shape of *Chaetura* species. *Aeronautes* swifts have more pointed wings and distinctively broad inner secondaries that taper sharply to much shorter outer secondaries and inner primaries. *Streptoprocne* swifts are considerably larger (even than Great Dusky Swift) and much more powerful and impressive.

The flight of *Cypseloides* species is perhaps most distinctive; the wing beats are slower than the smaller genera (with the exception of the rapid and rather bat-like White-chinned Swift), but lack the rather relaxed, deep wing beats of *Streptoprocne* species. The flight appears rather unsteady, and the wings are very stiff, held out at right angles to the body and often down-tilted, lacking the manoeuvrability of other genera.

Nesting behaviour

All known nest sites are close to water. The use of saliva in nest building has not been proved and mud is used as an adhesive. Clutches are notably smaller than other genera. Importantly the research of Marin and Stiles (1992) suggests that the nest shape of *Cypseloides* species depends mainly on the nature of the nest substrate rather than the species. Therefore, although the nest descriptions may be of typical nests, great variation should be expected. Similarly Marin and Stiles noted a positive correlation between the wetness of the nest and the amount of mud used proportional to the amount of plant matter (figure 9).

Figure 9. Nest of Spot-fronted Swift. (Based on Marin and Stiles 1992.)

In-the-hand identification

Navarro *et al.* (1992) reinforced the importance of tarsus measurements in separating various *Cypseloides* species. Table 6 based on Navarro *et al.* (1992), compares tarsus length with wing length. This graph shows the morphological closeness of the Black Swift superspecies (3, 4, 5 and 6), the similarity of the species pairs of White-chinned and White-fronted Swifts (8 and 9) and Tepui and Chestnut-collared Swifts (1 and 2), and the large size of Great Dusky Swift (10).

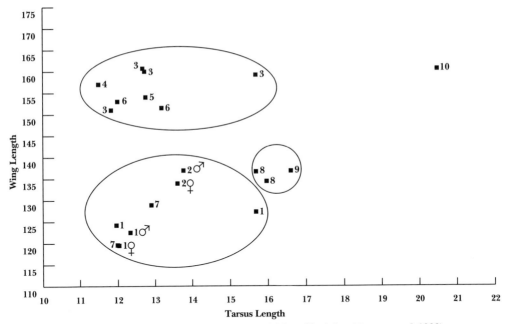

Table 6. Comparison of tarsus and wing lengths of *Cypseloides* swifts (after Navarro *et al.* 1992).

Calculation of the tarsus:tail ratio, by dividing the tarsus length by the tail length and multiplying this figure by 100, is a useful means of separating the *Cypseloides* species. This technique should be used in conjunction with other measurements and plumage features described in the species accounts.

Great Dusky Swift	44.50	Navarro *et al.* (1992)
White-fronted Swift	36.46	Navarro *et al.* (1992)[1]
White-chinned Swift	36.04	Zimmer (1945)
White-chinned Swift	35.64	Marin and Stiles (1992)
White-chinned Swift	35.23	Eisenmann and Lehmann (1962)
White-chinned Swift	34.05	Navarro *et al.* (1992)
Spot-fronted Swift	29.17	Marin and Stiles (1992)
Tepui Swift (Female)	28.19	Collins (1972)
Spot-fronted Swift	27.62	Navarro *et al.* (1992)
Tepui Swift (Male)	27.50	Collins (1972)
Chestnut-collared Swift	27.06	Navarro *et al.* (1992)
Sooty/Rothschild's Swift[2]	25.59	Navarro *et al.* (1992)
Black Swift (Costa Rica)	24.99	Marin and Stiles (1992)
Chestnut-collared Swift	24.89	Marin and Stiles (1992)
Rothschild's Swift	24.85	Eisenmann and Lehmann (1962)
Sooty Swift	24.49	Eisenmann and Lehmann (1962)
Black Swift (USA)	23.78	Navarro *et al.* (1992)
White-chested Swift	21.58	Eisenmann and Lehmann (1962)
Black Swift (Mexico)	20.91	Navarro *et al.* (1992)
Black Swift (Costa Rica)	19.39	Navarro *et al.* (1992)

[1] Figures amended slightly from Navarro *et al.* (1992) to include measurements of the fifth specimen described in Navarro *et al.* (1993).

[2] Navarro *et al.* (1992) treated these two species as conspecific.

Table 7. Average tarsus:tail ratios from a variety of sources.

1 CHESTNUT-COLLARED SWIFT
Cypseloides rutilus   **Plate 1**

Other name: *Streptoprocne rutilus*

IDENTIFICATION Length 13cm. This small *Cypseloides* swift is perhaps the most frequently encountered member of the genus over much of its neotropical range. The adult male is distinctively plumaged with a bright chestnut collar (females take longer to acquire full collar), a feature only shared by the Tepui Swift, which has only been recorded once in the range of this species. Separation from Tepui Swift is dealt with under that species. The chestnut collar is surprisingly easy to see, especially in sunlight. Immatures can show a partial collar on the nape. Individuals lacking the chestnut collar are very dull plumaged without any striking features. Full-collared birds have a rather black plumage, but non-collared individuals are rather paler and browner than the larger White-chinned and White-fronted Swifts, and particularly Spot-fronted Swift, especially on the wings which can appear paler than the upperbody. The face of collarless individuals appears particularly uniform and slightly capped compared to the other species which have varying amounts of pale areas on the head. Structurally has a proportionally longer tail than these species and usually shows a shallow but distinct cleft, although often appearing rather square (figure 10).

Figure 10. Chestnut-collared Swifts with different tail postures, showing effect on tail shape.

Black Swift is clearly larger than this species with blacker plumage (with a distinctly paler head), very distinct scaling on the juvenile and in most cases a more distinctly cleft tail. Flight is quite typical for the genus with much veering and twisting on rather stiff wings that are invariably held below the horizontal; often involves alternate bursts of rapid flaps on shallow wing beats and short glides. Marin and Stiles (1992) noted a great similarity between the flight of this species and Spot-fronted Swift.

DISTRIBUTION Rather broken Central and South American range. In Central America found in Mexico from E Sinaloa, Durango, Zacatecas, Hidalgo and Veracruz in the north, south into Guatemala in three separate highland areas, El Salvador, Honduras, Costa Rica (especially in the Cordillera Central and Cordillera de Talamanca but apparently only a rare visitor to the lower N Cordilleras) and into Panama where it has not been recorded east of Cerro Azul in E Panama province. Throughout Central America found mainly on the Pacific slope cordilleras and apparently not recorded from Nicaragua. In South America closely associated with the Andes and related cordilleras. Found in Guyana, and Venezuela from Sucre in the east, westwards to Carabobo and into the Andes to

Tachira. In Colombia found throughout all Andean ranges and Sierra de San Lucas, but not present in Santa Marta or Perija Mountains. In Ecuador found in the E and W Andes. South of Ecuador the range is rather broken and restricted to the E Andes through Peru south to Yungas of Cochabamba, Bolivia. Also found on Trinidad.

A rather local species that appears to be scarcer in Central America than in South America where it can be abundant in certain Andean localities and is certainly encountered with regularity. However, thought to be locally common in Costa Rica. Apparently recorded with greater regularity in recent years in Panama where up to 200 have been recorded. In Ecuador the species is said to be commonest on the western temperate slope of the Andes.

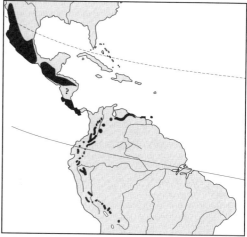

MOVEMENTS Populations in South America believed to be resident, but migration occurs in the northern areas of the species' Central American range. Probably some migration from N Mexico in winter as flocks of apparent migrants noted from mid March to May in W Mexico. However, most movement involves withdrawal from the high altitude interior, from at least October to February. In Guatemala the situation seems to be similar with specimens having been taken from 12 July to October (Land 1970). Previously it was believed that all Central American populations were only summer migrants (Phillips 1962), but Monroe (1968) states that the species is present in Honduras in the winter, although it is not certain whether these are birds that have bred in Honduras or are wintering after having bred further north. From Honduras south the species is present in the winter and is believed to be strictly a resident in Costa Rica and Panama. There is no evidence that Central American breeding populations winter in South America.

HABITAT Primarily a montane species. In Mexico and Costa Rica occurs between 1500-2450m, but in Costa Rica, at least, feeds regularly down to 300m or even sea-level on occasions. Similarly in Guatemala it has been recorded from close to sea-level to 2050m, and in Honduras it occurs mainly over 1000m, occasionaly feeding down to 400m in valleys. In the Andes it has been recorded to 3400m, although a range of 1000-2600m is more normal. Occurs over a variety of highland habitats, both cleared and forested and is equally likely to be found over rugged ravines as more gentle upland valleys and level ground.

Often feeds over lowlands especially when there is rain in the highlands. Often seen in the vicinity of waterfalls. Recorded in Ecuador (Marin 1993) from 500-3000m on the western slope and 500-2000m on the eastern slope.

DESCRIPTION *C. r. rutilus*
Adult male. Head Upper head from forehead to rear-crown sooty black-brown. The forehead and the narrow line of feathers over the eye appear a little paler due to grey-brown fringing. Black eye-patch. Chin to mid-throat and ear-coverts grey-brown with a variable admixture of dull red feathers. The red feathers can appear strikingly bright, almost orange, when caught in sunlight. Mid-throat to mid-breast orange and forming a wide collar over the nape. A variable amount of brown mottling can be seen in the collar. This mottling is mainly age-related. **Body** Below the nape the mantle is similarly black-brown. The upperparts are quite uniform but the rump and upper-tail-coverts appear a little paler. The underbody is a little paler than the mantle and is fairly uniform below the collar to the undertail-coverts. When fresh the feathers are narrowly fringed grey-white, most prominently on the undertail-coverts which are likely to retain these fringes longest. Plumage becomes slightly browner when worn. **Upperwing** Pattern very similar to typical *Apus* swift (especially Common Swift). Secondaries appear as palest area in upperwing, with outer primaries and lesser coverts appearing blackest. However, the whole pattern is very weak. The lesser coverts appear slightly paler than the mantle and the innerwebs of the remiges are paler than the black-brown outerwebs. **Underwing** Typically the remiges appear greyer and paler than on the upperwing and quite similar to the greater coverts with the median and lesser coverts appearing clearly darker. **Tail** The sooty black-brown tail is similar to the remiges and appears highly translucent when spread in sunlight.

Adult female. Females take longer to acquire the full collar (Marin and Stiles 1992). Non-collared females resemble the male in plumage, but are paler below. The face pattern appears very plain but with an impression of a darker cap. Those individuals with a full collar appear blacker in plumage and closer to the adult male. Individuals with partial collars are common.

Juvenile Either lacks the reddish collar or has it only partially. Narrow white fringing to the remiges and smallest wing coverts. Paler than full-collared adults.

Measurements Wing: male (22) 116-130.5 (122.5); female (20) 112-124.5 (119.1). Tail: male (22) 39.5-48.5 (44.80); female (20) 37.5-47.0 (42.7). Tail-fork: male (22) 1.0-3.0 (2.8); female (20) 0.0-4.5 (1.4). Tarsus: male (22) 11.5-13.0 (12.33); female (20) 11.3-13.0 (12.03) (Collins 1972).

GEOGRAPHICAL VARIATION Three races.
  C. r. rutilus (NE part of the South American range, from Venezuela to the Guianas, and on Trinidad.)
  C. r. griseifrons (W Mexico in Nayarit, W Jalisco, W Zacatecas and S Durango) Paler than *brunnitorques* with upperparts appearing sooty as opposed to almost black; the underparts are sooty grey-brown and the feathers of the forehead, the lores and just above the eye are fringed with pale grey. Wing: male 119 (worn at tip); female 121.5-123 (122.2). Tail: male 41 (worn at tip); female 40-42.5 (41.2) (Ridgway 1911).
  C. r. brunnitorques (SE Mexico to Peru) Collins (1972) considered that this subspecies may be synonymous with *rutilus* to which it is extremely similar. Mean

measurements: wing (131) 127.33; tail (136) 45.2; mass 21.32; tarsus 11.25 (Marin and Stiles 1992).

VOICE A vocal species from which a variety of calls can be heard; most calls have a rather insect-like quality. A chattering *kri-kik-kik-kik* rather parakeet-like but higher, a rather wheezing *t-t-t-teee-e* or *t-t-t-tsssss*, a staccato *chi chi chichi* and a very insect-like and accelerating *chi chi chuehu*. Other authors have noted calls that may simply be different transcriptions of the above calls: a rasping *zee-zee-zee* (Edwards 1989) and Stiles and Skutch (1989) describe the calls as 'sharp, scratchy, dry sputtering notes and high-pitched chatters; voice in general drier, more buzzy and metallic than that of *Chaetura*'.

HABITS Gregarious species that can be seen with a variety of sympatric swift species, especially White-collared Swifts with which it presumably sometimes roosts as it can often be seen associating with the large pre-roost flights of that species. Smaller species, notably *Chaetura*, are more loosely associated with. Flocks in the hundreds are not uncommon in the Andes. Often seen at great height over the ground. Appears to be faithful to particular areas where it may be seen day after day. Marin (1993) noted this species primarily foraged in the upper stratifications of mixed swift flocks in Ecuador.

BREEDING Nests between 50cm and 3m above water, in damp (humidity never below c. 95%) shaded sites. The nest is usually sited on a very small ledge, knob or niche, on a vertical or slightly overhanging rock face or similar situation. Marin and Stiles (1992) also noted a nest attached to a buttress under a log. Nests vary in morphology according to 'substrate'. Generally they are half-cupped or half an inverted cone with a shallow central depression. On wider ledges they are more bowl shaped or cup shaped. The nests are built almost entirely of mosses and liverworts, with relatively little mud which is mainly in the base or at the point of attachment. During incubation and/or whilst raising young the adults add dry mosses and liverworts as lining. The mean measurements of the Marin and Stiles sample were: 126mm height, 104mm width, 98.3mm front to back, 31mm depth of depression, 48mm back to front of depression, 66.2mm width of depression. Clutch always two (one egg might sometimes be lost due to egg rolling etc. Marin pers. comm.). Eggs measured (12): 23.62 x 15.49 (Marin and Stiles 1992). Collins (1968a) found nests in sea caves on Trinidad. Recorded breeding at 2300m in Ecuador (Kiff *et al.* 1989).

REFERENCES Collins (1968a and 1972), Fjeldså and Krabbe (1990), Howell & Webb (1995), Kiff *et al.* (1989), Land (1970), Marin and Stiles (1992), Marin (1993), Monroe (1968), Phillips (1962), Ridgway (1911) and Stiles and Skutch (1989).

---

2   TEPUI SWIFT
Cypseloides phelpsi   Plate 1

Other name: *Streptoprocne phelpsi*

IDENTIFICATION Length 16.5cm. Similar to the closely related, but rarely sympatric, Chestnut-collared Swift from which it differs primarily in its larger size, clearly more deeply forked tail (both sexes), brighter orange-chestnut collar that spreads to the ear-coverts and throat (figure 11), a narrow white supercilium and generally blacker plumage.

Figure 11. Heads of adult male Chestnut-collared Swift (left) and adult male Tepui Swift (right). The former species has a duller collar, darker nape and mottled throat patch and ear-coverts, whilst the latter has more a extensive and brighter collar.

The immatures, like Chestnut-collared Swift, lack full collars, but have more extensive pale scalloping on the underparts (not just restricted to the belly). The deeply forked tail excludes all sympatric *Cypseloides*, as does the collar of the adults. The immatures should still show some degree of tail-fork.

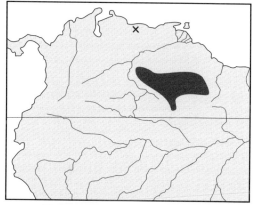

DISTRIBUTION Restricted NC South American range mainly in Venezuela: on Cerros Yapacana, Duida and Sierra Parima, in Amazonas and in the Gran Sabana of Bolivar in the upper Rio Caroni, Cerros Auyan-tepui and Jaua. Also found in the Merume mountains of NW Guyana and in the Brazilian mountains that border Venezuela. One record from N Venezuela where an individual was collected from the Rancho Grande Biological Station in Aragua on 16 February 1960.

There are no population data for this species and it is possibly quite scarce judging by the scarcity of museum specimens.

MOVEMENTS Believed to be resident in the Pantepui region, although the record from Aragua suggests at least occasional wandering.

HABITAT Believed to be restricted to the Pantepui region. The type was collected at an altitude of 1100m.

DESCRIPTION *C. phelpsi*
Adult male. Head Upper head from forehead to rear crown sooty black, except a distinct white supraloral streak (missing in some individuals). Black eye-patch. Rest of head orange-chestnut across whole of throat (including chin) and nape causing collar that is narrowest in the mid-nape and extends down onto the upper breast. **Body** From below the nape the upperparts are sooty black, uniform with the upper head. The underparts from the upper breast are also uniformly sooty black. **Upperwing** Deep black wing but the remiges are slightly paler on the inner-

webs and the outer primaries appear blackest. The coverts appear slightly blacker in flight, especially the lesser coverts and alula. **Underwing** The remiges appear slightly paler and greyer than on the upperwing. **Tail** The tail is very similar to the remiges in that the black appears paler below than above.

Adult female The female is very similar to the male in plumage but the breast can appear a little paler and is mixed with brown in some individuals. Females also take longer to acquire the full collar (Marin pers. comm.).

Juvenile Differs from the adults in lacking the collar. The dark underparts are extensively fringed pale grey (from the throat to the undertail-coverts).

Measurements Wing: male (12) 133-140.5 (136.9); female (18) 129.5-138 (133.9). Tail: male (12) 56.5-66 (61.3); female (18) 56.5-61.5 (58.9). Tail fork: male (12) 7.0-11.5 (9.6); female (18) 5.5-13.0 (9.7). Tarsus: male (12) 12.7-14.5 (13.76); female (18) 12.7-14.3 (Collins 1972). Weight: adult (5) 19-23 (21); juvenile (1) 21 (Dickerman and Phelps 1982).

GEOGRAPHICAL VARIATION None. Monotypic.

VOICE Not known.

HABITS Little is known of the biology and ecology of this species, although it probably resembles other *Cypseloides* species.

BREEDING No information.

REFERENCES Collins (1972), Dickerman and Phelps (1982), Meyer de Schauensee and Phelps (1978).

---

3   BLACK SWIFT
Cypseloides niger   Plates 1 & 2

Other name: *Nephoecetes niger*

IDENTIFICATION Length 18cm. This is the single representative of the genus over much of its range, indeed it is the only *Cypseloides* that is known to occur north of Sinaloa, Mexico. In its northern range it is the only largely dark swift with a forked tail. This tail-fork serves to separate the other members of the genus with which it is sympatric, except the considerably smaller and, in the male, strikingly plumaged Chestnut-collared Swift. However it should be noted that Black Swift shows considerable variation in tail-fork, some individuals (mainly females) showing little or no fork at all. White-chinned and White-fronted Swifts however, have obviously shorter tails, and Spot-fronted Swift is considerably smaller (mean wing 119.6 as opposed to mean wing 151 for the sympatric *costaricensis*). Although often hard to discern in the field, Black Swift further differs from all other *Cypseloides* in showing the head, down to the nape and lower throat, clearly paler than the rest of the plumage. Juvenile Black Swift is very extensively white-fringed on the remiges, rectrices and the body. Marin and Stiles (1992) state that the flight of Black Swift is not unlike White-collared Swift with 'much fast gliding, usually on wings held somewhat below the horizontal', but involving 'more veering and twisting' than that species.

DISTRIBUTION Extensive Nearctic, Neotropical and West Indian range. Occurs in North America south from SE Alaska, where it reaches its most northerly point at the Stikine River, through NW and C British Columbia and SW Alberta south through the western seaboard states to S California, NW Montana, Colorado, C Utah, NC New Mexico. The Central American range is rather broken. In Mexico on Pacific slope from WC Chihuahua to Oaxaca and on Atlantic slope from Hildgo to N Oaxaca. Also to southern interior, then south to Costa Rica through Guatemala and Honduras. In the West Indies it occurs on Cuba, Jamaica, Hispaniola, Puerto Rico, St Kitts, Guadeloupe, Dominica, Martinique, St Lucia and St Vincent. It has occurred as a vagrant on St Croix, Grenada and Barbados. The status on the Isle of Pines is apparently doubtful. Also recorded once 135km off the Pacific coast of Guatemala. It has occurred twice on Bermuda in May, with two birds together on one occasion.

Abundance varies considerably throughout this species' large range. Locally frequent to common in Mexico. In British Columbia the species is fairly common to locally very abundant on the coast and somewhat more local inland, although locally abundant south of the 56°N latitude (Campbell *et al.* 1990). In California the species is considered to be a rare and very localised breeder.

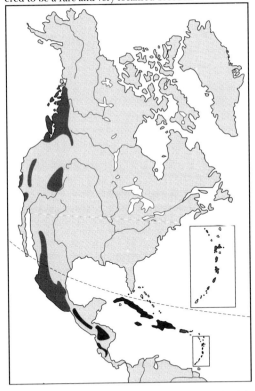

MOVEMENTS In the Lesser Antilles and Puerto Rico it is apparently only a summer visitor between March and September, and some authors consider these birds to winter in Guyana. The North American populations are migratory and are believed to winter in Central America perhaps as far south as S America. In S California the earliest spring record was on 20 April with the majority arriving in mid May and migration continuing until mid June. Concentrations of up to 400 have occasionally been noted. In British Columbia the earliest record is remarkably 1 March. Normally it arrives from mid to late May and con-

tinues into June. Flocks of less than 20 are most common but occasionally thousands have been reported, especially in times of stormy weather. Autumn migration can begin as early as late August, in British Columbia, continuing through September. Few remain by mid September and October, but it has recorded as late as 3 November on the coast. In S California migration occurs from late August, with the majority moving through in September. Numbers in the autumn tend to be smaller than in the spring. The latest record from California is 2 November. Migratory status in Central America not clear. Marin and Stiles note winter absence from Costa Rican breeding grounds and Howell and Webb note that most, if not all, leave Mexico north of the Isthmus from October to February.

HABITAT Primarily a mountainous species in its continental range. Occurs over a variety of open highland habitats, although often seen over rather rugged terrain. In Mexico found between 1500-3300m. In the West Indies tends to prefer forested uplands but it is sometimes recorded in the lowlands especially in poor weather.

DESCRIPTION *C. n. borealis*
Adult Sexes similar (strong white fringing on the underparts has been attributed to adult females, but Marin and Stiles agree with earlier work by Drew and Brooks in showing that there is no sexual dimorphism in adults). **Head** Dark grey upper head with extensive pale grey fringes from forehead to nape. The forehead, lores and the thin line of feathers over the eye are the palest areas. Black eye-patch. Sooty grey-brown ear-coverts and throat uniform, whitest on the chin, lacking the extensive fringing of the upper head. **Body** Upperbody black below nape on mantle and back, with the rump and uppertail-coverts appearing slightly paler, though not as pale as the head and nape. The sooty grey-brown throat extends to the mid breast. Below the mid breast the plumage is black-brown. When fresh the underparts are lightly grey-fringed, most extensively on the undertail-coverts (which appear slightly paler than the belly), and the belly. Even when worn some fringes will remain. **Upperwing** The wings are very typical of *Cypseloides* species, with the innerwing quite noticeably paler than the outerwing. The remiges are blackish, paler on the innerwebs especially the outer margins. The outer primaries appear darker than the inner primaries and secondaries. The greater coverts appear uniform with their respective remiges and the other coverts appear blacker and similar to the saddle. **Underwing** The remiges appear paler and greyer than above though still darkest on the outer primaries. The greater coverts are similar to the remiges with the median and lesser coverts appearing blacker and uniform with the underbody. **Tail** Black above like the remiges and similarly paler below. Marin and Stiles (1992) note that this species shows a slight tendency towards a spine-tipped tail but that the rachises do not protrude and are only slightly stiffened.

Juvenile Differs from the adult in that its head, remiges and particularly its underbody are more extensively white-fringed. On the underbody, below the paler throat area, the pattern is very spotty with broad white tips, black subterminal bars and paler bases.

Measurements Wing: adult male 157.5-175 (165.8); adult female 156-164 (160). Tail: adult male 53-66 (61); adult female 47-58.5 (52.5) (Ridgway 1911). Mean measurements (2): wing 160; tail 53.5; tarsus 12.72 (Navarro *et al.* 1992).

GEOGRAPHICAL VARIATION Three races.

C. n. borealis (North America as far south as SW USA) Described above.

C. n. niger (West Indies and Trinidad) Smaller than *borealis*. In adults the white tips on the underparts are less distinct than in the juveniles of the other subspecies and in some cases they are entirely missing. Within the West Indies some variation in darkness occurs. Previously, *jamaicensis* was recognised south of the mainly Greater Antillean range of *niger*. Wing: male (18) 148-158; female (8) 142-154 (147) (Eisenmann and Lehmann 1962).

C. n. costaricensis (C Mexico to Costa Rica) Distinctly darker and slightly smaller than *borealis*. The adult female has broader white fringing on the underparts. Mean measurements: wing (20) 159.15; tail (17) 53.55; tarsus (18) 13.38; weight (16) 35.71 (Marin and Stiles 1992).

VOICE Described as a harsh, rapid *chi-chi-chi-chik* (Edwards 1972) and as a soft *chip-chip* (Bond 1985).

HABITS This is a gregarious species that has been seen in flocks of thousands in British Columbia in both the breeding season and on migration.

BREEDING Nests on steep canyon walls near to water and in particular waterfalls. Six of seven recorded British Columbian nests were found on shallow ledges under overhanging moss. The nests are pads of moss (some have twigs or pine needles incorporated) bound together with mud, and one measured 140 x 100 and was 20mm deep. A clutch of one or two eggs is normal (Campbell 1990).

REFERENCES Amos (1991), Bond (1985), Campbell *et al.* (1990), Edwards (1972), Eisenmann and Lehmann (1962), Howell and Webb (1995), Marin and Stiles (1992), Navarro *et al.* (1992), Ridgway (1911).

---

4   WHITE-CHESTED SWIFT
     Cypseloides lemosi   Plate 2

IDENTIFICATION Length 14cm. This geographically restricted swift may be identified in its adult plumage from all other members of the genus by its large white pectoral patch. Immatures also show some white feathers in this area but the patch is generally smaller and less conspicuous. It is also unique amongst sympatric *Cypseloides* (indeed amongst all of the genus except for Tepui and Black Swifts) in having a distinctly forked tail in the adults (although this is significantly reduced to a shallow fork in immatures and females). The sympatric Chestnut-collared Swift has only a shallowly forked or occasionally square tail.

DISTRIBUTION Very restricted SW Colombian range. Known only from the Colombian departments of Valle and Cauca in the Upper Cauca Valley. All of the specimens collected were from near Santander. Sight records from Cauca have been made between Cali and Popayan on the right bank of the Rio Ovejas, some 30km south of the Cerro Coronado and 8km south of the village of Pescador. In Valle they were seen on the Cerro de los Cristales, above Cali. In March 1990 two swifts thought to be of this species were observed in a flock of c.150 *Cypseloides* swifts near Archidona, Napo Province in E Ecuador at an altitude of 1060m (Collar *et al.* 1992). Collar *et al.* 1994 note that

several subsequent sightings have been made in this area.

One of the world's rarest swifts. Listed as Vulnerable in the World List of Threatened Birds (Collar *et al.* 1994). In 1962 Eisenmann and Lehmann noted that what appeared to be the preferred habitat was expanding, presumably as a result of deforestation. If the species is closely linked to the habitat that it has been observed over then it is possible that it is under no threat. However, Collar *et al.* (1992) note that practically nothing is known of this species' ecological requirements and as long as these remain unclear the effect of the expanding area of eroded ground within its Colombian range is unquantifiable. This species should certainly be looked for in adjacent areas of Ecuador.

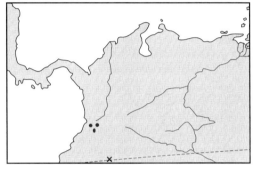

MOVEMENTS Believed to be resident, with records from February, April, May, and October.

HABITAT The species' preferred habitat appears to be highly eroded grassy hills that are sparsely covered with scrubby bushes and trees. On one occasion a flock was observed over flat pasture. All records have been at altitudes between 1050-1300m.

DESCRIPTION *C. lemosi*

Adult Male. Head Upper head blackish with the forehead and forecrown appearing slightly paler as a result of the paler grey-brown fringes. The ear-coverts appear black-brown and rather uniform with the crown. Black eye-patch. The chin and lower throat are lighter than the crown, appearing sooty black-brown. Body The upperparts from the crown to the uppertail-coverts are very uniform being blackish with a faint green gloss. The underparts are characterised by the large white patch covering most of the breast. In one of the specimens, a female, the white extends in a narrow, but clearly discernible, line to just above the legs. In this same individual there is some white fringing on the vent and undertail-coverts. Usually the remainder of the underparts of the adult is uniformly sooty black-brown. Upperwing These are said to be very uniformly blackish with a bluish gloss. Underwing The underwing pattern is presumably very typical as the remiges appear paler below and should contrast with the darker wing coverts. These coverts have grey-brown fringes.

Adult Female Some females may have much reduced or possibly missing chest patches. Like other members of the genus females may show extensive white fringing to body feathers and the marginal coverts.

Juvenile Similar to female in that the patch is much reduced.

Measurements Wing: male (3) 156-160 (158.3); female (4) 155-157 (156). Tail: male (3) 54-62 (57.3); female (4) 47-

55 (50.2). Tail-fork: male (1) 14; female (3) 4-8 (5.7); immature male (2) 5 + 7; immature female (1) 4. Tarsus: male (3) 11.5-12 (11.7); female (4) 11-12 (11.6) (Eisenmann and Lehmann 1962).

GEOGRAPHICAL VARIATION None. Monotypic.

VOICE Not recorded.

HABITS On the few occasions that this species has been encountered it has been in flocks of between 20-25 individuals. The birds have either been alone or with one of the following swift species: White-collared, Chestnut-collared or White-chinned.

BREEDING No information.

REFERENCES Collar *et al.* (1992 and 1994), Eisenmann and Lehmann (1962), Hilty and Brown (1986).

---

5   ROTHSCHILD'S SWIFT
Cypseloides rothschildi   Plate 2

Other name: Giant Swift

IDENTIFICATION Length 15cm. Very typical member of the genus. It is not sympatric with other members of the *niger* superspecies and only the rather different *Aeronautes*, *Chaetura* and *Streptoprocne* swifts might be expected in the same region. This is a very uniformly mid brown species with a square tail.

The very similarly plumaged Sooty Swift can be distinguished by its rather soft tail that lacks any sign of protruding, stiffened shafts (in this respect Rothschild's is unique amongst the genus together with Black and White-chested Swifts), longer wings on average (148-157, mean 154, compared with 135-153, mean 146.8) and its much paler plumage that is greyer anteriorly and lacking any suggestion of whitish areas on the face.

DISTRIBUTION CS South American range. Occurs in NW Argentina in the provinces of Salta, Tucuman and Santiago del Estero. It has also been recorded in S Peru at Cuzco and in S Bolivia. Zimmer (1945) was of the opin-

ion that the species probably occurred at other sites in Peru and therefore the Cuzco record was not a vagrant.

Little is published about the status of this species. Listed by Collar *et al.* (1994) as Near-threatened.

MOVEMENTS Resident throughout the range.

HABITAT Little is known of the behaviour of this species. It is likely, as with other *Cypseloides*, that it is commonest in wet mountainous areas in ravines and gorges with numerous waterfalls for breeding sites.

DESCRIPTION *C. rothschildi*
Adult Sexes similar. **Head** Grey-brown upper head, with black eye-patch and slightly paler forehead. The whole of the forehead and crown are distinctly pale fringed. Slightly paler ear-coverts and throat. The throat is uniformly light grey-brown. **Body** Dark brown mantle, darker than nape and head and marginally darker than rump. Underparts very uniform brown from below the throat to the undertail-coverts. When fresh the plumage can show some pale grey fringes (as with other *Cypseloides* these are probably most pronounced in the female). **Wing** Identical in plumage to *fumigatus*. **Tail** Black-brown above and paler and greyer below. Tail less stiff and lacking the slight shaft projections of *fumigatus*.

Juvenile Close to fresh adult. Some indistinct grey fringing on body and more distinct grey-white fringing on remiges, particularly secondaries.

Measurements Wing: male (6) 148-157 (154); female (1) 147. Tail: male (6) 48-56 (51.3); female (1) 147. Tarsus: 12 (Eisenmann and Lehmann 1962).

GEOGRAPHICAL VARIATION None. Monotypic.

VOICE Not known.

HABITS No information.

BREEDING No information.

REFERENCES Collar *et al.* (1994), Eisenmann and Lehmann (1962), Short (1975).

---

6   SOOTY SWIFT
Cypseloides fumigatus   Plate 2

IDENTIFICATION Length 15cm. Very typical member of the genus, although its sooty black-brown plumage is rather different from the black White-chinned and Spot-fronted Swifts, and lacks the distinctive plumage features of most White-chested, adult Tepui and adult male Chestnut-collared Swifts. It has a square tail and appears smaller and shorter-winged than the sympatric Great Dusky Swift, which is, however, similar in plumage. The very similar, but geographically separated, Rothschild's Swift is slightly larger, clearly paler sooty brown in plumage but without the tendency to show a contrasting paler forehead.

Can be separated from Rothschild's Swift not only by virtue of its darker plumage but also as a result of its smaller size, stiffer tail with a greater tendency to show protruding tail shafts.

DISTRIBUTION Rather small SC South American range. Occurs in E Bolivia in the province of Santa Cruz and in SE Brazil from the provinces of Espirito Santo and Rio de Janeiro south to Rio Grande do Sul. Also occurs in N Argentina in Misiones, and recently discovered in Paraguay where it was found at four different sites in the Oriental region. The species may have occurred in the N Brazilian province of Para, but this is considered to be erroneous by most authors.

This is probably not a rare species but is frequently overlooked as a result of confusion over its identification. Found to be relatively common in those areas in which it occurs in Paraguay.

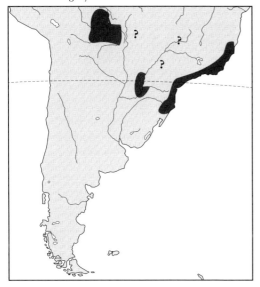

MOVEMENTS Believed to be resident throughout its range.

HABITAT As is typical of the genus, this species is closely associated with waterfalls. It could be found over any habitat in the vicinity of these features.

DESCRIPTION *C. fumigatus*
Adult Sexes similar. **Head** Brown upper head fairly uniform, but with black eye-patch and slightly paler forehead (especially supralorally). Ear-coverts slightly paler and uniform with very uniform throat. **Body** Mantle dark brown, darker than nape and head, and slightly darker than rump. Underparts very uniform brown from throat to undertail-coverts. In fresh plumage some pale grey fringes. **Upperwing** Innerwing paler than outerwing. Outer primaries, alula, lesser coverts and median coverts blackish-brown, fairly uniform with saddle. Inner primaries, secondaries and greater coverts paler brown. Innerwebs of remiges paler than outerwebs especially the outer margins. Very indistinct pale fringing on coverts and remiges. **Underwing** Remiges paler than above and similar to the greater coverts. The median coverts and lesser coverts are much darker and appear uniform with the body. **Tail** Black-brown above and paler and greyer below.

Juvenile Similar to fresh adult with some indistinct grey fringing on body and more distinct grey-white fringing on remiges, particularly secondaries.

Measurements SE Brazil. Wing: male (1) 153. Tail: male (1) 49. Tarsus: (1) 12 (Eisenmann and Lehmann 1962). Brazil. Wing: (4) 142.5-146.5 (146.8). Paraguay. Wing: (4) 135-147. Weight: 40-44 (Belton 1984).

GEOGRAPHICAL VARIATION None. Monotypic.

VOICE Not known.

HABITS Not a very gregarious species. Usually found in small groups of 3 - 6 only, but up to 100 noted at a roost in Paraguay (Belton 1984). It sometimes associates with *Streptoprocne* and *Chaetura* swifts.

BREEDING Breeds on vertical canyon walls, building a conical nest out of moss, pebbles and mud, lined with ferns. The clutch is of one egg (Reboratti 1918, Whitacre 1989).

REFERENCES Belton (1984), Brooks *et al.* (1992), Eisenmann and Lehmann (1962), Sick (1993), Short (1975), Reboratti (1918), Rothschild (1931), Whitacre (1989).

7   SPOT-FRONTED SWIFT
Cypseloides cherriei   Plate 1

Other name: Cherrie's Swift

IDENTIFICATION Length 14cm. A rather small but typical *Cypseloides* swift with short tail, and very striking facial pattern. Once identified as a *Cypseloides* the wholly sooty-black plumage (browner at times) and short square, or slightly rounded, tail will help to exclude the usually rather notch-tailed (and slightly longer-tailed) Chestnut-collared Swift (as will absence of a chestnut collar with regard to collared individuals of that species), and the distinctly fork-tailed Tepui Swift which also has a chestnut collar. In Costa Rica, where this species is sympatric with Black Swift, the latter can be excluded by its considerably larger size, distinctly forked tail (normally), lack of white facial markings and wholly paler head. The main confusion species is the similarly shaped but larger White-chinned Swift. In a good view the prominent white loral and post-ocular spots are diagnostic. The white chin patch is variable in extent and almost missing in some individuals. The flight of this species should prove a useful feature. Marin and Styles (1992) noted its similarity to Chestnut-collared Swift with 'twisting, veering glides on wings usually held below the horizontal and bursts of rapid flapping accompanying changes of speed or direction'. This is in contrast to the direct, rapid flight of White-chinned Swift on typically continuous wing beats which, as Marin and Styles note, gives the bird a rather 'bat-like' style.

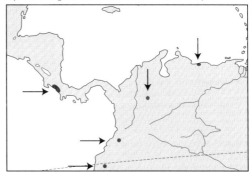

DISTRIBUTION Known from Central America and N South America, where the range is imperfectly known, with few specimens or sightings and breeding only known from three sites. In Central America it is known only from Costa Rica on the Pacific slope of the Cordillera Central, breeding on the upper Rio Tiribi, above Tres Rios, and the Cordillera de Talamanca, at Helechales. In South America the species is known from Aragua in Venezuela where breeding has been proved at Rancho Grande, in the Henri Pittier National Park. Two sites are known from Colombia: one specimen was taken near San Gil to the south of Bucaramanga, in Santander and a pair was collected from Charguayaco, north of Cerro Munchique, Departamento del Cauca, on the western Andean slope. In Ecuador two occupied nests were found and three adults collected as recently as 1989 from the west slope of the Andes at Las Palmas, near Chiriboga in western Pichincha.

This is one of the most enigmatic of the New World swifts. Despite the difficulties in identification it is safe to assume that this is one of the rarer swifts, as indicated by the scarcity of museum specimens and field sightings. However it has been suggested that the species may be considerably more widespread and less scarce than currently thought.

MOVEMENTS In Venezuela and Costa Rica there are records from most months of the year and it is probable that it is a breeding resident in both countries. The Colombian record is from January and it is perhaps more likely that it represents an undiscovered population rather than a vagrant from any of the known populations.

HABITAT All records of this species are from mountainous regions. The Colombian specimen was collected at 1100m, and the species has been caught on several occasions at around 1000m at Rancho Grande. The Costa Rican and Venezuelan sites are both forested mountain ranges with deep valleys. In Costa Rica at least the species shows a preference for deep gorges.

DESCRIPTION *C. cherriei*

Adult Sexes similar. **Head** Upper head sooty-black, except for pure white supraloral spot (does not reach bill), pure white post-ocular streak, distinct pale fringes over black eye-patch (continuous with supraloral patch) and indistinct forehead fringing. Ear-coverts very similar to upper head with throat very slightly paler. Central chin palest. Feathers of throat have broad dark brown terminal bars and paler bases, and accordingly when worn the throat can appear slightly paler and a little mottled. **Body** Head uniform with sooty-black saddle, with rump appearing marginally paler. Underbody is slightly paler with some light fringing when fresh that quickly abrades. **Upperwing** Closer in pattern to the *Apus* swifts with the outerwing appearing clearly darker than the innerwing. The outer remiges are blacker than the inner primaries and secondaries but like the other remiges they have paler innerwebs (especially the outer margins). The greater coverts are fairly uniform with the remiges. The median and lesser coverts and the alula appear blacker. The remiges and wing coverts are very lightly pale grey-fringed. **Underwing** The remiges appear paler and greyer than above and the outer primaries are even darker than the other remiges. The greater coverts are uniform with the remiges. The median and lesser coverts appear clearly blacker and quite uniform with the body. The underwing-coverts are not fringed, but may appear darker at the tip in the field. The leading edge coverts are extensively paler grey-fringed. **Tail** As remiges. Marin and Stiles (1992) note that tail spines are moderately developed.

Adult (brown form) Marin and Stiles (1992) note that a brownish plumage exists in which the feathers of the lower body and vent retain some pale tips. It is not certain whether this is a first summer plumage or a retained and therefore worn juvenile plumage.

Juvenile Like the fresh adult, but with more pronounced underbody fringing that persists longest around the vent and undertail-coverts. The remiges are narrowly but distinctly fringed grey-white. The post-ocular spot is smaller than in the adults.

Measurements Mean: wing (11) 128.90; tail (11) 44.22; weight (11) 22.90; tarsus (8) 12.90 (Marin and Stiles 1992).

GEOGRAPHICAL VARIATION None. Monotypic.

VOICE Described as a rarely heard high, thin chipping. Marin and Stiles (1993) note a continuous stream of harsh wittering clicking notes reminiscent of calls given by White-chinned Swift that may be a kind of rudimentary echolocation.

HABITS Difficulties in identification make this species' habits hard to describe, but it is certainly known to associate with Chestnut-collared Swift in Costa Rica, where it is thought to occur mainly singly or in pairs. The Rancho Grande records are particularly notable in that many concern individuals attracted to the lights at the research station between 19.30 and 21.30 during dense fog. Collins (1980) postulates that this is because *Cypseloides* swifts feed later than other Neotropical genera and therefore encounter problems in finding normal roost sites if adverse weather conditions occur. Marin (1993) notes that this species forages in the higher stratifications when with other swift species.

BREEDING Nests down to 1m above water but often higher (up to 5m). Of the species studied by Marin and Stiles (1992) this one showed widest variation in nest site. Nests were found to be moderately to fully saturated with water, but a few were as dry as the average Chestnut-collared Swift nest. Nests are typically sited on a narrow ledge or knob sloping outward at an angle of c. 45° and protected above by an overhang. Nest morphology varies with nest site. Depending on support (width and angle) and substrate nests are either cup or half-cup shaped. In wettest locations nests more substantial with considerable mud in the base. Normally, however, less substantial with little more than the minimum material necessary to form the nest cup. Nest consists of mosses and liverworts and a lining of dry bamboo leaves and ferns. Two nests found in Ecuador were only 3m from each other (and also in close proximity to the nest of an Andean Cock-of-the-rock *Rupicola peruviana*). Mean dimensions: height 56mm, width 104mm, front to back 68.2mm, depth of depression 18mm, width of depression 63.7mm, depression front to back 39.6mm. Clutch is a single egg. Egg dimensions (5): 25.44 x 16.63 (Marin and Stiles 1992 and 1993).

REFERENCES Collins (1980), Marin (1993), Marin and Stiles (1992 and 1993).

8   WHITE-CHINNED SWIFT
Cypseloides cryptus   Plate 1

Other name: Zimmer's Swift

IDENTIFICATION Length 15cm. This infrequently recorded swift is one of the most troublesome members of a confusing genus, as far as identification is concerned. The plumage is primarily blackish with its most distinctive feature, the small white chin, very hard to discern in the field except in the most exceptional circumstances. With this in mind it is best identified, once the observer is certain that it is a *Cypseloides* swift, by its short square tail and blackish plumage. Black, White-chested, Chestnut-collared and Tepui Swifts all have variably forked tails that are proportionally longer than this species. The first of these species, Black Swift, has blackish plumage with the entire head paler and greyer. In addition, the latter three can all show striking plumage features lacking in this species. With regard to the more troublesome immature or female Chestnut-collared Swifts, which have less apparent tail-forks and often lack any sign of the adult male's bright collar, it has been noted that this species has heavy, direct and more bat-like wing beats than its smaller relatives, and blacker less matt-brown plumage than non-collared Chestnut-collared Swifts. More troublesome, however, is the slightly smaller Spot-fronted Swift which is sympatric with this species in parts of N Venezuela, NE Colombia and the pacific slope of Costa Rica. This species typically has very black plumage and a square tail. However, the tail is longer (and can appear slightly rounded) and if seen in good conditions the prominent white spotting in front of and behind the eye would be diagnostic. Of the allopatric *Cypseloides* species, both Sooty and Rothschild's Swifts are paler and browner and the closely related, and even less well-known, White-fronted Swift is not separable in the field.

In the hand this species can be excluded from Rothschild's, Sooty and Black Swifts by a combination of a proportionally short tail and a long tarsus. Black Swift, at least, also differs in the nostril shape (figure 12): round in this species and ellipitical or narrowly oval in shape and very large in Black (Marin and Stiles 1992).

Figure 12. Nostril shape of Black Swift (left) and White-chinned Swift (right). (Based on Marin and Stiles 1992).

Great Dusky has a different ratio and is considerably larger (mean wing length 160.5 as opposed to 134.4). Tepui Swift has a long tail with a deep fork, even in the female (male 7.0-11.5, female 5.5-13.0) and Chestnut-collared Swift, as well as differing in the tail:tarsus ratio, is considerably smaller (mean wing length 124.4). The greatest problem is in separating White-fronted Swift which is morphologically very close to this species (and may prove to be a subspecies of this taxon). This is best done by examination of the head plumage pattern. White-fronted shows white on the lores and, importantly, the forehead as well as the chin. White-chinned always shows a dark forehead and tends to have sooty post-orbital feathers (as opposed to whitish in White-fronted) and a more abrupt, less tapering, face shape.

DISTRIBUTION Known from N South America and Central America. The type of this species was collected from Inca Mine on the Rio Tavara in Peru and is the only certain specimen from that country. A single specimen was collected in Ecuador but the precise details are lacking. There have also been two field sightings in both the E and W lowlands of Ecuador and it is presumed that this species occurs in the lower altitudes of both Andean slopes. In Colombia the species has been recorded (and probably breeds)from Cordoba in May and Cauca in January and April. In Venezuela recorded both in the northern Cordilleras from Tachira at Burgua and Aragua at Rancho Grande, and in the tepuis south of the Orinoco in Bolivar on Cerros Auyan-tepui and Sororopan-tepui. One record, a specimen collected at the Kaietur Falls, is the only one for Guyana. In Central America north of Costa Rica it occurs mainly on the Caribbean slope lowlands. It has been recorded from Belize at Manatee Lagoon, Honduras at San Esteban and Nicaragua at El Recreo. In Costa Rica it occurs in the San Jose and Terraba regions with definite breeding sites including Zapote de Upala in the Cordillera de Guanacaste; also the Rio Tiribi gorge above Tres Rios and the waterfall on the Rio Sardinal, Rara Avis and in Panama at San Blas, Isla Coiba.

Although almost certainly widely overlooked due to the problems in identification this is one of the rarest Latin American swifts.

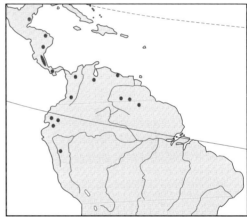

MOVEMENTS Not migratory (Marin pers. comm.).

HABITAT Typical of the genus in that it shows a preference for mountainous areas with gorges and waterfalls, behind or beside which the species roosts and nests.

DESCRIPTION *C. cryptus*
Adult Sexes similar. **Head** Crown to forehead dark brown. Slightly paler on sides of forehead, lores and narrow line over eye due to off-white fringes. Ear-coverts very similar to crown and slightly darker than the throat. Dark brown throat palest, sometimes white, on chin. Equally this pale area may be missing. Some individuals have clearly darker shaft streaks which can make the throat appear rather patchy; this is particularly enhanced by the paler feather bases which are revealed when worn. Black eye-patch. The area immediately behind the eye can be slightly paler than the adjacent areas of the head. **Body** Head and nape appear slightly paler than the very black-brown remainder of the upperbody. Underparts slightly paler than the upperparts. The whole of the throat looks paler and more mottled than the black-brown remainder of the under-

99

body. **Upperwing** Rather uniform and a little blacker than the mantle. The wing is blackest on the outerwebs of the remiges and the median coverts, lesser coverts and alula. The secondary greater coverts are very slightly paler and fairly uniform with the secondaries. The innerwebs of the remiges, and in particular the tertials, are clearly paler (medium brown) than the outerwebs. When fresh some pale fringes can be seen especially on the tips of the remiges. **Underwing** Remiges paler than above strongly contrasting with the darker median and lesser coverts, but similar to the greater coverts. **Tail** Similar, above and below, to the remiges. Tail spines well developed (Marin and Stiles 1992).

As with Spot-fronted Swift adults have a black or brown plumage. The white chin patch is smaller in this brown plumage. The brown plumage may be retained juvenile plumage.

Juvenile Dark brown below and slightly blacker above. Underparts white-tipped from the breast, the tips being particularly broad on the abdomen and vent.

Measurements Wing: 131-140 (135.4). Tail: 41.5-48 (44.4). Tarsus: 16 (Zimmer 1945). Colombia. Wing: male 142 + 143; female 136 + 140. Tail: male 43 + 49; female 40 + 44. Tarsus: male 15 + 16; female 15 + 16 (Eisenmann and Lehmann 1962). Costa Rica. Wing (mean): (14) 136.64. Tail (mean): (10) 44.02. Mass (mean): (13) 35.27 (Marin and Stiles 1992).

GEOGRAPHICAL VARIATION Monotypic, but White-fronted Swift may be a race of this species.

VOICE Noted as making sharp chips and more melodious chirping notes; also explosive, staccato clicking notes (Stiles and Skutch 1989). Marin and Stiles (1992) noted rapid strings of sharp clicking notes from a captive individual and speculated that these might be used in echolocation.

HABITS It has been recorded in Colombia associating with White-chested Swift and it is likely that, like other *Cypseloides* species, it occurs with other members of its genus, *Streptoprocne* swifts and, less frequently, other genera. In Costa Rica not believed to form large flocks. Occurs at altitudes from sea-level to 2000m. Noted (Marin 1993) to fly in the higher stratifications when present with other swift species.

BREEDING White-chinned Swift places its nests either in small cracks or on rock ledges or knobs beneath overhangs. The site is generally near falling water, protected by permanently wet *Pilea* sp. or less frequently mosses or liverworts. Nests remain permanently saturated. Typically the nest is a shallow bowl or half-cup with the following mean dimensions: 76.7mm from front to back, 106.9mm at its greatest width, 48mm in height; cup, 49.6mm from front to back and 65.4mm wide with a depth of 13.5mm. It is made from mosses and liverworts with the greatest amount of mud in the base. Clutch is a single egg. Egg dimensions (1): 28.33 x 19.41 (Marin and Stiles 1992).

REFERENCES Hilty and Brown (1986), Eisenmann and Lehmann (1962), Marin (1993), Marin and Stiles (1992), Zimmer (1945).

9   WHITE-FRONTED SWIFT
Cypseloides storeri   Plate 1

IDENTIFICATION Length 14cm. This is one of the least known of all swift species. Within the genus it closely resembles the allopatric White-chinned Swift and is probably only sympatric with two members of the genus: Black and Chestnut-collared Swifts. The very uniform plumage of this species, which only varies from sooty-brown on the chin, lores and forehead where the pale grey-white fringes of the feathers produce a frosted effect, is different from full-collared Chestnut-collared Swifts and the heavily white-fringed juveniles of Black Swift, and slightly less from the adults with their distinctly paler grey heads. Most importantly this species has a proportionally shorter tail than either of these species, which is square unless fully spread when it would appear more rounded.

Like White-chinned Swift this species can be diagnostically identified in the hand as one of this pair by calculating the tail:tarsus ratio. In this species this averages 36.46, in White-chinned it averages 34.05 - 36.04.

DISTRIBUTION Very restricted SW Mexican range. Currently known only from the mountains of Michoacan at Tacambaro and Guerrero in the Sierra de Atoyac and from the Sierra de Manantlan Biosphere Reserve, Autlan, Jalisco.

Because of the lack of information about this species it is hard to judge its status. However the lack of specimens and sightings of claimed dark *Cypseloides* from the known locations suggest that unless it has been mistaken for Chestnut-collared Swift in the past this is a genuinely rare species. Only five specimens are known (Peterson and Navarro-Siguenza 1993). Considered to be a Data Deficient Species on the World List of Threatened Birds (Collar *et al.* 1994).

MOVEMENTS Believed to be resident by Navarro *et al.* (1992) and Peterson and Navarro-Siguenza (1993). This viewpoint is considered to be without foundation by Howell (1993b). From the available data White-fronted Swift may be a migrant present in Mexico from July (or May?) to September.

HABITAT Only recorded or collected from mountainous regions between 1500-2500m. Both of the areas in which specimens were collected were forested with numerous ravines and many waterfalls. At Guerrero the forest is montane cloud forest; at Michoacan it is drier and the specimens were collected in a transitional zone between pine-oak and dry, tropical deciduous forest.

DESCRIPTION *C. storeri*

Within the small sample sexual plumage differences appear to be negligible. **Head** Sooty-brown with extensive white or buff-white tips to the feathers of the lores, forehead and chin, giving the appearance of a frosty-white face. The ear-coverts immediately behind the eye are similarly white-tipped. Black eye-patch. The upper head is a little darker than the throat. **Body** Whole of upperparts as crown. The underparts are a little paler than the upperparts, being generally less blackish. **Upperwing** Very typical of the genus with very uniform blackish wings. The outerwebs of both the remiges and the greater coverts are blacker than the innerwebs. The lesser and median coverts are more uniformly black and appear a little darker than both the innerwing and the mantle. **Underwing** Presumably typically paler on the remiges than above with contrast between the uniform greyish remiges and greater coverts and the blacker median and lesser coverts. **Tail** As remiges.

Measurements Wing: (5) 132.0-140.0 (136.8). Tail: (5) 42.6-48.5 (45.5). Tarsus: (5) 15.25-17.29 (16.6). Weight (holotype) 39.5 (Navarro *et al.* 1992).

GEOGRAPHICAL VARIATION None. Monotypic.

VOICE Not recorded.

HABITS Navarro *et al.* (1992) believed that they observed 4 individuals of this species in October 1990 entering a waterfall roost site with large flocks of White-naped and smaller numbers of Chestnut-collared Swifts.

BREEDING Not known.

REFERENCES Collar *et al.* (1994), Howell (1993b), Navarro *et al.* (1992), Peterson and Navarro-Siguenza (1993).

---

10   GREAT DUSKY SWIFT
Cypseloides senex   Plate 3

Other name: *Aerornis senex*

IDENTIFICATION Length 18cm. A rather atypical member of the genus. This is a large species which has a matt brown plumage with the head appearing rather pale, even whitish, on the forehead, lores and chin. The flight is rather fluttery and the bird, although large, does not have the rather graceful aspect of the *Streptoprocne* swifts due to its shorter wings and bulky body. Further differences from the sympatric *Streptoprocne* swifts are the much browner rather than blackish plumage, and from most White-collared Swifts the square tail. Within its genus the large size and bulky appearance are diagnostic. The plumage can be very similar to the much smaller Sooty Swift but that species tends not to be so heavily fringed or appear so pale on the head.

DISTRIBUTION Restricted to SE South America. Occurs in C and S Brazil in the Upper Madeira; Aripuana, Dardanelos and Mato Grosso, the Sierra do Cachimbo; Para, to Sao Paulo; Salto do Itapura. It is also found in the Oriental region of Paraguay and in the NE Argentine province of Misiones. A common species that is locally abundant.

MOVEMENTS Although generally thought to be resident within range Hilty and Brown (1986) suggest that it should be looked for in Colombia east of the Andes and is a possible trans-Amazonian migrant. They quote probable sightings

from Iquitos, Peru in June by Peter Alden. However, Marin (pers. comm.) states that this is not a migratory species.

HABITAT Associated with waterfalls around or behind which it breeds and roosts, often in great abundance. During the day the species feeds high over the forest canopy.

DESCRIPTION *C. senex*

Adult Sexes similar. **Head** Feathers of upper head are chocolate brown, with blackish shaft streaks and distinct pale grey fringes. The throat is similar but the shafts are much blacker than the slightly paler chocolate centres, and the pale grey fringes are similar to the upper head. Across the throat and the ear-coverts the shaft streaks overlap resulting in faint streaking across the feathers. The whole effect of the head is very pale especially on the dense feathers of the lores, forehead and chin where little can be seen of the darker centres and shafts. The nape is similarly pale. **Body** The mantle and upper back are black-brown and clearly darker than the head and nape. The lower back, rump and uppertail-coverts are slightly paler (but still clearly darker than the head) and very lightly pale-fringed. The underparts are clearly darker from the lower throat and are only very slightly paler than the mantle. The undertail-coverts appear very slightly paler and have indistinct paler margins. **Upperwing** The wing is very similar in tone to the mantle and is very uniformly dark. The outer primaries appear slightly darker than the rest of the wing. Innerwebs of the remiges are slightly paler and the inner primaries and secondaries are tipped, especially on the innerwebs. The greater coverts are also paler tipped. The coverts are generally black-brown and uniform. **Underwing** The remiges appear only marginally paler beneath and are slightly paler than the greater coverts. Median and lesser coverts are darker and browner. The leading edge coverts are grey-brown with clearly paler fringes. **Tail** The rectrices are similar above and below to the remiges.

Juvenile Similar to adult. Very narrow pale fringes to remiges. On the underwing, the median coverts are paler than the greater coverts and the outerwebs of the underwing primary coverts are proximally white.

MEASUREMENTS Mean (2): wing 160.5; tail 46.0; tarsus (2) 20.47 (Navarro *et al.* 1992).

GEOGRAPHICAL VARIATION None. Monotypic.

VOICE Distinctive call a *ti-ti-ti* followed by a buzzing *tirr-tshaarr* (Sick 1993).

HABITS A gregarious species often associating with White-collared Swift.

BREEDING Great Dusky Swift is a colonial breeder that nests on horizontal ledges next to or behind waterfalls. The nest is a disc-shaped cone constructed of moss and pebbles held together by mud. The nests are often fully exposed to sunlight (Whitacre 1989, Sick 1993).

REFERENCES Brooks *et al.* (1992), Hilty and Brown (1986), Navarro *et al.* (1992), Sick (1993), Whitacre (1989).

STREPTOPROCNE

Three large species are placed in this Neotropical group. One is endemic to S Mexico, one to SE South America and the other has a wide range from Mexico to the Caribbean and south to Argentina.

Separation from sympatric genera

The large size and distinctive plumages of this genus make recognition straight forward. Smaller sympatric genera can all be easily separated both on plumage features and, most easily, on jizz. *Chaetura* swifts lack the white collar and are less deeply black; with the exception of Short-tailed they all have square tails and lack the power and impression of bulk and large size. *Tachornis* swifts appear tiny and have delicate shimmering wing beats, quite unlike the powerful beats of these species. *Panyptila* swifts, although sharing some similar plumage traits, are far more streamlined with thin wings and long deeply forked tails that are usually held closed, appearing very spike-like. The *Aeronautes* swifts are more similar in tail shape but are obviously far slighter with very rapid shimmering wing beats and thin wings, especially the outer primaries.

Figure 13. Fresh (left), slightly worn (centre) and worn (right) tails of White-collared Swift showing formation of tail-spines and effect of wear on tail shape. (Based on Orr 1963).

Cypseloides swifts offer the greatest area of confusion for inexperienced observers. If seen well, when plumage tones and patterns can be seen, then separation is straightforward. Only White-chested Swift shares with this genus areas of white around the upper chest. At height the *Streptoprocne* swifts appear longer and broader winged and broader-tailed than any of the *Cypseloides*. Additionally, the tail can typically be seen to be forked in White-collared Swift which excludes Sooty, Rothschild's and Great Dusky Swifts and the shorter-tailed White-chinned and Spot-fronted Swifts. The typically square-tailed White-naped and Biscutate Swifts have large broad tails that can appear slightly notched. Most importantly the flight of the two genera is quite different. *Streptoprocne* swifts have a leisurely loose flight action often employing deep wing beats. Furthermore the wings appear rather swept back. In *Cypseloides* swifts the wings appear very stiff and rather down-tilted, held out at right angles to the body. The flight appears rather unsteady with frequent rocking from side to side, and awkward sudden turns interspersed with rapid flickering wing beats. *Cypseloides* swifts can fly more directly, but the impression is still less graceful and relaxed, with more flickering wing beats.

11   WHITE-COLLARED SWIFT
Streptoprocne zonaris   Plate 3

Other names: Antillean Cloud Swift; Ringed Swift

IDENTIFICATION Length 22cm. Throughout most of this species' range it is an easily identified and familiar species. Its large size and distinctive plumage make recognition straightforward. However, the two other members of the genus, Biscutate and White-naped Swifts, are far more taxing, although fortunately their ranges are very restricted. Separation of these species is dealt with in de-

tail under their accounts. It is sufficient to say here that this is the only *Streptoprocne* that can show a complete white collar and an obviously forked tail.

DISTRIBUTION Extensive Neotropical range. In the Greater Antilles it occurs on Cuba (including the Isle of Pines), Hispaniola, Jamaica, Grenada, Grenadines, and Saba. Also found on Trinidad. Has occurred on one occasion on Vieques Island, Puerto Rico. In Central America it occurs in Mexico from S Tamaulipas on the Gulf coast and from Guerrero on the Pacific coast, and throughout Guatemala, Belize, Honduras, Nicaragua, Costa Rica and Panama. Found throughout Colombia west of the Andes, with most sightings east of the Andes in the foothills or the adjacent lowlands. Found throughout Venezuela north of the Orinoco, and more locally in the cerros of Bolivar and Amazonas south of it. Found in Guyana, and Brazil in NW Amazonia, Roraima and the Mato Grosso. In the east of Brazil it is found from Minas Gerais to Rio Grande. South of Colombia the species occurs in the Andes and adjacent lowlands of Ecuador and Peru, into E Bolivia and as far south as Mendoza in WC Argentina. May occur (or may have occurred) in Uruguay, as an individual was said to have been collected near Montevideo in 1934 (Cuello and Gerzenstein 1962).

Up to 1987 there had been a handful of records from the southern USA from Texas west to Florida. A sight record from Point St. George, Del Norte County, California on 21 May 1982 (Erickson *et al.* 1989) was accepted by the California Birds Records Committee (Parkes 1993). A well documented individual was photographed in December 1987 at Freeport, Texas and it was noted that during this period a number of large high-flying swifts was seen and it was thought possible that a flock of White-collared Swifts might be present. All N American records are assumed to relate to *mexicana*.

Throughout most of its range a locally common to abundant species, frequently outnumbering all other swifts. However, the race *pallidifrons* of the West Indies is rather scarce in places. Described as a casual visitor to the Isle of Pines, Vieques and Saba and a rare seasonal visitor to Grenada. In parts of Central America it is rare on the Pacific coast. Marin (1993) noted this species was almost equally abundant at all elevations in Ecuador.

MOVEMENTS In many parts of this species' range seasonal movements occur, especially altitudinal. In Peru it is always present in the Andes, but also moves into the coastal lowlands from May to August. This behaviour is common throughout the lowlands adjacent to the Andes. Rare visitor during May to October to Grenada. Otherwise the species can be said to be a resident, though with some dispersal outside the breeding season.

HABITAT Occurs over a variety of habitats in its large range, both lowland and highland, coastal and interior, forested and agricultural. In many parts of the range the species is fundamentally a montane and submontane species occurring in the lowlands only seasonally or as part of large daily feeding movements. Occurs in the Andes to 4350m. Tends to be far less abundant in arid areas. This is no doubt as a consequence of the importance of waterfalls for nesting sites. Only breeds on Caribbean islands with mountains higher than 2000m. It has been postulated that on islands without a great altitudinal range this species is outcompeted by Black Swift. Found to have the widest elevational range of Ecuadorian swifts: western slope 300-4000m, eastern slope 300-4200m (Marin 1993). The same study revealed that this species fed higher than other Ecuadorian swift species when feeding in the same vicinity.

DESCRIPTION *S. z. subtropicalis*
Adult Sexes similar (but see below). **Head** Deeply black with a slight gloss on the upper head from forehead to nape. Lores and forehead can appear slightly paler. Throat entirely black to mid or lower throat. Black eye-patch. **Body** Broad white collar encircles body without break. It is widest across the breast narrowing over side of neck and nape, contrasting highly with black plumage. Rest of body plumage is, like the head, deeply black with a slight blue gloss most apparent on the mantle and the breast. The feather bases are dusky and therefore, when worn, the plumage may not look so deeply black; this can be particularly apparent on the underparts. **Upperwing** Remiges deeply black on the outerwebs, paler grey-brown on the innerwebs. The outer primaries appear blacker than the inner primaries. The coverts are also deeply black, appearing slightly darker on the lesser coverts and alula. Marginal coverts white-fringed. **Underwing** Remiges appear distinctly paler and contrast with the slightly darker greater coverts and greatly with the median and lesser coverts. The leading edge coverts are black-brown with a greyer fringe. **Tail** As remiges, black-brown and slightly paler below. Strongly stiffened rachises can protrude when rectrices are worn (see figure 13).

Female It is possible that the collar in the female may be incomplete or missing in some individuals. This was not found to be the case in the study by Marin and Stiles (1992).

Juvenile The plumage of the juvenile is generally duller, a rather sooty-black lacking the gloss and deepness of the adult. Furthermore the remiges and body feathers when fresh have narrow grey or white fringes (most apparent on the underparts, head and lower back/rump). The collar is reduced in width and clarity, due to the more extensive dark feather bases and centres. In some the collar is incomplete, mottled or, in extreme examples, absent. The tips of the rectrices are often white-fringed.

Measurements Wing: 194-207 (Parkes 1993).

GEOGRAPHICAL VARIATION Russell (1964) and Monroe (1968) considered that the Honduran populations were intermediate in features between *mexicana* and *albicincta*, being closer to the former in size but closer to the latter in blackness of plumage. However, Parkes (1993) notes that both authors were mistaken as to the racial identity of the 'southern element' as they had accepted the wide ranging *albicincta* proposed by Peters (1940). Parkes also notes intergradation in Belize and possibly part of Guatemala. Nine races are recognised here.

S. z. subtropicalis (mid elevations from Santa Marta, Colombia and Merida, Venezuela south to Peru) Described above.

S. z. albicincta (Venezuela south of the cordilleras and Andes and Guyana) Broad breast band, but very restricted fringing (even in immatures) which tends to be brownish as opposed to white. Small race, but blacker than *minor*. Wing: 182-197 (Parkes 1993).

S. z. mexicana (From S Mexico, to Belize where it is believed to intergrade with *bouchellii*) Large race. Underparts greyer than other races. Broad white breast band. Entire underparts lightly fringed in fresh immatures. Wing: (20) 204-217.5 (Parkes 1993).

S. z. bouchellii (Nicaragua to Panama) Blacker than *mexicana* with narrower breast band (especially in immatures, which also have fringed throats). Wing: (14) 195-210 (Parkes 1993). Mean measurements (40): wing 201.86; tail 69.10: mass (44) 103.75 (Marin and Stiles 1992).

S. z. minor (Mountainous regions of coastal Venezuela; frequently strays to Trinidad) Narrow breast band (white restricted to tips and can appear mottled even in the adult). Greyish throat. Immatures extensively white-fringed on underparts, uppertail-coverts and marginal coverts. Wing: 180-197 (Parkes 1993).

S. z. pallidifrons (Greater and Lesser Antilles) Chin, forehead and streak above eye contrastingly pale. Pale feathers also sometimes on throat. Marginal coverts buff-fringed, most conspicuous in immatures which lack other fringing. Wing: (17) 193-206.5 (Parkes 1993).

S. z. altissima (Colombia and Ecuador at high elevations) Very large. Broad white collar not narrowing on nape or sides of neck. Blackish plumage, slightly paler throat. White-fringed marginal coverts. Well defined fringing on underparts in immatures. Wing: (5) 223-232 (226.5) (Parkes 1993).

S. z. zonaris (Lowlands in S Brazil, Bolivia and W Argentina) Large, deep black race with no fringing to marginal coverts and white on nape restricted to tips and therefore distinct narrowing or mottling can be apparent in this area. Less deeply black than *subtropicalis*. Immature practically unfringed. Wing: (18) 204-221 (Parkes 1993).

S. z. kuenzeli (Andean Bolivia and NW Argentina at high elevations) Underparts and throat brownish with marginal coverts distinctly fringed and often with a narrow white line of feathers over eye. Whole of plumage in immatures heavily white-fringed, with white on lores and over eye. Wing: (4) 206-211 (207) (Parkes 1993).

VOICE At times very vocal, especially when it occurs in large numbers. The presence of a large flock at altitude gathering to roost is often revealed as a result of large numbers calling simultaneously. In these cases, as Fjeldså and Krabbe (1990) state, they may sound like a distant parakeet flock. A wide variety of insect-like trills and buzzes, chattering and screeches can be heard; a nasal twitter *chee chee chee*, *whiss whiss*, and scratchy but not shrill *tseet tchee*, and *chirrio* (Fjeldså and Krabbe 1990), a loud somewhat musical *cleek*, *cleek*, *cleek* (Edwards 1972), and harsh

mini-scream *shra-shra-shra...* repeated several times that lacks the harsh piercing tone of the Common Swift. Bond (1985) transcribes the last call which he finds reminiscent of Common Swifts, as *screee-screee*.

HABITS Highly social species that can be observed in flocks numbering many hundreds, especially in the evening when they gather to roost communally in caverns (including sea caves) and behind waterfalls. Other swift species, *Cypseloides*, *Chaetura*, *Panyptila* and *Aeronautes*, often associate with this species. Daily foraging takes the bird over a very large range, and local movements linked to the weather are a well recognised behavioural feature. In the northern cordilleras of Venezuela this can be observed very readily. Preferred feeding in this area appears to be over cloud forest at around 1000m, but with onset of poor weather a very easily observed and marked movement occurs into the lowlands, which because of the orographic nature of the rainfall often remain dry. Once in the lowlands the species will feed even at low level over major cities. Feeding occurs from low level, especially in poor weather, to great height.

Figure 14. Nest of White-collared Swift. (Based on Marin and Stiles 1992).

BREEDING Nests mainly in caves next to or behind waterfalls. Caves vary from holes of 30cm deep and 30cm wide at the entrance to over 1m at the entrance and several metres deep. Hanging vegetation often screens entrances. Nests singularly or in colonies of up to 12 pairs. Marin and Stiles (1992) found four nests 1-3m apart and 3m above the water level in the wall of a dark river gorge. Marin and Stiles (1992) note two nest site requirements: darkness and a more or less horizontal, solid surface such as a shelf, ledge, flat rock or cave floor upon which the nest is placed. Nests are often placed in the zone of fine spray from falling water but not where water directly falls. Usually nests are disc-shaped pads of mud, rootlets, mosses, liverworts and other vegetable matter with a shallow, saucer-shaped central depression. Mosses and liverworts compose the nest bulk with, on occasions, a lining of fine mosses or plant fibres. Base of nest usually wet, upper nest just moist. One clutch observed by Marin and Stiles (1992) was laid directly onto the soft sand of a narrow shelf below an overhang, in a narrow, dark gorge only 1m above the river. Average dimensions: height 95, depth of depression 10, width of depression 147. Clutch 2. Egg dimensions (4): 35.90 x 23.14 (Marin and Stiles 1992).

REFERENCES Anon. (1988), Beebe (1947), Belton (1984), Bond (1985), Cuello and Gerzenstein (1962), Edwards (1989), Erickson *et al.* (1989), Fjeldså and Krabbe (1990), Hilty and Brown (1986), Kepler (1972), Land (1970), Marin and Stiles (1992), Monroe (1968), National Geographic Society (1987), Parkes (1993), Ridgely (1976), Ridgway (1910), Rowley and Orr (1965), Russell

(1964), Meyer de Schauensee and Phelps (1978), Wetmore (1968), Zimmer (1953).

12 BISCUTATE SWIFT
Streptoprocne biscutata   Plate 3

IDENTIFICATION Length 22cm. This typical *Streptoprocne* swift differs from the sympatric White-collared Swift in its slightly smaller size, square tail and the broken collar with two white patches on the lower throat/upper breast and the nape separated by obvious gaps. The mark on the nape is rather diamond shaped. The complete collar may be lacking in some individuals of White-collared Swift when identification will have to be based upon the difference in tail shape. In White-collared Swift sustained views are often required to discern the often rather shallow tail-fork. Worn, or moulting, White-collared Swifts can show rather square tails. However, these birds will have undergone post-juvenile moult and will show white collars.

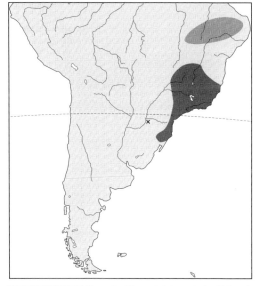

DISTRIBUTION Found with certainty only in C South America. Occurs in E Brazil from Rio Grande do Norte, Piaui, Minas Gerais, Espirito Santo and Rio de Janeiro south to Rio Grande do Sul. Also observed in Misiones, N Argentina and apparently in Paraguay (Olrog 1979). Hilty and Brown (1986) considered that this species might have occurred in the Amazonas region of Colombia but they do not supply any evidence to support this theory.

Traditionally this was considered to be an uncommon species, but large numbers roost in Serido and in 1984 approximately 1200 individuals were present at fifteen cave colonies in the Ibitipoca Mountains of Minas Gerais. Considered to be common to abundant in Rio Grande do Sul.

MOVEMENTS This species' movements are not fully understood. The breeding grounds of those birds roosting in the caverns of Serido from February to October are not known. *Biscutata* breeds in Minas Gerais from October to the end of December, but they arrive as early as August with some, at least, staying until February. The wintering grounds of this population are unknown. It is

possible that *biscutata* is a transequatorial migrant as has been suggested for Great Dusky Swift by Hilty and Brown (1986).

HABITAT A species associated with dry caverns for breeding and roosting, but which frequents waterfalls to feed. At Ibitipoca Park in Minas Gerais it breeds at 1600m.

DESCRIPTION *S. b. biscutata*
Adult Sexes similar. **Head** Sooty black-brown. Distinctly paler on the forehead, lores and line of feathers over eye. Throat black-brown, sometimes appearing slightly paler and closer to forehead in colour. Black eye-patch. **Body** Broken white collar which is broadest on the upper breast and lower throat, extending to sides of head, where it is black across the rear ear-coverts and sides of neck and then white on the nape from the back of the upper ear-coverts. Otherwise the remainder of the body plumage is uniformly sooty black-brown, with some pale grey-white fringing when fresh. **Upperwing** Remiges blackest on the outerwebs, pale grey-brown on the innerwebs. The outerwing appears blacker than the innerwing, but generally the whole effect is rather uniform. **Underwing** Distinctly paler remiges that are fairly uniform with the greater coverts and distinctly paler than the darker median and lesser coverts. Leading edge coverts are black-brown with a greyer fringe. **Tail** Similar to the remiges, above and below.

Juvenile The plumage when fresh is fringed grey-white, most noticeable on the belly and also on the tips of the rectrices.

Measurements Wing: male (6) 200-211 (206.8); female (5) 198-210 (206.0). Tail: (6) male 64-70 (67.3); female (5) 64-69 (65.8) (Sick 1991).

GEOGRAPHICAL VARIATION Two races.
   S. b. biscutata (Resident, south of the range of *seridoensis*). Described above.
   S. b. seridoensis (A migrant in NE Brazil at Rio Grande do Norte and Piaui) Very similar to *biscutata*, but is smaller. Wing: male (8) 185-196 (188.6); female (15) 172-1932 (182.8). Tail: male (8) 53-74 (62.4); female (15) 53-72 (59.8). Most of 78 *seridoensis* weighed less than 100g, whereas 165 *biscutata* were above 100g (Sick 1991).

VOICE As White-collared Swift with similarly strident calls. Belton (1984) noted the calls of a flock that he considered to be composed primarily of this species to be *chee chee chee chee chee...* which is how he transcribes the call of White-collared Swift.

HABITS An exceptionally gregarious species that can be found roosting communally during the winter in vast numbers in dry caverns. In August 1978 in Serra do Bico da Arara, in the Serido region of Rio Grande do Norte in Brazil 90,000 - 100,000 roosted in one of what the local people describe as 'swallow grottos'. Readily associates with other swift species, but not apparently White-collared Swift.

BREEDING Nests in large caverns, on horizontal ledges. The nest is composed of lichens, mosses and dry leaves which are formed into a shallow bowl-shaped disc (Whitacre 1989).

REFERENCES Belton (1984), Hilty and Brown (1986), Olrog (1979), Sick (1991 and 1993), Whitacre (1989).

13   WHITE-NAPED SWIFT
Streptoprocne semicollaris   **Plate 3**

IDENTIFICATION Length 22cm. The great size of this species with its powerful flight and strong dimensions, creates a jizz found only in the New World amongst the *Streptoprocne* swifts. For a more detailed analysis of the characters that separate this genus from the smaller genera see genus introduction. Often considered to be the largest of the world's swifts, but in wing length at least, it overlaps with some individuals of the larger races of White-collared Swift. It is best separated from that species by the lack of any white on the underside. Even in those White-collared Swifts that lack a complete collar there is always some white mottling. However, the vast majority of White-collared Swifts will show a complete collar. This can be hard to see when the bird is high overhead, but is usually very distinct especially if the bird is seen against a background. The tail of White-naped Swift usually appears square, or at times very slightly notched. When soaring with the tail fully spread it can appear quite rounded. In most cases White-collared Swift shows a distinct tail-fork, even when the tail is tightly closed. Unlike some species, White-collared has the convenient habit of frequently flexing its tail which allows for easy checking of the tail shape. However some worn individuals have reduced tail-forks that in extreme cases can appear quite square ended. Fortunately juveniles lacking white collars will have forked tails. It is possible that there is some difference in the calls of the two species, but this must be used with caution as it may be the result of differences in transcription only. Indeed Rowley and Orr (1962) noted a difference only in that the calls of White-naped Swift were quieter and shorter. It is also possible that this species has a greater tendency towards soaring in flight than White-collared Swift.

DISTRIBUTION Near-endemic to W Mexico. Found locally in the highlands of C Mexico from Chihuahua and Sinaloa in the north, south through Nayarit, Hidalgo, Morelos and Mexico. One record from Chiapas, on the Guatemalan border. There has been some speculation that this species may occur to the south of Mexico but there is no such evidence from Guatemala. However, recently recorded in groups of seven and 12, during April-May 1993, from Upper Rio Raspaculo, Belize. An interesting series of records is detailed by Monroe (1968) concerning 'White-naped Swifts' in the Caribbean lowlands of Honduras. These records involve sightings of large flocks of *Streptoprocne* swifts showing white napes on many dates and by several observers between 1952 and 1960. Monroe con-

cludes his discussion of this by stating that the definite identity of the Honduran birds will have to remain a mystery until a specimen can be secured. It is interesting to note that the flocks appeared to be homogeneous and that the observers not only noted the plumage features but also an apparent difference in flight pattern from White-collared Swift.

In the few widely separated sites that this species is regularly known from it is relatively common. However, it appears to be entirely absent from the intervening areas. It is possible that the species goes largely unnoticed due to a combination of the lack of observers in remote areas of highland Mexico and the problems of identification caused by the species rapid high-flying behaviour. If the species were proved to occur in Honduras, it seems likely that it also occurs in intervening Guatemala, and may be rather less scarce than is currently thought.

MOVEMENTS There is no evidence of movement other than in localised feeding groups. The records from Honduras, if they do refer to this species, were from June to February. Observations from the Cuernavaca colony in Morelos, Mexico by Rowley and Orr (1962), indicate that eggs were found in late May, and the birds were presumed to be feeding young in mid June. When it is considered that the Morelos colony is towards the north of the species' range the dates in Honduras do not preclude some form of post-breeding dispersal.

HABITAT Although this species has been recorded at sea-level it is essentially a bird of wild highland landscapes, where it favours cliff faces, deep river gorges and high crags. It has been recorded to 3600m. It has on occasions been observed feeding over towns, but more usually occurs over forested slopes, brushy flatlands and open countryside.

DESCRIPTION *S. semicollaris*
Adult Sexes similar. **Head** Sooty black-brown, appearing distinctly paler on the forehead and slightly paler on the lores, throat and chin. Black eye-patch. **Body** A pure white band on the nape extends from the back of the upper ear-coverts. These feathers have sooty bases. The remainder of the upperparts are uniformly sooty black-brown like the crown. The underparts are slightly paler and more brown than the upperparts. It is likely that the body feathers are lightly grey-white fringed when fresh, as in White-collared Swift. **Upperwing** Black-brown remiges darker on the outerwebs and pale grey-brown on the innerwebs. The outer primaries are slightly blacker than the inner primaries. The coverts are darker towards the leading edge of the wing. **Underwing** The remiges are paler and greyer beneath contrasting more with the distinctly darker lesser and median coverts though less so with the similarly coloured greater coverts. **Tail** Similar to the remiges. The rectrix spines are very strong and when the surrounding feather webs become worn, tail spines are apparent.

Juvenile It is likely that the plumage of the juvenile like other *Streptoprocne* swifts is lightly white-fringed on the body with less on the rectrice tips and coverts. The white nape band is also possibly reduced.

Measurements Wing: 228-233 (222.3). Tail: 73 (Ridgway 1911).

GEOGRAPHICAL VARIATION None. Monotypic.

VOICE Like White-collared Swift this species is generally silent when single and very vocal when encountered in a group. Rowley and Orr (1962) describe the call as *cree-cree-cree*, delivered both in flight and when calling from the walls of breeding caves. Edwards (1972) states that the call is a rather musical *cleek, cleek, cleek*, and later (1989) as a harsh extended *chi-ik, chi-ik*.

HABITS A gregarious species which has been observed in colonies of up to 200 individuals. It has been observed roosting with Chestnut-collared Swift, and possibly the little known White-fronted Swift, at a waterfall near Tacambaro, Michoacan. In the daily course of feeding, a breeding colony observed in Mexico, was seen several miles away from the nesting site.

BREEDING White-naped Swift is a cave nester. The main study (Rowley and Orr 1962) of nesting habits describes a colony of sixteen pairs in a cave whose floor, mainly around the entrance, was partly flooded by a river where the nests were sited on ledges in small cavities off of the main chamber. The nest is typically a shallow depression (made by the bird) in dry sand, lacking any material or saliva. However, Whitacre (1989) notes that nests similar to those of White-collared Swift are sometimes made. Of 32 nests studied by Whitacre (1989) 26 had nest structures. These were composed mainly of mud with substantial plant material and were either entirely circular with shallow depressions or 40-75% circular with truncation towards the slope of the substrate. 18 out of 27 nests were lined with a variety of plant matter. Normal clutches consist of two eggs. Most nests were on firm horizontal substrates.

REFERENCES Edwards (1972, 1989), Howell (1993a), Howell and Webb (1995), Land (1970), Monroe (1968), Ridgway (1911), Rowley and Orr (1962), Whitacre (1989).

COLLOCALIA

A wide ranging genus of many tiny to small species whose taxonomy is still poorly understood.

Field identification of swiftlets is notoriously difficult and many species are not safely identifiable in the field. However, the following accounts give a number of pointers, including some of a rather subjective nature. In certain areas such as Borneo, the Philippines and New Guinea field identification of some species is not possible.

Despite this pessimistic introduction many species can be safely identified in the field. Fortunately some species either occur alone on oceanic islands or with just one other species. The wide ranging Glossy Swiftlet is always easily identified, except in the few areas where it occurs in sympatry with Linchi Swiftlet. At present we understand too little about the gloss of these two closely related species and the normally blue-glossed Glossy Swiftlet can sometimes show greenish-glossed plumage.

In other areas such as the south Pacific, White-rumped Swiftlets are easily separated from the dark-rumped Uniform Swiftlets. Significantly larger species such as Whitehead's and Giant Swiftlets should also be apparent amongst their smaller relatives. By highlighting potential differences this guide will hopefully contribute towards our understanding of this most difficult group.

Field observers should concentrate on plumage features such as rump and throat colour, and ideally, if more than one species is present, depth of tail-fork, size and structure. Larger species will typically have a more laboured or relaxed flight compared with smaller species. However, as with all swifts, a number of variables can lead to inconsistencies in flight patterns.

14 GIANT SWIFTLET
Collocalia gigas   Plate 4

Other names: Waterfall Swift; *Hydrochous gigas*

IDENTIFICATION Length 16cm. This rare swiftlet's atypically large size and distinctive flight action make it comparatively distinctive. One of the darkest and most uniformly plumaged swiftlets, it is wholly blackish above and dusky brown below. Considerably larger than all other swiftlets and when seen in silhouette is more likely to be mistaken for House Swift. Wings appear broader, more triangular, than other swiftlets and the body bulkier so that even when it is not seen in direct comparison with smaller relatives it is likely to appear clearly larger. Flight is very smooth and direct with deep wing beats and the wings held only just below the horizontal (unlike other members of the genus). Lacks the unsteady rocking typical of other swiftlets, and flight has been likened to that of House Swift. The tail can be held tightly closed for long periods without showing any sign of the fork.

DISTRIBUTION Poorly understood. Known with certainty to breed on Java. On Java, restricted to the west in Salak, Pelabuhanratu, Pangrango and Patuha. Has also been recorded from Sumatra, West Malaysia and Borneo. Recorded in West Malaysia on at least 8 occasions since the type was collected at Semangkok Pass in 1900; at Fraser's Hill on five occasions 1968-1973, at Genting Simpah in March 1969 and in the Negri Simbilan lowlands in October 1965. In Sumatra two specimens were collected from Solok, Padang Highlands in 1914 and although it has been said that there are no reliable sight records the species has been recorded from Gunung Kerinci, west Sumatra (Higgins *et al.* 1989). In Borneo, Smythies (1981), cites records from Mount Sepali, Kanowit district May 1952; Santubong, August 1956; Cape Ngosong, July 1960; Klias, September 1960 (5 individuals) and Galisa'an Island,

Sandakan, July 1965 (2 individuals). Since these records the slopes of Mount Kinabalu have become known as a site for the species.

Rare and local. Listed by BirdLife International as a near-threatened species (Collar *et al.* 1994). Little is known about its exact status, but it is known only from a handful of sites. No accurate population counts have been made, even from well known sites. However 60-80 were noted together at the Cibeureum waterfall in Gunung Gede-Pangrango National Park, West Java in August 1985 (King 1987).

MOVEMENTS This is an area of considerable mystery. All records from Fraser's Hill have occurred during autumn and at night, trapped with certain migrants of other species. Although Wells (1992) stated that these were probably disturbed from nearby roosts, the species has not apparently been recorded during the day at the site. Likewise, records from both Borneo and Sumatra do not include certain breeding records.

HABITAT Primarily associated with waterfalls in mountainous rainforest regions.

DESCRIPTION *C. gigas*
Adult Sexes similar. **Head** Upper head black-brown slightly glossed. Forehead and narrow line of feathers above eye are greyer-fringed, appearing slightly, but not contrastingly, paler. Ear-coverts become slightly paler further from crown. The whole of the throat is mid brown or grey-brown and slightly mottled due to the appearance of some feather bases. Black eye-patch is whitish-based. **Body** Upperparts are completely glossy black-brown and uniform with head. Rami are whitish. Underparts appear fairly uniform and mid brown, in most individuals slightly paler-throated. Undertail-coverts have distinct greyish-white margins. **Upperwing** Uniformly black-brown like upperparts and showing little contrast between tracts. **Underwing** Remiges appear paler and greyer than above. Some contrast between the dark glossy grey remiges and the dark grey coverts, especially the rather blackish lesser coverts. Leading edge coverts appear slightly pale grey-fringed. **Tail** Very similar to remiges above and below showing little or no contrast with tail-coverts.

Juvenile/First winter Differs from adult as a result of the less prominent grey-white margins to the undertail-coverts and the more pronounced white on the concealed feather barbs on the body feathers. Structurally, first year birds differ as the central pair of rectrices have asymmetrical outer and innerwebs at the tips (which are symmetrical in adults; see figure 15).

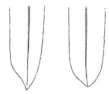

Figure 15. Tail feathers of Giant Swiftlet. Note more rounded profile of adult (right) than juvenile (left). (Based on Somadikarta 1968).

Measurements Males. Java (25): wing 142.5-158 (150.1); tail 58-64 (61.5); tail-fork (20) 6.5-13.0 (9.6). Sumatra (1): wing 155; tail 64.5; tail-fork 9.0. Females. Peninsular Malaysia (type): wing 153.5; tail 62; tail-fork 11. Java (21): wing 145.0-155.5 (150.3); tail 59.0-66.0 (61.9); tail-fork 6.0-14.0 (9.2). Sumatra (1): wing 159.0 (Somadikarta 1968). No statistically significant differences between male and female measurements. Weight: 35-39 (Cranbrook 1984).

GEOGRAPHICAL VARIATION None. Monotypic.

VOICE Recorded as 'sharp wicker' (Medway and Wells 1976). King (1987) recorded a loud twittering... readily audible above the roar of the waterfall. Does not produce an echolocating call.

HABITS Gregarious with 60-80 birds noted together. On Java regularly occurs with Brown Needletail and less frequently with Linchi Swiftlet. At a West Javan breeding site roosting occurs both behind, and on cliffs next to, the waterfall. When breeding has ended, the site is rapidly vacated at dawn and returned to after sunset. However,

on occasions the birds can be seen close to the site after dawn and high over adjacent forest at any time during the day, as is the case at other regular sites. It has been noted that the species is more likely to be seen during the day on Java during the rainy season (October-May). Studies proving that this species does not echolocate revealed that it has acute vision in poor light conditions and is therefore most likely mainly a crepuscular feeder. These habits lead to occasional associations with bat flocks.

BREEDING Medway and Wells (1976) describe a nest from peninsular Malaysia attributed to this species as a truncated cone of liverworts and small amounts of mosses, bound together with feathers and saliva, with which it was also attached to the rock substrate. It measured 9cm high, the egg chamber 3.7cm across x 1.2cm deep, and was sited on a small rock ledge within the spray zone of a waterfall. The single egg is described as glossless white, measuring 29 x 19mm, comparable with 26-31.5 x 17.7-18.1mm for eight eggs from Java. Hoogerwerf (1949b) describes the nest as a sturdy bowl with a rather thick wall and a shallow cup constructed mainly of dark roots and fibrous material, covered with moss on the outside. Parts of the nest had glue-like spots of hardened saliva. The inside measurements were 7 x 7.5cm in width and 2cm in depth. The outside measurements were 9cm at its widest (upper) part and about 6.5cm in height. Both of the eggs were blunt-ended, thin-shelled, rough-textured, and dull white in colour. The two eggs measured 30 x 19.1mm and 27 x 17.7mm.

REFERENCES Becking (1971), Collar *et al.* (1994), Cranbrook (1984), Higgins *et al.* (1989), Hoogerwerf (1949), King (1987), Medway and Wells (1976), Somadikarta (1968), Smythies (1981), Wells (1992).

15   GLOSSY SWIFTLET
Collocalia esculenta   Plate 4

Other names: White-bellied Swiftlet; Philippine Swiftlet, Grey-rumped Swiftlet (races *septentrionalis* and *marginata*)

IDENTIFICATION Length 9-10cm. This tiny species (although in some areas of its range it is medium-sized, such as over 3000m in New Guinea) is one of the easiest swiftlets to identify.

Figure 16. Glossy Swiftlets of the white-rumped race *uropygialis*.

109

Although there is some variation in plumage glossiness (both geographically and with wear) and some races show variable amounts of white in the tail and rump (*marginata, septentrionalis, stresemanni, desiderata, tametamele, natalis* and especially *uropygialis* show extensive white around the rump - figure 16), the underpart pattern is diagnostic (away from areas of sympatry with Linchi and Pygmy Swiftlets): blackish throat, sometimes onto upper breast, and white remainder of underparts except for the black undertail-coverts. The tail appears square, sometimes with a slight cleft when tightly closed. The races *manadensis*, (N Sulawesi) and *spilura* (Bacan), are very dark below (figure 17) and the white abdomen does not serve as a useful feature. However, the small size, square tail, glossy plumage on the upperparts and mottled plumage below serve to separate them from brown swiftlets.

Figure 17. Glossy Swiftlets of the nominate race (left) and the dark-bellied race *spilura* (right).

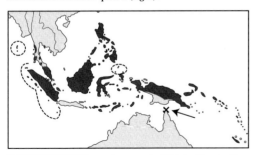

DISTRIBUTION Extensive range in SE Asia, Greater Sundas, Philippines, E Indonesia and New Guinea. Occurs throughout the Andaman and Nicobar islands; Burma, in S Tenasserim; peninsular Thailand from the west coast to the north of Phuket east to the mountains around Khao Luang, and in the far south the coast in Satun province and inland in Yala province; throughout peninsular Malaysia, including the major islands and Singapore; Sumatra, throughout the mainland and satellite islands, including Nias, Batu and Mentawi; widely distributed on Borneo; Christmas Island; Philippines, on Luzon, Cebu, Mindanao, Masbate, Mindoro, Calayan, Sibuyan, Banton, Tablas, Bohol, Camiguin Norte, Babuyan Claro, Polillo, Panay, Palawan, Mactan, Samar, Camiguin Sur, Dinagat, Siargao, and Bongao; Wallacea, on Sumbawa, Flores, Sumba, Alor, Wetar, Romang, Kisar, Damar, Sawu, Timor, Semau, Dao, Roti, Tanahjampea, Kalao, Talaud, Sangihe, Talisei, near Lembeh, throughout Sulawesi, Banggai, Sula, Obi, Buru, Ambon, Seram, Manawoka, Gorong, Watubela, Tayandu, Kai, Bacan, Halmahera, Ternate, Tidore, Rau and Morotai; throughout New Guinea, except the southern Trans-Fly region and the plain of the Idenburg River, in N Irian Jaya; on many of the New Guinea satellite is-

lands: Trobriand Islands, D'Entrecasteaux Archipelago, Louisiade Archipelago; most of the Bismarck Archipelago (not St Matthias group); Admiralty Islands; Bougainville and Buka. A flock of 40 Glossy Swiftlets was recorded in 1967 in the Iron Range, Cape York Peninsula in Queensland, Australia.

Often abundant in suitable areas. In New Guinea tends to be scarce or absent in areas receiving low rainfall.

MOVEMENTS Resident. The Australian record may imply some seasonal wandering but Dickinson (pers. comm.) suggests that it may be more likely to represent a previously undiscovered population.

HABITAT New Guinea from sea-level to 3600m, and even to 4500m in the Carstenz Massif, Irian Jaya, and over a great variety of habitats. In parts of the range, such as the Andamans and Nicobars, closely associated with human dwellings and noted for remarkable tameness. One individual was noted flying into a house at night on Fraser's Hill, Malaysia where it became entangled in a large spider's web (Nach, verbally). Ali and Ripley (1970) state that it flies freely in and out of residential bungalows and office buildings in settlements while hawking, unmindful of the human inmates.

DESCRIPTION *C. e. cyanoptila*
Adult Sexes similar. **Head** Upper head uniformly black-blue with green-blue gloss. Black eye-patch. Line of feathers over eye fringed paler grey. Ear-coverts grey, appearing paler than upper head, giving slightly capped appearance but merging into nape with less cut-off. Ear-coverts at times appear quite rusty in shade. Throat dark grey and quite uniform with ear-coverts though with greater tendency to pale fringing. **Body** Upperbody uniform with head, black-blue with green-blue gloss, from nape to uppertail-coverts. Feathers of upperbody have broad black-blue tips and paler grey bases and off-white basal shaft streaks. Underparts uniform from throat to upper breast, becoming more heavily fringed on lower breast, with solid grey breaking up and blotched pattern emerging on upper belly. Blotched pattern produced by white feathers with dark grey shaft streaks and bases extending midway up shafts. Below blotching there is an area of white extending to vent, although the shaft streaks are grey. Undertail-coverts are black-blue with some narrow white fringing, clearly the darkest tract on the underbody. Extent of solid grey and area of blotching and pure white varies individually. **Upperwing** Uniform, with black-blue wing coverts heavily glossed as mantle. Remiges very uniform (slightly browner than wing coverts) though less glossed than coverts. Outer vanes of innerwebs palest. Some contrast between coverts and remiges at times, with coverts appearing slightly darker. No fringing in upperwing. **Underwing** Remiges paler than above and uniform with greater coverts (greater coverts slightly darker towards tips). Median and lesser coverts clearly blacker than remainder of underwing and adjoining grey breast patch. Leading edge coverts similar to lesser coverts but with indistinct pale fringes. **Tail** Black-blue, paler below with conspicuous gloss.

Juvenile Possibly more heavily fringed on underparts than adult. Pale grey/buff fringes to remiges. Gloss appears more greenish than on adults.

Measurements Wing: (9 Borneo, BMNH) 92.25-103 (98.6). Natunas and Linga Island (4): wing 102-107 (104.6); tail 40-43 (41.3) (Oberholser 1906). Weight: 8-11 (Cranbrook 1984).

GEOGRAPHICAL VARIATION Many races have been described. Thirty-one are recognised here. It is likely that considerable intergration occurs between races. This is certainly the case with the Philippine races (Dickinson 1989b). Two forms, *septentrionalis* and *marginata*, are sometimes considered to constitute a separate species.

C. e. cyanoptila (Malay Peninsula, E Sumatra and the lowlands of Borneo) Described above.

C. e. affinis (Andaman and Nicobar Islands) Paler-throated than *cyanoptila*, and gloss more purple. Narrow pale fringes to rump. Extensive white rami on upperparts especially rump and nape. Wing: 91-100. Tail 36-39 (Baker, in Ali and Ripley 1970). Wing: (5 BMNH) 86-101 (94.5).

C. e. elachyptera (Mergui Archipelago) Similar to *affinis* with rather greenish gloss. Wing: (5) 97-101 (99.4). Tail: 39-42.5 (40.5) (Oberholser 1906). Wing (1 BMNH): 99.8.

C. e. oberholseri (W Sumatra, Nias Island and Mentawi Island) Similar to *cyanoptila*.

C. e. marginata (Philippines, from C Luzon south to Bohol and Leyte) Pale-rumped due to white-fringed rump feathers. From similarly pale-rumped *septentrionalis* by its duller underparts. Dickinson (1989b) did not place birds from Palawan in *marginata*, stating they were an unnamed form said to be darker on the vent than other races and lacking white fringes to rump. Wing: (4 BMNH) 80.6-104 (89.3).

C. e. septentrionalis (Philippines, on Babuyan, Calayan and N Camiguin) Similar to *marginata* but underparts whiter.

C. e. isonota (N Luzon, Philippines) Generally rather more steel-blue in plumage than latter Philippine races. Rump only faintly white-fringed (if at all). In last character similar to *bagobo* from which it differs by whiter underparts. Wing: (6 BMNH) 95-102 (98.4).

C. e. bagobo (Philippines, on Mindanao, Mindoro and Sula Archipelago) Only in alpine regions. Similar to *isonota*, but differs in browner underparts.

The following races have white tail-spots on innerwebs of lateral rectrices and differ from preceding plain-tailed races. However, Salomonsen (1983) noted that individuals of some races can sometimes have plain tails and that many show considerable variation in the size of this feature.

C. e. esculenta (S Sulawesi, north to Gulf of Tomini, Banggai and Sula Islands, Obi, S Moluccas, Kai and Ambon) Highly blue-green glossed upperparts, broad white fringes to grey throat and chest. Flanks, with grey bases and white fringes, can appear pure white. Rectrix spots small, but distinct. Wing: male (17) 94-105 (99.4); female (5) 97-105 (99.8) (Salomonsen 1983). Weight: (2) 7.3-8.0 (Diamond 1972).

C. e. erwini (Upper altitudes, from c.1600m, of mountains in the W New Guinea highlands: Carstensz peaks, Nassau Range, Snow Mountains and Oranje Range) From *nitens* only in size. Wing: male (4) 113-120 (116.3); female (5) 107-118 (112.8); unsexed (7) 108-115 (111.6) (Salomonsen 1983). Limits between *nitens* and this form not well-defined as Salomonsen shows size positively correlates with altitude. Birds with wing lengths between 102-110 said to be intermediate. Mountains peripheral to those described above have *nitens* to high altitudes or birds only approaching *erwini* in size.

C. e. stresemanni (Manus, Admiralty Islands and surrounding small islands; also Rambutyo, Nauna and Los Negros) From other forms in this group in having extensively white rump feathers (dark bases give an untidy appearance). Upperparts duller than *kalili*. Pale grey throat and chest sharply defined from white underparts. Undertail-coverts as *tametamele*. White fringing to innerwebs of secondaries and often greater underwing-coverts. Large, distinct rectrix spots. Wing: male (13) 97-103 (99.6); female (15) 97-103 (99.8) (Salomonsen 1983).

C. e. becki (C Solomon Islands and once Malaita Island) Upperparts as *nitens*, usually lacks white rami on neck. Underparts vary; grey can extend across abdomen with broad white fringing. Small rectrix spots. Wing: (28) 98-107 (101.1) (Mayr 1931).

C. e. natalis (Christmas Island) Resembles Linchi in that the plumage gloss is basically greenish. Pale grey chest and very white abdomen. Broad white fringes to rump, especially on the sides. Wing: (7 BMNH) 96-105.5 (100.2).

C. e. neglecta (Timor and Roti Islands) Although Somadikarta comments (1986) that this race has greenish plumage, Salomonsen (1983) states that the upperparts differ from all other races in being grey-blue without gloss. Further diagnostic features are distinct white fringes to innerwebs of secondaries (except four outer ones), white fringes to some scapulars and narrow white fringes on rump. Breast and lower throat extensively white-fringed giving appearance of contiguous white area with abdomen. Small. Wing: male (7) 95-101 (98.1); female (4) 95-105 (99.5) (Mayr 1944).

C. e. sumbawe (Sumba, Flores and Sumbawa, Lesser Sundas) Small race similar to *esculenta*, differing in duller darker blue (not green-blue) gloss to upperparts and whiter underparts (white flanks and broader white fringing to grey feathers). Wing: (8) 92-95 (93.3) (Salomonsen 1983).

C. e. nitens (Lowland New Guinea, almost to 2500m in some ranges, and Yapen, Karkar Island, Long Island, Misool and Batanta. Has occurred, as a vagrant, to northern Australia and probably also found on Salawati) Similar to *esculenta* but more brilliantly blue-glossed, dark bases to flank feathers more restricted and variably sized rectrix spots (lacking in some). Wing: male (19) 97-110 (101.8); female (15) 97-108 (101.1); 15 unsexed birds 96-104 (99.3) (Salomonsen 1983).

C. e. numforensis (Numfor Island in Geelvink Bay, W New Guinea) From *nitens* by grey flanks and breast sides with broad white fringes broken at tips by distinct dark shaft streaks. Slightly smaller than *nitens*. Wing: male (7) 95-99 (97.0); females (7) 96-100 (98.1) (Salomonsen 1983).

C. e. amethystina (Waigeo Island) Known only from one specimen which is brilliantly dark violet-blue glossed on upperparts, upperwing-coverts and undertail-coverts. Similar below to *numforensis*. Wing: (male) 96 (Salomonsen 1983).

C. e. misimae (Misima and Rossel Islands, Louisiade Archipelago) Similar to *nitens*, duller above (more so than *numforensis*) more bluish sheen and narrow white fringes to rump feathers. Wing: male (6) 99-103 (100.6); female (8) 96-101 (98.4) (Salomonsen 1983).

C. e. tametamele (New Britain and Witu Islands, Bismarck Archipelago, and Bougainville Island, Solomon Islands) From *nitens* by extensive white on lateral rump feathers, dark on chest not well demarcated from sides of breast and grey flanks with extensive white fringes, large white fringes to undertail-coverts which are not glossy. Very large rectrix spots, in some extending to outerwebs. Wing: male (20) 95-103 (98.3); female (12) 97-104 (101.0) (Salomonsen 1983).

C. e. kalili (New Ireland, New Hanover, and Djaul Island, C Bismarck Archipelago) Brighter and deeper blue-glossed than *nitens*, with paler grey chest and throat, rest of underparts as *tametamele*. Rectrix spots very small or absent. Some concealed white rami on neck. Wing: male (8) 95-102 (97.6); female (7) 97-104 (99.6) (Salomonsen 1983).

C. e. perneglecta (Alor, Wetar, Roma, Kisar, Damar and Sawu Islands, Lesser Sundas). Intermediate between *sumbawae* and *neglecta*. Considerably duller and slightly paler above than former (though underparts similar). More glossy and bluer (as opposed to greyer) than latter. Wing: male 91-97 (94.7); female 90-96 (94.2) (Mayr 1944).

C. e. minuta (Tanahjampea and Kalao Islands, Flores Sea) Similar to, but smaller than, *esculenta*. Wing: (3) 93 (Salomonsen 1983).

C. e. manadensis (N Sulawesi) Very dark race with throat, upper breast and flanks black (dull green gloss and quickly abraded narrow white fringes). Upperparts and undertail-coverts have strong green gloss (more so than *esculenta*) becoming more bluish when worn. Rectrix spots large and obvious. Wing: male (4) 99-102 (100.3); female (7) 97-101 (98.6) (Salomonsen 1983).

C. e. spilura (N Moluccas. Birds from Halmahera have been described as *nubila*) From all other races by dark grey (with broad white fringes, broken by dark shaft streaks) as opposed to white abdomen. Throat, chest and flanks brownish-grey, with narrow white fringes, lacking gloss. Strongly dark blue-glossed upperparts and undertail-coverts. Small, indistinct rectrix spots. Wing: (5) 92-97 (Salomonsen 1983).

C. e. spilogaster (Lihir Group and Tatau Island, Bismarck Archipelago) Similar to *kalili*. Differs in more extensive white rami on neck, duller more greenish upperparts, lower breast and abdomen have needle-like shaft streaks. Small but distinct rectrix spots. Wing: male (18) 93-100 (97.5); female (21) 94-102 (97.3) (Salomonsen 1983).

C. e. hypogrammica (Nissan and Green Islands, Bismarck Archipelago) Similar to *spilogaster*, upperparts duller and bluer, lacks white rami on neck. Wing: male (7) 92-100 (96.3); female (6) 96-99 (97.3) (Salomonsen 1983).

C. e. makirensis (San Cristobal Island, Solomons) Similar to *becki*, but duller and darker blue above and abdomen white with flanks having little grey. Wing: (5) 94-96 (94.4) (Salomonsen 1983).

C. e. desiderata (Rennell Island, Solomons) Similar to *makirensis*, duller above with broad white fringes to rump (less than *stresemanni*) and darker chest and throat. Usually lacks white rami on neck. Wing: male (4) 95-97 (96.5); female (7) 96-101 (99.1) (Mayr 1931).

C. e. uropygialis (Santa Cruz, Torres and Banks Islands, New Hebrides) Distinct blackish, slightly glossed, upperparts and distinct white rump band, lacks white rami on neck, dark throat and chest lacking white fringes, clear-cut from white abdomen. Rectrix spots missing or small and indistinct. Wing: male (31) 94-100 (96.5); female (22) 92-102 (97.9) (Salomonsen 1983).

C. e. albidior (Loyalty Islands and New Caledonia) Similar to *uropygialis*, differs in paler grey throat and chest with broad white fringes and large distinct rectrix spots. Wing: male (7) 95-101 (98.0); female (7) 96-102 (98.3) (Salomonsen 1983).

VOICE A rather silent species that does not echolocate. Coates (1985) describes the call as a sharp twitter.

HABITS Gregarious. Often in large, rather loose groups, although not in the hundreds as in some species. Frequently observed with sympatric swiftlets and loosely with other swifts and hirundines. Sometimes observed flying within forests when the canopy and understorey are quite open.

BREEDING According to Coates (1985) a crude cup of moss, rootlets, lichen and other material (obtained from trees) attached with saliva to a vertical or steeply sloping surface. Nest sites include the walls of shallow caves, sheltered rock faces, holes in rocks, under overhanging boulders beside streams, against overhanging banks along streams, under the hanging foliage of pandanuses in beech forest, in a wartime concrete bunker and in an aircraft hangar. Nesting occurs singly and in small colonies. Ali and Ripley (1970) describe the nest as 'black' inedible, and of no commercial value; made entirely of vegetable matter, e.g. moss lichens, casuarina needles, coconut fibre, agglutinated with the birds' saliva. Shallow, flat-bottomed half-cups (c. 7x5cm externally and c. 2cm deep) attached like brackets to the substrate with brownish saliva; usually in vast colonies in sawmill sheds etc., clustered densely along wall near angle with ceiling, the nests often touching and built partly upon one another in a disorderly jumble. Two eggs form the clutch. The egg is blue-white to white. Baker (in Ali and Ripley 1970) states that the average size of 100 eggs was 17.5 x 11.2mm. Nests only in well lit sites, as unable to echolocate.

REFERENCES Ali and Ripley (1971), Coates (1985), Cranbrook (1984), Diamond (1972), Dickinson (1989a and 1989b), Lekagul and Round (1991), McKean (1967), Mayr (1931 and 1944), Medway and Pye (1977), Oberholser (1906), Salomonsen (1983), Somadikarta (1986).

16   LINCHI SWIFTLET
Collocalia linchi   Plate 4

Other name: Cave Swiftlet

IDENTIFICATION Length 10cm. As with Glossy Swiftlet easily separated from the all other swiftlets by its small size, highly glossed plumage and, most importantly, the white abdomen contrasting with the dark chest, throat and undertail-coverts. Identification from Glossy Swiftlet is more problematic and is based upon Linchi Swiftlet having green gloss to the dark feathers whilst Glossy Swiftlet has blue gloss. Fortunately the two species are probably only sympatric in the highlands of Sumatra and on the upper slopes of Mount Kinabalu in Sabah.

In the hand, the hind toe is naked in Linchi Swiftlet and has a feather tuft in Glossy Swiftlet (figure 18).

Figure 18. Legs of Linchi Swiftlet (left) and Glossy Swiftlet (right), showing absence and presence of hind-claw tuft respectively.

However, these seemingly diagnostic features should be used with some caution. Browning (*in litt.*) noted that the type of *dodgei* has a hind-claw tuft and that there is evidence from at least one specimen of *linchi* that it had 'lost a feather from its hind toe'. Additionally Browning notes that some individuals of one of the Philippine races of Glossy Swiftlet, *septentrionalis*, lack the hind-claw tuft. With regard to the plumage colour wide variation in the relative blue/green gloss of Glossy Swiftlet has been noted (Salomonsen 1983) states that *manadensis* is green-glossed and Somadikarta (1986) notes that *natalis* and *neglecta* have greenish plumage. Gloss in all swift species has a tendency to vary with wear and those species that are initially bluish-glossed tend to go through a period of the gloss becoming greenish before it is lost altogether in extreme cases. Additionally, at least some juvenile Glossy Swiftlets are clearly more green in plumage than fresh adults. Linchi Swiftlet has a plain tail lacking the white spots present on the innerwebs of the lateral rectrices on most Glossy Swiftlets to the south and east of 'Stresemann's line' (this stretches between Christmas Island and Java, then north between Sumbawa and Lombok, between the west coast of Sulawesi and Borneo and between the north coast of Sulawesi and Mindanao).

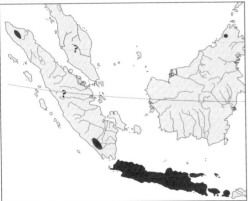

DISTRIBUTION Endemic to the Sundaic region. Throughout Java and the nearby islands of Madura, Bawean, Kangean, Nusa Penida, Bali and Lombok. In Sumatra confined to the Bukit Barisan mountains. Somadikarta (1986) shows the distribution in this long range as the extreme north and south. Despite this apparent discontinuity it seems likely that the species may well be found throughout the mountain chain if white-bellied swiftlets from this island were studied in detail. In Borneo only from the W and SW slopes of Mount Kinabalu, Sabah. Evidence that the species occurs in the Malayan Peninsula is based upon one specimen with an attached label stating 'Malacca' purchased in 1856 by the British Museum from Maison Verreaux. Somadikarta believes the bird would only be found at higher altitudes in the peninsula and presumably this opinion is based upon the species'

preferences in Sumatra.

Common throughout Java and its satellite islands. Appears to be far rarer on Mount Kinabalu and on Sumatra. Somadikarta considered it likely that the Malayan Peninsula population could be extinct.

MOVEMENTS Resident.

HABITAT On Java and surrounding islands it is found from the lowlands to high altitude over both forest and more open wooded country. On Mount Kinabalu found only at high altitudes as on Sumatra.

DESCRIPTION *C. l. linchi*
Adult Sexes similar. **Head** Upper head very uniformly black-brown with a pronounced green gloss. Black eye-patch. Narrow line of feathers over eye has pale grey fringes. Ear-coverts dark grey, paler than upper head (slightly capped appearance) but merging into nape. Dark grey throat uniform with ear-coverts though usually shows more pale grey fringing. **Body** Mantle very slightly paler than head or rump but all are obviously green-glossed and essentially black-brown. Feather pattern of upperparts is broad black-brown highly glossed tips and paler grey bases. Underparts are uniformly dark grey with paler grey fringes from the throat to the lower breast where the fringes become more extensive and the dark grey more blotchy. On the belly the feathers are white with dark grey bases and shaft streaks. The undertail-coverts are dark grey, distinctly green-glossed and with some white fringes. The exact pattern of the underparts, is individually variable. **Upperwing** Very uniform, with black-brown wing coverts heavily green-glossed. Remiges are very uniform, slightly browner than wing coverts and less glossed. Remiges palest on the outer vanes of the innerwebs. Coverts can appear darker than remiges. No fringing on wings. **Underwing** Remiges paler and greyer than on upperwing and uniform with greater coverts. Greater coverts can appear darker towards the tips though they are not fringed. Median and lesser coverts clearly blacker than the rest of the underwing with which they contrast, as they do with the grey breast patch. Black-brown leading edge coverts have indistinct pale fringes. **Tail** Plumage as remiges, and similarly glossed and paler below.

Juvenile Possibly more heavily fringed on underparts than adult.

Measurements Wing: average (77) 94.50. Tail average: (58) 41.48 (Somadikarta 1986).

GEOGRAPHICAL VARIATION Four races.
C. l. linchi (Throughout Java and nearby islands of Madura, Bawean and Nusa Penida) Described above.
C. l. dodgei (Only at high altitude on Mount Kinabalu, Sabah) Similar to *linchi* but wing and tail lengths significantly shorter. Wing: average (3) 88.67. Tail: average (3) 34.50 (Somadikarta 1986).
C. l. ripleyi (Bukit Barisan Mountains, Sumatra) The Malacca specimen (and therefore the Malayan population) is thought to be of this race. Differs from *linchi* in that dark shafts on the abdomen feathers are pronounced. Average tail length much shorter. Wing: average (25) 93.02. Tail: average (25) 38.24 (Somadikarta 1986).
C. l. dedii (Bali and Lombok) Has much the longest wings and tail on average. Significantly blacker grey on feathers of neck, chin, throat, breast and flanks than *linchi*. Wing: average (45) 97.19. Tail: average (40) 44.25 (Somadikarta 1986).

113

VOICE The call has been noted to be a high-pitched *cheer-cheer* (Mackinnon 1988). No echolocation ability.

HABITS Highly gregarious. Found in large flocks and at times it has been recorded with all of the sympatric swift species. In Java, often found tightly circling, or even flying through the crowns of emergent trees, especially fruiting figs.

BREEDING Similar to that of Glossy Swiftlet and like that species the nest is placed in well lit areas of the caves or in buildings or rock crevices. Nests are bracket-shaped and self-supporting. A network of fine threads of cement is bound together with the chosen nest material which is very dependent upon what materials, usually thin items of vegetable matter, are available. Two white elongated eggs are laid.

REFERENCES Dickinson (1989b), MacKinnon (1988), Medway (1966), Salomonsen (1983), Somadikarta (1986).

---

17 PYGMY SWIFTLET
Collocalia troglodytes   Plate 4

IDENTIFICATION Length 9cm. A distinctive tiny swiftlet, black above with a narrow but distinct white rump band and glossy wings and tail. Underparts largely mottled whitish grey-brown throat and upper breast and between the black-brown undertail-coverts. Within range distinguished from main confusion species, Glossy Swiftlet, by white rump band. Although some subspecies of Glossy Swiftlet occurring on the Philippines have pale rumps they are rather mottled and grey in appearance and quite unlike the very neat, well-defined rump band of Pygmy Swiftlet. Very black upperparts and whitish abdomen exclude brown swiftlets with whitish rumps.

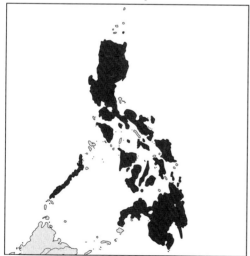

DISTRIBUTION Philippine endemic. Found on Banton, Bohol, Bucas, Calauit, Camiguin Sur, Camotes, Catanduanes, Cebu, Dinagat, Gigantes, Guimaras, Leyte, Luzon, Marinduque, Masbate, Mindanao, Mindoro, Negros, Palawan, Panay, Romblon, Samar, Siargao, Sibuyan, Siquijor and Ticao. Fairly common to common in the foothills.

MOVEMENTS Resident.

HABITAT Often encountered over inland waters. Common away from water over or near forest.

DESCRIPTION *C. troglodytes*
Sexes similar. **Head** Upper head wholly black-brown with slight oily gloss. Some individuals appear a little browner on forehead. Feathers black with grey bases not visible, dark-tipped effect produced in area of overlap. Grey-brown throat with pale-grey fringes often appearing darkest on chin and paler on lower throat before becoming darker again on upper breast. Feathers have darker subterminal bands and grey bases. Ear-coverts are grey-brown, paler than crown and darker than throat and lacking fringes. Black eye-patch. **Body** Saddle black-brown, uniform with nape and either very slightly paler or uniform with head. The rump has a narrow though highly contrasting white band, formed by broad white bases to the rump feathers which have broad black tips and conspicuous dark shaft streaks. Uppertail-coverts are black-brown and uniform with saddle. Below the lower throat the breast is dark grey with some paler fringing. Between upper belly and vent pattern becomes progressively paler with dark feather bases getting smaller and white fringes becoming larger. Feather pattern of vent is very similar to that of rump, with a broad white base and a black-brown tip. The undertail-coverts are black-brown. **Upperwing** Very black, uniform wings. Remiges are black-brown with browner, paler innerwebs. Remige tips and outer vanes of innerwebs are markedly, but very narrowly, pale grey. Inner primaries and secondaries are a little browner than the outer primaries. Coverts and alula are black-brown with only very marginal contrast with remiges which can appear very slightly darker (especially outer primaries). **Underwing** Remiges appear paler than the upperwing and uniform with greater coverts. Median coverts and lesser coverts are markedly blacker than the rest of the underwing and underbody. Coverts show no fringes, although the effect of darker tips on the greater coverts can sometimes be discerned. Leading edge coverts are uniform with lesser coverts. **Tail** Uniform with the outer remiges and similarly paler below.

GEOGRAPHICAL VARIATION None. Monotypic.

MEASUREMENTS Wing: 86-96 (Dickinson 1989b).

VOICE It is not known whether or not this species can echolocate. However, nests have been reported up to 15m into caves.

HABITS A gregarious species usually found in small groups.

BREEDING Made of vegetable matter bound together with strands of saliva cement (Dickinson 1989b). The nest is a half-cup attached to the wall of a cave or 'water-tunnel' and the saliva is hardened, white and translucent. There has been debate in the past as to the edibility of the nests and this has not been resolved with certainty. Medway and Pye (1977) characterise the nest as self-supporting, bracket-shaped; vegetable materials: sparse with firm nest cement.

REFERENCES Dickinson (1989b), Medway and Pye (1977).

114

18   SEYCHELLES SWIFTLET
Collocalia elaphra   Plate 5

Other names: Seychelles Cave Swiftlet; *Aerodramus elaphra*

IDENTIFICATION Length 11cm. Very typical of its genus and somewhat larger than Mascarene Swiftlet from which it further differs through its darker upperparts, with its rump barely paler than the back. Underparts browner than on Mascarene Swiftlet. Tail forked. This is the only swift species that commonly occurs on the Seychelles.

Medway (1966) notes that the type specimen has much white on the concealed barbs of the plumage of the back and that the dark shafts of the pale grey abdomen are inconspicuous. In the hand, this species has proportionally longer wings and a larger bill than Mascarene Swiftlet.

DISTRIBUTION Endemic to the Seychelles on the islands of La Digue, Mahé and Praslin.

One of the rarest swifts. Listed by BirdLife International as Vulnerable (Collar *et al.* 1994). The population in 1988 was considered to be under 1,000. Availability of suitable caves for nesting seems to be a natural limit to both distribution and population (only 'some half-dozen' nest caves are known Collar *et al.* 1994) but the situation has been exacerbated by disturbance and in some cases deliberate vandalism. It has been recommended that metal grilles are placed across the mouths of these caves to limit human access in an attempt to halt the decline of the species. Formerly a colony existed on Félicité but this has disappeared. It may have moved to La Digue where a recent increase in numbers has occured although this may well have been caused by a change in water distribution patterns on the island (Collar *et al.* 1994, quoting J Stevenson verbally).

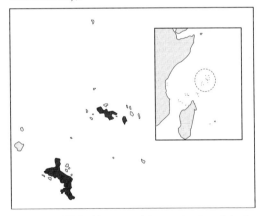

MOVEMENTS Resident, remaining predominantly in the neighbourhoods of the nesting colonies.

HABITAT Despite its sedentary behaviour this swiftlet can be seen over most habitats on the islands on which it breeds, from the lowlands up into the mountains.

DESCRIPTION *C. elaphra*
Adult Sexes similar. **Head** Dark grey-brown. Forehead very slightly browner. Slightly capped appearance as whole of ear-coverts mid-brown. Throat pale grey-brown with darker brown feather shafts. **Body** Upperparts uniform with crown, rump indistinctly paler. Underparts rather uniformly grey-brown, darkest on undertail-coverts which are

slightly darker grey than the remainder of the underparts and have a paler grey fringe. **Upperwing** Very uniformly black-brown with greenish gloss most apparent on the darker outerwebs. Gloss when worn becomes purplish and later a lacklustre black-brown. Paler innerwebs are most apparent on the tertials. Outerwing appears slightly darker than mantle. **Underwing** Slightly paler below than the upper surface with remiges rather uniform with the greater coverts but paler than the black-brown median and lesser coverts. **Tail** Similar to remiges.

Measurements Wing: 120-121 (Oberholser 1960). Medway (1966) measured the type at 124. Wing: (4 BMNH) 114-118 (116). Tail-fork: 5-6.5 (5.9).

GEOGRAPHICAL VARIATION None. Monotypic.

VOICE Procter (1972) believed that due to the nature of the caves in which the colonies were situated it was possible that it could echolocate. Penny (1974) describes what he believes to be the echolocating call as a regular low-pitched *pewterpewter*, extraordinarily metallic, almost electronic, in character.

HABITS Gregarious. Due no doubt to the relatively low population, does not occur in the large colonies typical of the genus.

BREEDING Typical bracket-shape formed by grey-green strands of lichen bound together by saliva. Strong nest but quite flexible despite being dry. The hinge of the nest is dark brown and has large quantities of saliva. Measures 5.5cm in height, 10.5cm in width and 3.5cm in cup depth (front to back). The clutch usually contains one egg. (Procter 1972). Medway and Pye (1977) describe the nest as self-supporting, bracket-shaped; vegetable materials: sparse, firm nest cement.

REFERENCES Collar *et al.* (1994), Medway (1966), Mountfort (1988), Oberholser (1960), Penny (1974), Procter (1972), Sibley and Monroe (1990).

19   MASCARENE SWIFTLET
Collocalia francica   Plate 5

Other names: Grey-rumped Swiftlet; Indian Ocean Swiftlet; Mascarene Cave Swiftlet; Mauritius Swiftlet; *Aerodramus francica*

IDENTIFICATION Length 10.5cm. This is the only member of the genus that occurs on the Mascarenes, which facilitates identification. A typical swiftlet with a grey-white rump band and forked tail.

In the hand, differs from Seychelles Swiftlet in proportionally shorter wings, smaller bill, slightly darker, browner underparts and distinctly paler rump.

DISTRIBUTION Endemic to the W Mascarene islands of Mauritius and Réunion.

Listed by BirdLife International as a Near-threatened species (Collar *et al.* 1994). Lack of suitable sites limits population size.

MOVEMENTS Resident. Has occurred once as a vagrant on Madagascar (Langrand 1991).

HABITAT Occurs over any habitat. Breeding caves are often the mouths of subterranean waterways and Staub (1976) noted that they often open in the middle of canefields.

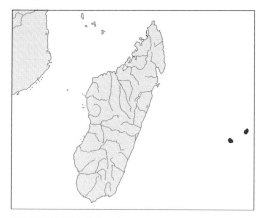

DESCRIPTION. *C. francica*

Adult Sexes similar. **Head** Black-brown crown, distinctly glossed. Ear-coverts mid brown creating a slightly capped appearance and darker than grey-brown throat. Throat feathers have clearly darker shaft streaks. **Body** Upperparts uniform with crown except for the rather indistinct grey-white rump band. Underparts grey-brown with least distinct shaft streaking on chest. The undertail-coverts are darker grey with grey-white fringes. **Wing** Very uniform above with blackish outerwebs that are distinctly glossed greenish although this becomes lacklustre when most worn after being purplish in the interim. Paler innerwebs most apparent on the tertials. Underneath, the remiges and greater coverts are paler than above and appear paler than the blacker median and lesser coverts. Upperwing appears a little darker than mantle, especially the outerwing. Darkest underwing-coverts appear darker than the underbody. **Tail** Similar above and below to the wing.

Measurements Wing: 112-114. Medway (1966). Tail fork: (4 BMNH) 4-6 (4.7).

GEOGRAPHICAL VARIATION Browning (*in litt.*) considered that four specimens at United States National Museum from Réunion are duller (browner, less green) above and darker-rumped than birds of similar museum age from Mauritius. The Réunion series also appears slightly darker on the underparts, especially the undertail-coverts. Browning considered that Réunion birds should be a separate subspecies but no name was available.

VOICE Staub (1976) states that in caves they emit a constant *tic-tic-tic* call to direct their flight, as with an echo-sounder. Medway and Pye (1977) state that the species can echolocate.

HABITS Typical swiftlet behaviour, though as with last species its relatively small population ensures that this species is not seen in massive flocks, nor does it form huge colonies like some other swiftlets.

BREEDING Typical bracket-shaped nest made from lichen filaments bound together with saliva, which is only copious at the base. Two measured nests were 6x5.5cm wide, 2.5cm deep and 6.5x5.5 wide, 3.5cm deep (Medway 1966). Usual clutch is two eggs. Medway and Pye (1977) describe the nest as self-supporting, bracket-shaped; vegetable materials: sparse, firm nest cement.

REFERENCES Collar *et al.* (1994), Langrand (1991), Medway (1966), Medway and Pye (1977), Procter (1972), Staub (1976).

20   INDIAN SWIFTLET
Collocalia unicolor   Plate 6

Other names: Indian Edible-nest Swiftlet; *Aerodramus unicolor*

IDENTIFICATION Length 12cm. A medium-sized brownish swiftlet with a shallowly forked tail, rump concolorous with the rest of the upperparts. Within range this species is the only regularly occurring swiftlet. The larger Himalayan Swiftlet may be encountered in the subcontinent during the winter and is separated from this species by its grey rump band that is clearly paler than the remainder of the upperparts.

In the hand, told from Himalayan Swiftlet by smaller size (wing 113-117 as opposed to minimum wing of 116 for the smallest race of Himalayan, *rogersi*, and most are considerably longer) and rump being uniform with rest of the upperparts. Upperparts are also slightly paler and less glossy than Himalayan Swiftlet (Ali and Ripley 1970). Tarsus unfeathered.

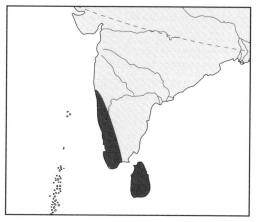

DISTRIBUTION Endemic to S India and Sri Lanka. Found from about 16°N at Ratnagiri in S Maharashtra and south into Goa, Karnataka W Tamil Nadu and Kerala. Occurs on small islets off the Malabar coast and throughout Sri Lanka.

Abundant throughout Sri Lanka, and occurring in huge colonies on islets off the Malabar coast and in cliffs of the Western Ghats.

MOVEMENTS Believed to be resident.

HABITAT Occurs over a variety of habitats from sea-level to 2200m. Most abundant on small rocky islets and around natural caves in ranges of the Western Ghats.

DESCRIPTION *C. unicolor*

Sexes similar. **Head** Upper head dark grey-brown. Extensively white (basal two-thirds) eye-patch, tipped black. White most visible from front. Ear-coverts paler than upper head, but not contrastingly so. Throat slightly paler than in *brevirostris*. **Body** Whole of upperparts uniform, dark grey-brown, with upper head, including the rump. However, some specimens show slight paling on rump, although this is mainly a result of disturbance to the feathers causing some white basal tufts to be revealed. White tufts obvious on nape. Underparts slightly darker pale grey-brown below the throat with brown shaft streaks. Undertail-coverts slightly darker grey and can show very diffused paler fringes. **Wing** Uniformly blackish-brown above with paler innerwebs. The underwing is paler than above with the lesser and median coverts contrastingly darker brown.

Measurements Wing: 113-117. Central tail: 41-45. Outer tail: 48-55. Weight: 11 (Ali and Ripley 1970).

GEOGRAPHICAL VARIATION None. Monotypic.

VOICE Said to resemble Himalayan Swiftlet. Roosting birds are said to keep up a feeble but shrill clicking or twittering *chit-chit*, with distinct livelier choruses intermittently (Ali and Ripley 1970). Bates (in Ali and Ripley) says it has a very harsh call note, reminiscent of the Whiskered Tern. Can echolocate.

HABITS Usually encountered during the day in small groups of up to six birds. Most active around colonies in early morning and late evening.

BREEDING According to Ali and Ripley similar to Himalayan Swiftlet 'but whitish and less mixed with extraneous matter such as grass, moss and feathers, the attachment practically of pure coagulated saliva; thus of some gastronomic and commercial value. Often hundreds of nests clustered densely, 5 to 20 cm from one another in patches on the rock wall or ceiling of dark grottos'. Average size of 80 eggs 20.9x13.5. Described by Medway and Pye (1977) as self-supporting, bracket-shaped; vegetable materials: sparse to moderate, firm nest cement.

REFERENCES Ali and Ripley (1970), Medway (1966), Medway and Pye (1977).

---

21 PHILIPPINE GREY SWIFTLET
Collocalia mearnsi   Plate 8

Other names: Philippine Swiftlet; *Aerodramus mearnsi*

IDENTIFICATION Length 10-10.5cm. One of the most troublesome of all swiftlets, with both identification and taxonomy being a matter of debate. Mainly submontane and may only infrequently be observed with the much commoner, essentially lowland Uniform Swiftlet. The two are very similar differing only in Philippine Grey having a shallower tail-fork and averaging smaller (wing 106-119.5 against 111-127) than the distinctly forked Uniform Swiftlet. They are perhaps inseparable in the field. On the Philippines, Edible-nest Swiftlet has a pale rump as opposed to Philippine Grey's dark-rump and a more heavily forked tail. Black-nest Swiftlet has occurred on Palawan and would be hard to separate in the field from this species although it is slightly larger and a little darker below. The rare montane Whitehead's Swiftlet with its very restricted highland range (Mount Apo and Mount Data on Mindanao and Luzon respectively) is a larger species with a massive head, deeply forked tail and distinctive pale frosting on the lores.

In the hand, as well as size and the structural differences quoted above can be told from *amelis* (the Philippine race of Uniform) by feathered tarsi, smaller more decurved bill (figure 19) and white tips to basal barbs of rump. Of the specimens at BMNH there was no overlap in body length between the two species (98-110.5 in 12 specimens of Philippine Grey Swiftlet versus 111.5-117 in three specimens of *C. v. amelis*). However, the size of this sample and the difficulty of comparing skin lengths leaves room

for caution.

Figure 19. Heads of Philippine Grey Swiftlet (left) and Uniform Swiftlet of the race *amelis* (right) showing apparent differences in bill shape.

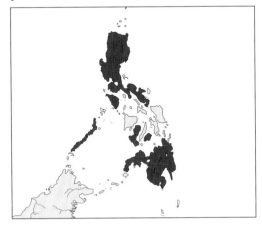

DISTRIBUTION Endemic to Philippines, where it is uncommon. Known from the islands of Bohol, Luzon, Mindanao, Mindoro, Negros and Palawan.

MOVEMENTS Apparently sedentary.

HABITAT Believed to be submontane (Dickinson 1989a).

DESCRIPTION *C. mearnsi*
Sexes similar. **Head** Crown glossy black-brown. Ear-coverts slightly paler and browner with throat appearing grey-brown. **Body** Upperparts, including rump, uniform with crown. White tips to basal barbs. Underparts mid grey-brown, with some darker brown shaft streaking, and appearing darkest on undertail-coverts. **Wing** Glossy black-brown upperwing, blackest on outerwebs and paler brown on innerwebs. Underwing is paler. Remiges appear uniform with the greater coverts and the median and lesser coverts appear blacker. **Tail** As remiges above and below.

Measurements Wing: 106-119.5. Tail: 45-52 (Oberholser 1912). Tail-fork: (3 BMNH) 5.8-6.3.

GEOGRAPHICAL VARIATION None. Monotypic.

VOICE No known evidence on echolocation. Rapid, excited chattering *Sh-sh-Shoo*, with emphasis on the first and third notes (transcribed from recording by C. Robson).

HABITS Presumably similar to other members of the *spodiopygius* superspecies.

BREEDING Builds moss nests held together with salival cement which hardens rather than remaining moist (Dickinson 1989a).

REFERENCES Dickinson (1989a), Oberholser (1912), du Pont (1971), Ripley and Rabor (1958).

117

22 MOLUCCAN SWIFTLET
Collocalia infuscata   Plate 8

Other name: *Aerodramus infuscatus*

IDENTIFICATION Length 10cm. This small species has a shallowly forked tail, uniformly pale grey underparts and, except in the northern Moluccas, a distinct white rump band. Apart from the easily separated Glossy Swiftlet, it need only be distinguished from Uniform Swiftlet which has a dark rump. In the northern Moluccas the race *infuscata* has paler more uniform underparts (the form of Uniform Swiftlet from the northern Moluccas *waiguensis* has a pale throat contrasting with the rest of the underparts).

In the hand, differs from Uniform Swiftlet most notably in having a feathered tarsus. However, it should be noted that the degree of tarsal feathering is variable and can be rather sparse (and in Uniform Swiftlet feathers are occasionally present in some populations and individuals). Most importantly *sorurum* and *ceramensis* have white rumps. *Infuscata* has a greyish rump and although this can be absent there are always extensive pale bases to the rump feathers and, as with the other two races, the pale grey underparts should differ from the more brownish underparts of the Wallacean races of Uniform Swiftlet.

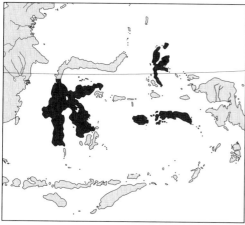

DISTRIBUTION Endemic to Wallacea, where it has an extensive range. Found in central, southern and south-eastern Sulawesi and on the Moluccan Islands of Morotai, Ternate, Halmahera, Buru and Seram. Apparently not uncommon.

MOVEMENTS Thought to be resident.

HABITAT Apparently a higher altitude species in central Sulawesi than Uniform. Stresemann found that it was not uncommon at around 700m on Seram and Heinrich collected many specimens in Sulawesi between 50-2500m (Stresemann 1940).

DESCRIPTION *C. i. sorurum*
Sexes similar. **Head** Black-brown upper head. No discernible fringing around eyes or lores. Grey supraloral patch. Throat pale grey below with extensive dark brown bases to feathers (can be dark on the chin). **Body** Upperparts as crown except for narrow, but distinct, white rump band with brown shaft streaks. Proximal uppertail-coverts greyish. Some white rami at bases of nape feathers. Underparts

as throat but undertail-coverts a little darker. **Wing** Uniform black-brown upperwing with only innerwebs to remiges paler. Underwing paler on remiges and somewhat greyer with median and lesser coverts appearing dark brown, contrasting with underbody. **Tail** As remiges.

Measurements Wing: male (17) 105-114 (109.3); female (8) 106-114 (110.9). Tail: 42-46. Tail-fork: 5-8 (Stresemann 1940). Wing: (5) 106-110 (108.6). Tail: 42-46. Tail-fork: 5-8 (Salomonsen 1983).

GEOGRAPHICAL VARIATION Three races.
    C. i. sororum (Sulawesi, except north) Upperparts blackish, faintly glossed, narrow (6-7) grey-white rump band with black shaft streaking. Underparts and eye-spot silvery grey, chin slightly darker, brownish. Tarsus feathered, but some variation in extent.
    C. i. infuscata (Halmahera, Morotai and Ternate, N Moluccas) Similar to *sororum*, differing in greyish or dark rump. Great variation in tarsal feathering. Wing: (5) 109-113 (110.6). Tail: 44-46. Tail-fork: 7-8 (Salomonsen 1983).
    C. i. ceramensis (Buru, Ambon, and Seram, S Moluccas) Similar to *sororum*, rump broader (10-12) and very white with contrastingly dark shaft streaks. Wing: (4) 104-108 (106.3). Tail: 41-43. Tail-fork: 5-6 (Salomonsen 1983).

VOICE It is not known whether or not this species can echolocate.

HABITS As other members of *spodiopygius* superspecies.

BREEDING White and Bruce (1986) consider that it is likely that the nests described, rather briefly, by Heinrich (in Stresemann 1940) belonged to this species and were made of vegetable matter and attached to a cave wall.

REFERENCES Browning (1993), Salomonsen (1983), Stresemann (1910 and 1940), Watling (1983), White and Bruce (1986).

---

23 MOUNTAIN SWIFTLET
Collocalia hirundinacea   Plate 9

Other name: *Aerodramus hirundinaceus*

IDENTIFICATION length 11-13cm. Medium-sized grey-brown swiftlet with moderately forked tail. Very hard to distinguish in the field from the *granti* form of Uniform Swiftlet which occurs in sympatry (except the significantly larger upland race *C. h. excelsa*). However the underparts of this species are more uniform and paler than in *granti* Uniform Swiftlet. More importantly some individuals show indistinct grey rump bands and as a result of extensive pale bases to the rump feathers most birds will show some paling in this area. The other two sympatric brown Swiftlets, Papuan and Bare-legged, are both significantly larger with proportionally shorter tails. Both are generally darker on the underparts and Three-toed shows considerable contrast between the dark underparts and the silvery-grey throat.

In the hand, distinguished from the very similar and sympatric Uniform Swiftlet by its tiny more strongly curved bill (figure 20), more extensive white to the bases of the back feathers, marginally darker more blue-sheened upperparts (the crown in particular is much darker) and the very pale white or silvery-grey underparts. The tarsus is feathered as opposed to, typically, unfeathered in Uniform Swiftlet.

Figure 20. Heads of Mountain Swiftlet (left) and Uniform Swiftlet of the race *granti* (right) showing differences in bill shape.

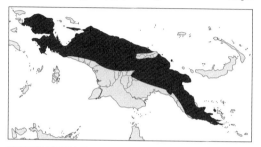

DISTRIBUTION Endemic to New Guinea and the islands of Yapen, Karkar and Goodenough (D'Entrecasteaux Archipelago). Found mainly in mountainous regions. Although it can be seen in the lowlands it is not present in the southern lowlands, or in the northern flood plain of the Sepik.

Coates (1985) recorded this species as locally common to abundant throughout the highlands, although apparently scarce in the lowlands.

MOVEMENTS Like many birds in New Guinea prone to wander. This explains periods of absence in areas where it is usually abundant.

HABITAT Essentially a highland swiftlet, found only in low density in the lowlands where there are nearby hills. In the highlands recorded to 4000m. On the island of Karkar only encountered above 240m, and believed to be confined to the mountains of Goodenough island. Feeds over a great variety of upland habitats, open and forested.

DESCRIPTION *C. h. hirundinacea*
Sexes similar. **Head** Upper head black-brown with bluish gloss. Extensive white to bases of supraloral patch. Throat can appear brownish on chin and centre as dark brown feather bases are extensive and are often revealed. **Body** Upperparts as crown. White rami liberally scattered about upperbody. However pale bases very extensive around rump, making it appear paler than surrounding tracts. In some even the tips to the rump feathers are grey and an indistinct band can be seen. Below the often dark central throat the underparts are mid or pale grey. **Wing** Very uniform above being somewhat similar to mantle in colour. Innerwebs paler and browner. Underwing pale on remiges but darker brown on medium and lesser coverts. **Tail** As remiges.

Measurements Wing: 114-117 (115.6) (Mayr 1937). Weight: male (7) 8.7-10.0 (9.2); female 8.0-9.3 (8.9) (Diamond 1972).

GEOGRAPHICAL VARIATION Three races.
   C. h. hirundinacea (New Guinea (except in range of *excelsa*), Goodenough Island in the D'Entrecasteaux Archipelago and Karkar Island) Lighter upperparts

and slightly lighter underparts (shaft streaking less obvious). Strongly feathered tarsus.
   C. h. baru (Yapen Island, Geelvink Bay) Very uniformly dark upperparts, underparts brown-grey with distinct shaft streaking. Well feathered tarsus. Wing: (6) 109-114.5 (111.2) (Mayr 1937).
   C. h. excelsa (Above 1600m in the central range of New Guinea in the Snow Mountains and the Carstensz Peaks) Similar to *hirundinacea* but much larger. Wing length increase with altitude (see table 1). Rand (1942) recorded the following measurements from varying altitudes: 1600m; male (3) 118-120 (119.0), female (6) 113-122 (1117.2), 2220m; male (3) 121-124 (122.0), female (3) 120-127 (123.7), 3225m; (4) 121-128 (125.0), female (7) 121-130 (125.0), 3600-4000m; male (8) 129-137 (133.4), female (6) 125-134 (129.8).

VOICE Has been proved to echolocate. Flocks occasionally emit a high-pitched repeated twitter (Coates 1985).

HABITS Typically gregarious, mixes freely with other swiftlets.

BREEDING The nest is composed of plant material such as mosses, filmy ferns, grasses and rootlets, and is a bulky structure with a depression in the top, attached without saliva to a small ledge or niche in the wall of a sink-hole in subdued light or in complete darkness (Coates 1985). Hadden (1975) recorded one egg per nest in a colony in the Tari district during late September. Colonial species. Medway and Pye (1977) characterise the nest as externally supported; vegetable materials; little or no nest cement.

REFERENCES Bravery (1972), Coates (1985), Diamond (1972), Hadden (1975), Mayr (1937), Medway and Pye (1977), Rand (1942), Salomonsen (1983), Schodde (1972).

24   WHITE-RUMPED SWIFTLET
Collocalia spodiopygius   Plate 9

Other names: Grey Swiftlet; Grey-rumped Swiftlet; Pacific White-rumped Swiftlet; *Aerodramus spodiopygius*

IDENTIFICATION Length 10-11.5cm. A medium-sized grey-brown swiftlet with a deeply forked tail and a pale rump band. The paleness of this band is geographically variable, but is pale enough throughout to separate the sympatric Uniform Swiftlet. The *leletensis* race of Mayr's Swiftlet has a pale rump band but it is a much larger species with a proportionally shorter tail. Underparts vary but in many subspecies are much paler than Uniform Swiftlet though often with a darker throat. Described in Coates (1985, by Lindgren), as having a more fluttering flight than Uniform Swiftlet.

DISTRIBUTION Extensive range mainly on islands in Papuasia, Melanesia and Polynesia. New Guinea; on islands and archipelagos of Manam, Long, New Britain, Watom, New Ireland, New Hanover, Tabar, Lihir, Mussau, Emira, Bougainville, Buka, and on the Admiralty Islands of Manus, Lou, Pak, Rambutyo and Horno, S and E Solomons, Duff Swallow and Santa Cruz Islands, Banks Islands, the N New Hebrides to Epi Island, the S New Hebrides, Loyalty Island, and the island groups of Fiji, Tonga and Samoa.

Coates (1985) describes it as fairly common to abundant

119

on the islands off E Papua New Guinea. Bregulla (1992) notes range contraction in the race *leucopygia* which by the 1960s was present only on the islands of Malo and the west coast of Santo and was considered to be uncommon. Tarburton's (1987) study of Fiji found 5 colonies with a total 2785 birds.

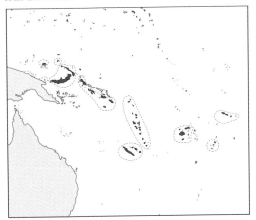

MOVEMENTS Sedentary.

HABITAT Essentially an island species occurring over any habitat. In highlands to at least 1890m on New Britain and at least 1600m on Bougainville. Breeds both in island caves and mainland gorges and hills.

DESCRIPTION *C. s. spodiopygius*
Sexes similar. **Head** Crown glossy black-brown, with forehead, lores and particularly small line of feathers over eye appearing slightly paler. Ear-coverts uniformly paler and browner, creating somewhat capped appearance. Chin and mid throat often grey-brown appearing darker than pale grey (almost grey-white) lower throat. **Body** Upperparts uniform with crown (pale feather bases particularly apparent on nape which can be slightly paler), except for well-defined pale grey rump band. Underparts lack streaking and are pale grey to grey-white, slightly darker grey on undertail-coverts which are pale grey-fringed. **Wings** Blackish above with purple gloss when fresh. Innerwebs paler and browner, most apparent on tertials. Below, the remiges and greater coverts appear paler and greyer, contrasting with blackish median and lesser coverts. **Tail** Uniform above and below with remiges.

Juvenile Clearly defined pale fringes to tips of remiges and more extensive white rami on upperparts.

Measurements Wing: (8 BMNH) 112-123 (118.06). Wing: (10) 113-124 (118.2) (Salomonsen 1983).

GEOGRAPHICAL VARIATION 11 races recognised here.
  C. s. delichon (Manus Island, Admiralty Archipelago) Dull glossed blackish upperparts, copious white rami on neck and mantle, strikingly broad white rump (13-17), broader than all other forms. Grey-white (silver tinge) underparts and eye patch, throat whitish, chin brownish, very dark undertail-coverts. Tarsus unfeathered or only thinly, proximally. Wing: male (22) 101-109 (105.2); female (15) 103-109 (105.4) (Salomonsen 1983).
  C. s. noonaedanae (New Ireland and New Britain) Rump band narrower and greyer than *eichorni* (4-7), upperparts blacker than *delichon*, underparts as *eichorni* but throat contrastingly grey-white as is eye patch. Bare

tarsus. Wing: male (14) 102-108 (104.5); female (16) 101-109 (104.8). Those from Lemkamin in New Ireland mountains (900m) are slightly larger averaging: male (6) 105.0; females (3) 108.3 (Salomonsen 1983).
  C. s. reichenowi (S and E Solomon Islands) Similar to *noonaedanae*, underparts more uniform and darker, rump uniformly sooty-grey and narrow. Tarsus unfeathered or only proximally. Wing: (8) 102-108 (104.4) (Salomonsen 1983).
  C. s. desolata (Duff, Swallow, and Santa Cruz Islands) As *noonaedanae* above but darker below than *reichenowi*, throat and ear-coverts contrastingly darker brown, white on eye patch only on concealed bases. Tarsus unfeathered. Wing: (10) 99-102 (101.0) (Salomonsen 1983).
  C. s. epiensis (Banks Islands, N New Hebrides to Epi Island) As *desolata*, but darker underparts, darker brown throat and rump usually lighter. Unique in Melanesian forms of the species in having whole length of tarsus densely feathered. Wing: (5) 100-104 (102.0) (Salomonsen 1983).
  C. s. eichorni (Mussau Island, St Matthias Group, Bismarck Archipelago) Similar to *delichon*, underparts grey-brown (clearly darker) upperparts paler, rump band narrower (7-14) and slightly greyish, and tarsus unfeathered. Wing (shorter than *delichon*): male (5) 100-1103 (102.2); female (7) 100-103 (101.4) (Salomonsen 1983).
  C. s. ingens (S New Hebrides) Deeply black slightly glossed above (no white rami on neck), pure white rump band (9-13), greyish underparts (intermediate between *epiensis* and *leucopygia*). Throat and face sides brown, contrasting with underparts. Only base of eye patch white. Tarsus feathered as *epiensis*. Wing: (4) 108-116 (110.8) (Salomonsen 1983).
  C. s. leucopygia (Loyalty Islands and New Caledonia) As *ingens*, but underparts below brown throat and sides of face almost pure white. Wing (Loyalty Islands): male (9) 109-115 (112.9); female (6) 111-118 (113.0). Wing (New Caledonia): male (6) 104-111 (108.0); female (1) 108. (Salomonsen 1983).

Due to the large distance between Polynesian forms of this species and all others, comparison between the following three races concentrates on their mutual differences and similarities only. All have unfeathered tarsi.
  C. s. assimilis (Fiji) Smaller than *spodiopygius* with whiter rump band, darker chin and paler greyer underparts (less contrast on undertail-coverts). Wing: (9) 111-119 (113.9) (Salomonsen 1983) (Stresemann 1912) gives the wing: (21) 107-117 (111.0). Seventeen measured at BMNH: 104-113 (110.3).
  C. s. townsendi (Tonga) Similar to *assimilis*, but sootier grey below with greyer rump band. Slightly smaller than *spodiopygius*. Wing: (12) 112-118 (114.7) (Salomonsen 1983). Seven from BMNH: 110.5-114 (112.1).
  C. s. spodiopygius (Samoa) Described above. Only separable on size from *townsendi*.

VOICE An echolocating species. Bregulla (1992) notes a variety of high pitched squealing and short twittering notes and a pleasant warble *tooweet-tooweet-ziweet*.

HABITS A gregarious species often encountered in large flocks, freely associating with Uniform and Glossy Swiftlets. Feeds at a variety of different levels.

BREEDING A substantial, but shallow cup, made from fine vegetation and mosses. This is attached to the walls of a cave. On Bougainville one colony uses the adits of an old gold mine (Hadden 1981). Silva (1975) described a colony in two sea-level caves on Long Island, and Lindgreen (in Coates 1985) describes a colony in a cave created by wave erosion on Lou Island. One or two white eggs (18-19 x 12-13). Tarburton (1993) records the average size of 36 nests of *A. s. assimilis:* 50mm x 49.7mm x 21.1 (average volume index 52.4cm³). Nest characterised by Medway and Pye (1977) as self-supporting, bracket-shaped; vegetable materials, copious; firm nest cement.

REFERENCES Bregulla (1992), Coates (1985), Hadden (1981), Medway and Pye (1977), Salomonsen (1983), Silva (1975), Stresemann (1912), Tarburton (1987 and 1993).

25   AUSTRALIAN SWIFTLET
Collocalia terraereginae   Plate 9

Other names: Grey Swiftlet; *Aerodramus terraereginae*

IDENTIFICATION Length 11cm. The only swiftlet breeding in Australia. Invariably, pale grey to grey-brown rump contrasts with rest of upperparts and underparts are pale grey-brown. Shallowly forked tail. Vagrant Uniform, Mountain and Glossy Swiftlets could theoretically occur within range. The latter species with its white abdomen and highly glossed plumage would be easily recognised. However, as some examples of this species show dark rumps individuals of the former two would be very hard to separate in the field. Uniform Swiftlet (Papuan *granti*) shows the throat paler than the rest of the underparts (whereas Australian is uniform below) and Mountain has generally paler underparts, but these features (especially for Mountain) would be very hard to confirm in the field.

In the hand, strongly feathered tarsus and in most examples pale rump exclude Uniform Swiftlet. Within the White-rumped superspecies this form has the longest tail (49-50, Salomonsen 1983), with the exception of the longest tailed individuals of the *excelsa* race of Mountain Swiftlet. Pale-rumped examples easily separated from Mountain (which can show some indistinct paling on the rump), otherwise considerably darker on underparts (indeed darker than all in White-rumped group except *desolata* and *epiensis*).

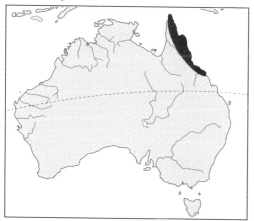

DISTRIBUTION NE Australia in coastal and highland

areas of NE Queensland from Claudie Ridge, Cape York Peninsula south to Eungella Range near Mackay. Occurs on suitable offshore islands (those with caves) south to Dunk, Family Group and Hinchinbrook Island.

Common below 500m in Australia. The largest known colony is 500 pairs. Tarburton (1993) recorded 34 active colonies between 1985-1987.

MOVEMENTS Sedentary.

HABITAT Mainly below 500m, occasionally to 1000m. Breeds both in island caves and mainland gorges and hills.

DESCRIPTION *C. t. terraereginae*
Adult Sexes similar. **Head** Crown brown to black-brown, forehead, lores and line of feathers over eye slightly paler. Ear-coverts quite uniform with upper head. Throat uniform sooty-brown. **Body** Upperparts uniform with crown. Distinct grey-white to pale grey-brown rump. Width of rump (10 BMNH) 4-10 (7.3). In some birds rump band can apparently be quite uniform with mantle. Underparts lack streaking and are uniform with throat. Undertail-coverts darker grey-brown with slightly paler fringes. **Wings** Uniformly black-brown above. Innerwebs paler and browner. **Tail** As remiges. Typically contrasts with paler uppertail-coverts.

Juvenile Pale fringes on remiges.

Measurements Wing: (2) 114 + 115. Tail: (2) 49 + 50. Tail-fork: 5 + 6 (Salomonsen 1983). Wing: (21 BMNH) 107-118.2 (112.9). Weight: (16) 10.5-12.5 (11.25).

GEOGRAPHICAL VARIATION Two races.
   C. t. terraereginae (N Queensland) Described above.
   C. t. chillagoensis (Queensland) Smaller and paler than nominate form. Wing: average (433) 107. Weight: average (519) 9.39 (Tarburton pers. comm.)

VOICE Pizzey (1980) describes the flight call as a high-pitched *cheep* and the echolocating call as a metallic incessant clicking.

HABITS Very similar to White-rumped Swiftlet.

BREEDING Medway (1966) examined a nest consisting entirely of casuarina twiglets interspersed between laminae of firm, translucent nest cement. This species has the smallest nests of any taxa in the *spodiopygius* group. Thirty *terraereginae* nests from Tully Falls averaged 56mm x 45.5mm x 3mm (volume index 7.6cm³), whilst 100 nests of *chillagoensis* averaged 49mm x 42.7mm x 11.9 (volume index 24.9cm³). Clutches of both subspecies always consist of one egg only (Tarburton 1993).

REFERENCES Medway (1966), Medway and Pye (1977), Pizzey (1980), Salomonsen (1983), Tarburton (1993).

26   HIMALAYAN SWIFTLET
Collocalia brevirostris   Plates 6 & 7

Other names: Volcano Swiftlet (race *vulcanorum*); Indochinese Swiftlet (race *rogersi*); *Aerodramus brevirostris*

IDENTIFICATION Length 13-14cm. A large grey-brown swiftlet usually with a pale grey rump (some forms show darker rumps) and obviously forked tail. Throughout most of range this species is identified with comparative ease as it is the only swiftlet. However in the south of its breeding range and in its wintering quarters it comes into contact with Edible-nest and Black-nest Swiftlets. Himalayan is less dark above and less pale below than Edible-nest Swiftlet and therefore appears more uniform. Edible-nest Swiftlet of the race *germani* also has a broader and considerably whiter rump. Black-nest Swiftlet is extremely hard to separate from Himalayan in the field. Himalayan has a deeper tail-fork and probably averages paler on the rump band. The isolated Javan race, *vulcanorum*, has rather dark underparts (especially the throat) and shows an indistinct pale rump patch.

In Malaysia, Medway and Wells (1976) separated this species in the hand from Black-nest Swiftlet on wing-length (121-136), tail furcation of 14-24% and light tarsus feathering (*rogersi* lacks tarsal feathering), compared with Black-nest's wing-length of 128-135, tail furcation of 9-16% and heavy tarsus feathering. However, see variation in wing lengths given below (extreme range for smallest and largest races 116-141). *Vulcanorum* shows some variation in tarsal feathering with males having bare tarsi and females having feathered tarsi. White rami present in plumage of all races (except *rogersi*) although it is sparse in *vulcanorum* (figure 21). For separation of Edible-nest Swiftlet in the hand, see under that species.

Figure 21. Typical upperpart feathers of Mossy-nest Swiftlet (left), Himalayan Swiftlet (centre) and Black-nest Swiftlet (right). Note differences in the colour of the basal barbs (rami).

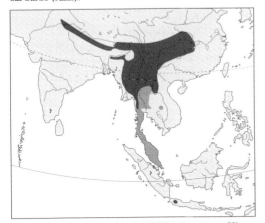

DISTRIBUTION Extensive S and SE Asian range. Himalayas from about 76°E in Himachal Pradesh (Kangra, Dalhousie) east through Nepal, Sikkim and Bhutan to NE India, in the Assam hills (both north and south of Brahmaptura), Nagaland, Manipur, Mizoram, and in upland areas of Bangladesh. Also in SW China, in NW Yunnan, in S and N Sichuan, Hubei and N Guizhou, and N Laos, Burma, W Thailand. Range in Sibley and Monroe (1990) includes C and S Vietnam, however Robson (pers. comm.) believes there may be no breeding records from Indochina. Status in Sumatra is uncertain. Van Marle and Voous (1988) mention that according to Medway (1966) the species is thought to be resident but it may be a winter visitor. Medway and Wells (1976) state that it occurs discontinu-

ously throughout SE Asia to Sumatra. This opinion is based on Medway (1966) from the description in Robinson and Kloss (1924) of a cave containing the nests of *brevirostris* in Buo, Padang Highlands. However, Wells (1975) states that the existence of a breeding race of *C. brevirostris* in Sumatra has not been acceptably demonstrated. In the winter, disperses south into the central plains of Thailand, the Thai border with Laos and in the far east of Thailand as well as the Malay Peninsula, south to Negiri Sembilan. One record from Port Mouat in S Andaman, although also observed on Narcondam where Osmaston (in Ali and Ripley 1970) considered they probably bred in caves on the south coast of the island. The isolated population on Java is known only from four volcanic peaks: Gunung Gede-Pangrango, Tangkubanprahu and Papandayan. It has been stated that *vulcanorum* occurs on volcanic peaks in Sumatra, but there is no firm evidence to support this theory.

Throughout breeding range tends to be the commonest swiftlet. However only relatively small numbers seem to move south to winter and it is an uncommon winterer in peninsular Thailand and Malaysia. The Javan population is considered by some authorities to be a separate species: Volcano Swiftlet *C. vulcanorum*. BirdLife International list this as a Vulnerable species and state that recent records are only known from Gunung Gede in Gunung Gede-Pangrango National Park (Collar *et al.* 1994). An additional problem is caused by the active nature of all the known craters around which it occurs with colonies vulnerable to 'periodic extinction' (MacKinnon 1988). These problems coupled with the limited breeding habitat suggest that this is a naturally rare form.

MOVEMENTS In the Malay Peninsula evidence suggests that this is a non-breeding passage migrant or winter visitor, with extreme dates of November and March. At Fraser's Hill, during 1966-69, three night-flying migrants netted from 1 to 16 November had pale rumps and winglengths of 128-135mm, and were considered to be *brevirostris* or *innominata*. Three individuals (unidentified subspecifically) that were believed to be migrants were collected in November in the Malacca Straits on One Fathom Bank lighthouse. Apparently *rogersi* has not be recorded south of Surat Thani. In Nepal the subspecies concerned, *brevirostris*, is mainly an altitudinal migrant. In summer it occurs up to 4575m, but winters between 915-2745m, and on occasions it is seen in the lowlands (tarai), for example after storms. *Vulcanorum* is apparently highly sedentary with no records away from the peaks. Due to identification problems this is perhaps not surprising.

HABITAT Mainly a highland species, feeding over a variety of habitats but particularly favouring forested river valleys. Especially in periods of bad weather or late in the evening the species will feed over a wider range of terrain such as paddy fields and other cultivated areas. In Java restricted to high peaks and ridges, and especially along crater rims.

DESCRIPTION *C. b. brevirostris*
Adult Sexes similar. **Head** Dark grey-brown upper head. Line of feathers immediately over front half of the eye are, on average, pale grey-fringed. The eye patch is largely white, best seen when viewed from the front. Ear-coverts are paler than upper head, but they merge without notable contrast. Throat is pale grey-brownish with slight brown shaft streaking. **Body** Head to lower back uniformly dark

grey-brown. Distinct, but narrow, greyish rump band with brown shaft streaks. Uppertail-coverts as mantle. The underparts are pale grey-brown with streaking as the throat. Undertail-coverts slightly darker grey, especially the longest, lacking whitish fringes although some show diffusely paler edges. White rami. **Wing** Upperwing uniformly blackish, with paler innerwebs. Underwing paler with darker brown median and lesser coverts. **Tail** As remiges.

Juvenile (applies to all races). More concealed white in the contour feathers, less defined rump band and more sparsely feathered tarsus.

Measurements Wing: 120-133. Central tail: 45-48. Outer tail: 54-59. Tail-fork: 8-10 (Ali and Ripley 1970). Wing: (13) 123-132 (Deignan 1955). Five measured by Deignan (1955) from NE Burma had a range of 128-132. Wing: 122-135 (Medway and Wells 1976). Weight: 12.5-13 (Ali and Ripley 1970).

GEOGRAPHICAL VARIATION Five races recognised here, two of which (*rogersi* and *vulcanorum*) are sometimes considered to be separate species.

C. b. brevirostris (Himalayas from Himachal Pradesh east through Nepal, Sikkim and Bhutan to NE India, Bangladesh and south into Burma and Thailand) Described above. The palest-rumped race.

C. b. innominata (C China migrating to SW Thailand, Malay Peninsula, and possibly Sumatra) Rump colour intermediate between *brevirostris* and *inopina*. Wing: 125-132. Tail: 52-55. Tail-fork 8-10 (Ali and Ripley 1970). Medway and Wells (1976) give the type wing as 137 and Deignan (1955) measured 15 individuals with a range of 132-141.

C. b. inopina (Sichuan, Yunnan and C China) This is a dark race, but variation in the rump colour seems to be apparent, and is responsible for some authors separating the form *pellos*. Wing: 135-136 (Medway and Wells 1976).

C. b. rogersi (E Burma, W Thailand and Laos) Similar to *brevirostris*; with typically pale rump (quite dark in some individuals) but lacking tarsal feathering and white rami. Smallest race. Wing: (19) 116-128 (Deignan 1955). Deignan considered that five birds with lightly feathered tarsi from NW Thailand and the S Shan States, Burma may represent a more northern atypical population of *rogersi* and that three adults taken in winter from Selangor may be migrants from this population.

C. b. vulcanorum (Volcanic peaks on Java) Dark underparts. Indistinct pale grey rump. Wing: 118-124.5. (Stresemann 1931). Two at BMNH measure: wing 120 + 123; tail 50 + 56; tail-fork 5 + 7.

VOICE Ali and Ripley (1970) describe a conversational twittering *chit-chit* uttered at a roost and also a low rattle-like call as of a knitting needle drawn across a few teeth of a wooden comb. *Vulcanorum* has been heard to utter a piercing *teeree-teeree-teeree* (MacKinnon and Phillips 1993) and can echolocate.

HABITS Gregarious, usually seen in small groups of up to 50, but sometimes up to 300 individuals may be seen together. Frequently seen with other swifts and hirundines.

BREEDING According to Ali and Ripley (1970) nest is a tiny, unlined cup of agglutinated moss with the rim slightly sloping down from the vertical rock wall to which it is attached like a bracket with the inspissated saliva. Built in colonies close to and often touching one another, haphazard and in rows in angle of ceiling, not clustered together in 'villages' (cf House Swift), the nests pockmarking large areas of the rock face or ceiling within dark grottos. Diameter of cup c. 6cm, depth inside c.2cm, with a slightly tapering pedestal below, c.5cm thick. The side of the nest adhering to the wall projects upwards giving increased purchase. Eggs, 2, white ellipsoid, narrowing slightly at one end. Average size of 8 eggs 21.8 x 14.6mm. Characterised by Medway and Pye (1977) as self-supporting, bracket-shaped; vegetable materials; sparse, firm nest cement.

REFERENCES Ali and Ripley (1970), Collar *et al.* (1994), Deignan (1955), MacKinnon and Phillipps (1993), Medway (1966), van Marle and Voous (1988), Medway and Pye (1977), Medway and Wells (1976), Somadikarta (1967) Stresemann (1931).

27   WHITEHEAD'S SWIFTLET
Collocalia whiteheadi   Plate 8

Other names: *Aerodramus whiteheadi*

IDENTIFICATION Length 14cm. This very rare species is unlikely to be encountered away from the summits of Mount Apo or Mount Data. The large-headed appearance, so typical in museum specimens, should create a distinctive jizz in the field. The deeply forked tail should also aid identification. Otherwise the plumage is typically indistinct, with no rump band and indistinct underparts, apart from extensive pale frosting on the lores (figure 22).

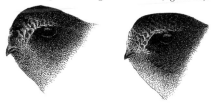

Figure 22. Heads of Whitehead's Swiftlet (right) showing frosting on lores and over eye typical also of the allopatric Bare-legged Swiftlet and Papuan Swiftlet (left).

In the hand, large (wing 129-140.5) with a massive skull and a distinctly forked tail with a length of 57-60. Naked tarsus. Upperparts uniformly brownish-black with some white rami. Distinct pale fringes on lores .

DISTRIBUTION Endemic to the Philippines. Known only from Luzon, where four specimens were taken near the summit of Mount Data by Whitehead in 1899, and from Mindanao where 3 specimens were taken on Mount Apo at 1,200m in 1904.

Clearly very rare with a restricted range. Previous authors have confused the issue by stating that the species has a wide distribution and altitudinal range. Dickinson (1989a) showed that this is not the case. Only seven museum specimens are known and it has never been recorded with certainty in the field. It is critically in need of further study. Listed as Vulnerable by BirdLife International and possibly at risk from habitat loss (Collar *et al.* 1994).

MOVEMENTS Presumably resident.

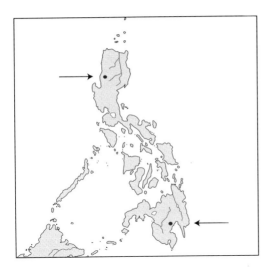

HABITAT Strictly montane. Both sites are naturally forested. Otherwise nothing is known of its behaviour.

DESCRIPTION *C. w. whiteheadi*
Sexes similar. **Head** Upper head uniformly dark grey-brown. Lores and line over eye are extensively pale grey-fringed and appear frosty. Forehead as upper head separating pale grey frosting of lores. Eye patch with whitish bases but mainly black. Throat mid or pale grey with indistinct browner shaft streaks. Ear-coverts darker than throat and slightly paler than upper head, although lacks any capped appearance. **Body** Whole of upperparts, including rump, dark grey-brown. Underparts pale or mid grey-brown, with brown shaft streaking, becoming slightly darker on undertail-coverts, which have the darkest shaft streaks. Some white tufts on plumage. **Wing** Dark blackish-brown, with innerwing slightly browner. The innerwebs of all remiges are paler and browner. Distinct, but narrow, whitish tip to the longest tertial extends along the tips of all secondaries. The underwing is paler with the lesser and median coverts darker brown. **Tail** As remiges.

Measurements Wing: 133.5-140.5 (Stresemann 1914). Tail: 64 (Ogilvie-Grant 1895).

GEOGRAPHICAL VARIATION Two races.
    C. w. whiteheadi (Mount Data, Luzon) Described above.
    C. w. origenis (Mount Apo, Mindanao) Similar to *whiteheadi*, but more blackish upperparts and underparts darker. Wing: 129-138. Tail: 53-60 (Oberholser 1906).

VOICE It is not known whether this species can echolocate.

HABITS No information.

BREEDING Medway (1966) describes four nests thought to have been made by *origenis* that were collected on Mearns' expedition to Mount Apo (1904) from a hollow tree as 'rounded vegetable nests constructed of green bryophytes together with some fibrous plant material, and apparently not incorporating nest cement'.

REFERENCES Collar *et al.* (1994), Dickinson (1989a), Oberholser (1906), Ogilvie-Grant (1895), Salomonsen (1983), Somadikarta (1967), Stresemann (1914).

28   BARE-LEGGED SWIFTLET
Collocalia nuditarsus   Plate 9

Other names: Schrader Mountains Swiftlet; New Guinea Swiftlet; Naked-legged Swiftlet; *Aerodramus nuditarsus*

IDENTIFICATION Length 14cm. A rather rare, large, highland species with dark rump and shallowly forked tail. The two noticeably smaller brown sympatric swiftlets, Uniform and Mountain, have paler underparts. The similarly sized Papuan Swiftlet is largely allopatric and can be separated in the field by its darker underparts - particularly the throat which is uniform with the underparts as opposed to contrastingly silvery-grey.

In the hand, closest in appearance to the allopatric Whitehead's Swiftlet from the Philippines, from which it differs, most notably, in the bare tarsus. Differs from Papuan Swiftlet in having four toes, bare tarsus, matt not glossy upperparts, throat uniform with (generally darker) underparts and dark bases to nape feathers not contrasting with tips. Browning (*in litt.*), however, noted that there are white rami on the nape and upper back.

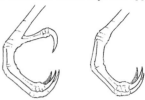

Figure 23. Feet of Bare-legged Swiftlet (left) and Papuan Swiftlet (right). (Based on Somadikarta 1967).

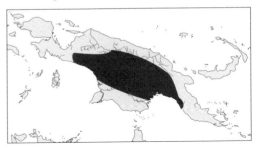

DISTRIBUTION Known from S and SE New Guinea, from Mimika River, the Snow Mountains from Mount Goliath, Kubor Mountains and from near Hall Sound at Baroka. All these sites are south of the central watershed, but it is believed to breed in the western highlands north of the watershed in the Schrader Range.

The scarcity of museum specimens suggests that it may be uncommon.

MOVEMENTS Presumably resident.

HABITAT Mainly a highland species as suggested by the high altitude (over 1500m) collection of four of the five specimens discussed by Salomonsen (1963). However, the remaining specimen was collected at 30m.

DESCRIPTION *C. nuditarsus*
Head Upper head matt black (only very slight gloss). Nape feathers very slightly paler, black-grey, than tips. Feathers immediately over eye with whitish fringes forming slight supercilium as in *papuensis*. Throat sooty-grey. **Body** Upperparts uniform with crown. Base of nape feathers

black-grey with little or no contrast with tips. Underparts uniformly sooty-grey as throat. **Wing** Very uniformly blackish above (paler innerwebs). Underwing paler on rectrices than above. **Tail** As remiges. Tarsus completely bare.

Measurements Wing: male (1) 138; female (2) 137 + 140. Tail: 49-51. Tail-fork 4-5 (Salomonsen 1983).

GEOGRAPHICAL VARIATION None. Monotypic.

VOICE Not recorded.

HABITS Little is known about its behaviour.

BREEDING No information.

REFERENCES Beehler *et al.* (1986), Salomonsen (1962 and 1983), Somadikarta (1967).

---

29   MAYR'S SWIFTLET
Collocalia orientalis   Plate 9

Other names: Guadalcanal Swiftlet (race *orientalis); Aerodramus orientalis*

IDENTIFICATION Length 13-14cm. Probably one of the rarest swiftlets, only known from two specimens. Differs from allopatric members of the same superspecies in having a pale grey rump band. Differs from the two sympatric brown species in being larger with a proportionally shorter tail with a very shallow tail-fork; from Uniform Swiftlet in having a pale rump band and from White-rumped Swiftlet in having an indistinct greyish as opposed to white (or grey-white) rump band.

In the hand, differs from Uniform Swiftlet in larger size (wing 127-134), from Papuan Swiftlet by having four toes, naked tarsus, dull upperparts lacking iridescence (only in respect of *orientalis*) and darker underparts including throat, and from the similar Bare-legged Swiftlet by its pale rump band.

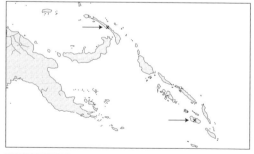

DISTRIBUTION Only recorded on two widely separated Melanesian Islands. One specimen was collected at 900m on the Lelet Plateau in central New Ireland in 1962, whilst the other was collected in 1935 from Guadalcanal in the Solomons.

With only the type specimens for this species it seems likely that it is indeed a very rare species. However, studies in the region may lead to further clues as to its true status. Listed by BirdLife International as a Data Deficient Species (Collar *et al.* 1994).

MOVEMENTS Not known.

HABITAT Never knowingly observed in the field. However, Salomonsen (1983) considered it undoubtedly a mountain bird.

DESCRIPTION *C. o. leletensis*
Upperparts Wholly glossed bluish-black with dark grey bases to the nape feathers that do not contrast with tips. Extensive frosting over eye and sides of forehead caused by pale grey tips to the feathers. **Underparts** Sooty-brown underparts, lacking contrast between throat and abdomen. **Wings** Very uniform with mantle above, with only the innerwebs appearing paler and browner. Underwing paler grey with some contrast with coverts. **Tail** As remiges.

Measurements Type (adult female): wing (worn) 134; tail 52; tail-fork 4 (Salomonsen 1983).

GEOGRAPHICAL VARIATION
  C. o. leletensis (C New Ireland) Described above.
  C. o. orientalis (Guadalcanal Island, Solomons) Tarsus thinly feathered. Similar to *leletensis* but slightly paler and much broader grey rump. Upperparts lack dark bluish gloss of *leletensis*. Wing: 127 (Salomonsen 1983).

VOICE Not known.

HABITS Nothing is known about its behaviour.

BREEDING Not known.

REFERENCES Beehler *et al.* (1986), Coates (1985), Collar *et al.* (1994), Salomonsen (1963 and 1983), Somadikarta (1967).

---

30   MOSSY-NEST SWIFTLET
Collocalia salangana   Plate 7

Other names: Mossy Swiftlet; Sunda Swiftlet; Thunberg's Swiftlet; *Aerodramus salangana*

IDENTIFICATION Length 12cm. Dark-rumped medium sized swiftlet with very shallowly notched tail (some individuals have deeper forks). Within usual range (it has occurred on one occasion on Basilan in the Philippines), needs to be separated from Edible-nest, Himalayan and Black-nest Swiftlets. The race of Edible-nest that occurs on Sumatra and Borneo, *vestita*, is a dark-rumped form of that species, although extensive pale bases to the rump feathers cause the impression of a paler rump in extended views. The Javan race of Edible-nest, *fuciphaga*, shows variation in rump colour but it is always at least slightly greyer than the mantle and often considerably paler. Mossy-nest Swiftlet also has slightly darker underparts and upperparts and typically a much shallower tail-fork than Edible-nest. These last differences also apply to the much whiter-rumped *germani*, which is sympatric with this species in coastal northern Borneo and offshore islands. Himalayan Swiftlet, which may occur as a wintering species in Sumatra, is paler grey below, has a more heavily forked tail (though less so than Edible-nest) and a typically a clearly pale grey rump band. The pale-rumped forms of the larger Black-nest Swiftlet *tichelmani* (possibly never sympatric with this species as it occurs only in southeast Borneo) and *maxima* with which it is sympatric in parts of Java, can be easily separated. The dark-rumped race *lowi* is more problematic. It is sympatric with Mossy-nest Swiftlet in Java (where it intergrades with *maxima* in W Java making it possible to have examples of Black-nest in W Java with pale or dark rumps) and N Borneo. These dark-rumped forms still have the slightly deeper tail-forks of the species and are a little greyer below and not so black above as Mossy-nest. In areas of sympatry with dark-rumped Black-nest Swiftlets, identification is best clinched by observing the species on the nest at one of the many accessible cave colonies where both species (and invariably Edible-nest) are present.

In the hand, tarsal feathering variable. Uniform blackish-brown upperparts, with basal barbs not white but greyish-brown. This latter character excludes Uniform Swiftlet and the Edible-nest superspecies (except some individuals of the nominate *fuciphaga* and *vestita*). Underparts lack shaft streaking in some individuals (very light in all).

DISTRIBUTION Endemic to Greater Sundas. In Sumatra, colonies near Buo and Ngarai Sianok and also Padang Highlands, Barat. Possibly widespread in the Padang Highlands. Also on islands off W Sumatra. On Java found throughout, wherever there is suitable habitat. In Borneo, restricted to the north and the Natuna Islands. A specimen from the Philippines was taken in 1906 by Mearns from Basilan.

Locally abundant. Often occurs in vast numbers in optimum cave colonies. Uncommon on Java, although this may be the result of it not being recognised.

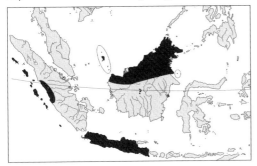

MOVEMENTS Thought to be sedentary, although the record from the Philippines suggests some vagrancy.

HABITAT Breeds in caves and feeds over adjacent habitats. Particularly abundant in areas such as Niah caves in Sarawak where a suitable cave is situated amongst primary forest.

DESCRIPTION *C. s. salangana*
Adult Sexes similar. **Head** Very black upper head with purple gloss. The line of feathers over the eye is slightly greyer. The ear-coverts are paler brown, but the contrast with the darker crown and the paler grey throat is not striking. Supraloral spot white-based, black-tipped. **Body** The whole of the upperparts are uniform with the crown (no rump band). The underparts are dark with the chocolate grey-brown throat just paler. The underparts become progressively darker from the chest in many individuals, with the undertail-coverts darkest grey with paler fringes. The shaft streaking is very light (lacking in many). **Wings** The upperwing is very black with a purple gloss, with paler brown innerwebs, most apparent on the tertials. The underwing is paler, with contrast between the dark brown lesser and median coverts and the dark grey remiges and greater coverts. **Tail** Like the remiges above and below.

Juvenile Differs in having pale fringes to the tips of the remiges.

Measurements (BMNH). Wing: (10) 113-123 (118). Tail:

(5) 49-52 (50.6).

GEOGRAPHICAL VARIATION Four races recognised. Those occurring in the Sumatran Padang Highlands, at least at Bukit Tinggi (Wells 1975), are thought to be closest to *natunae*.
  C. s. salangana (Java) Described above. Tarsal feathering weak or lacking.
  C. s. natunae (Natuna Islands, N Borneo and probably Sumatra) Apparently averages longer-winged than *salangana* otherwise similar. Feathered tarsus. Wing (Sumatra, 5) 121-128 (124), tail 52-56 (53.8), tail-fork 4%-7% (Wells 1975). BMNH Wing: (10) 113.5-124.5 (119.4). Weight: ca.23 (Cranbrook 1984).
  C. s. maratua (Maratua archipelago off the NE Bornean coast) Smaller than *salangana* or *natunae*. Generally paler than *aerophila*. Type measurements: wing 118; tail 48 (Riley 1927).
  C. s. aerophila (Nias and other islands west of Sumatra) Similar in size to *maratua*, thus smaller than *salangana* or *natunae*. Darker than *maratua*. Tarsus naked. Most deeply forked tail, with two measuring 11% and 13%. Wing: (1) 118 (Wells 1975).

VOICE Both *salangana* and *natunae* echolocate.

HABITS Like most other swiftlets, most active at dawn and dusk when it can be seen forming large swarming flocks around cave mouths and over adjacent forest. Like other echolocating species it can be seen feeding amongst forests in twilight amongst the canopy and at times at much lower levels in a rather bat-like fashion.

BREEDING Built of vegetable materials, held together by a sparse amount of nest cement which is transparent when fresh, and does not harden but remains permanently soft and moist to the touch. Apparently, because this nest cement is structurally weak, the nests always depend, at least for partial support, on a ledge or on some small irregularity in the cave wall, and the truly self-supporting bracket-shaped type is not found (Medway 1966). The clutch is two eggs according to Smythies.

REFERENCES Chapman (1985), Cranbrook (1984), Dickinson (1989), Medway (1961 and 1966), Riley (1927), Smythies (1981), Wells (1975).

31   UNIFORM SWIFTLET
Collocalia vanikorensis   Plates 8 & 9

Other names: Island Swiftlet; Lowland Swiftlet; Vanikoro Swiftlet; Grey Swiftlet (race *amelis*); Palawan Swiftlet (race *palawanensis*); *Aerodramus vanikorensis*

IDENTIFICATION Length 13cm. This medium-sized grey-brown species which lacks a paler rump (except in the race *pallens* from the Bismarck Archipelago which has an indistinctly paler rump) has a wide range and due to its nondescript plumage is one of the main identification problem species in the genus. In the Philippines it can be separated in the field with some certainty from all species except Philippine Grey Swiftlet from which it differs in having a deeper tail-fork. Fortunately Philippine Grey Swiftlet is apparently uncommon and primarily submontane compared to this species which is mainly lowland where it is common. In Wallacea the primary confusion species is Moluccan Swiftlet which differs in having a whitish rump except in the northern Moluccas where the race *infuscata* either has an indistinct pale grey rump or has the band totally missing. The race of Uniform Swiftlet in the northern Moluccas *waigeuensis* is slightly less black than *infuscata*, is browner beneath with a paler throat (as opposed to uniformly grey) and always has a uniform rump. In Papua New Guinea the main confusion species is Mountain Swiftlet which is hard to separate as it has a dark rump (note that pale bases to the rump feathers can give the impression of a paler rump) and similar measurements. However *granti* (the Papuan race of Uniform) has a pale throat contrasting with the darker brown underparts compared to the more uniformly marked Mountain Swiftlet. In the Central Highlands the upland form of Mountain Swiftlet, *excelsa*, is recognised by its much larger size. In the Bismarck Archipelago Uniform Swiftlet can be distinguished from the only other grey-brown species occurring there, White-rumped Swiftlet, as it is at best slightly paler on the rump, with brownish or brownish-black upperparts, and grey-brown on the underparts with clearly paler throat patch.

Figure 24. *Coultasi* race of Uniform Swiftlet (left) and *eichorni* race of White-rumped Swiftlet (right). Both forms are present in the St Matthias group, Bismarck Archipelago.

White-rumped Swiftlet has a white rump band (that varies somewhat between islands but is always distinct) blackish upperparts and very uniformly pale grey underparts. This species is the only grey-brown swiftlet in the Louisade Archipelago. In the Solomon and Santa Cruz Islands it needs to be separated from White-rumped Swiftlet which in these island groups has a narrow, yet distinct, grey-white rump. The forms of White-rumped Swift in the New Hebrides and Loyalty Islands further differ in having whitish underparts except for a much darker throat. Typically slightly larger than White-rumped Swiftlet through most of range.

In the hand, this species is distinguished from Mountain Swiftlet by its wider, longer bill, less white on the rami of the back feathers, and browner plumage above and below. The tarsus is typically unfeathered as opposed to feathered in Mountain. From Philippine Grey Swiftlet the sympatric race of Uniform, *amelis*, differs in naked tarsus and measurements (see under that species). All races can be told from Mossy-nest Swiftlet by having white rami on back feathers. From most Edible-nest Swiftlets by uniform rump. From most White-rumped and all Moluccan Swiftlets by naked tarsus. Distinguished from Melanesian and Polynesian forms of White-rumped, that lack tarsal feathering, by uniform as opposed to white or greyish rump bands. Tarsus usually bare but Browning (1993) notes that feathers are present in some populations and individuals. Browning also notes the tail/wing ratio of the nominate form to be 0.45-0.48 (mean 0.46).

DISTRIBUTION Philippines, E Indonesia, New Guinea and Melanesia. Widespread in the Philippines; Catandu-

anes, Cebu, Gigantes, Dinagat, Luzon, Mindanao, Mindoro, Panay, Sibuyan and Palawan. Likely to occur on Bantayan, Batan, Bohol, Cagayancillo, Marinduque and Verde. Moluccas, on Morotai, Halmahera, Ambon, Banda and Seram Laut. Found throughout New Guinea in lowlands and hills, and the Aru islands, W Papuan Islands of Misool and Waigeo, Numfor, Biak, Yapen, Trobriand, Woodlark, Goodenough and Fergusson in the D'Entrecasteaux Archipelago, Misima and Tagula in the Louisiade Archipelago, most of the Bismarck group, Manus and Rambutyo in the Admiralty group, Nissan, Bougainville and Buka. According to Coates (1985) the presence of *hirundinacea* on Karkar requires confirmation of this species on Manam. In Melanesia, occurs on the Solomons, Santa Cruz Island, Vanuatu and New Caledonia. Recorded once from NE Queensland Australia.

Common to abundant throughout.

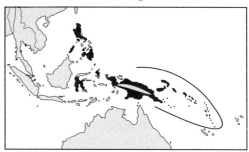

MOVEMENTS Essentially sedentary with only localised movements.

HABITAT In New Guinea throughout the lowlands to 1400m, but on Goodenough altitudinal range extends to 1600m. Although mainly seen over forest it will occur over all habitats including coastal types. Offshore islands often visited.

DESCRIPTION *C. v. vanikorensis*
Sexes similar. **Head** White supraloral spot obvious. Upper head dark grey-brown, with feathers above eye paler-fringed on average. Ear-coverts slightly paler, but contrasting enough with upper head to give capped appearance. Throat pale grey-brown with little shaft streaking. Dark brown bases to throat feathers can give mottled appearance especially on chin. **Body** Grey-brown upperbody, head often appearing slightly blacker. Rump uniform with upperparts, but slightly paler impression can be seen due to the exposure of the usually concealed paler feather bases and white feather tufts. Plumage with distinct sheen. Underbody pale grey-brown with shaft streaking being rather restricted. Slightly darker grey undertail-coverts lacking whitish fringe. **Wing** Blacker more glossy than upperparts, but with paler innerwebs. Underwing paler on remiges and greater coverts, but uniform with darker median and lesser coverts. **Tail** As remiges.

Measurements New Hebrides. Wing: male (16) 118-125 (120.4); female (11) 118-125 (121.7). Tail: 50-53. Tail-fork 7-11 (Salomonsen 1983).

GEOGRAPHICAL VARIATION 14 races recognised here. *Amelis* and *palawanensis* are sometimes considered to be separate species, amd *amelis* is sometimes included in Edible-nest Swiftlet *C. fuciphaga*. The two forms on Sulawesi, *aenigma* and *heinrichi* are sometimes included in Mossy-nest Swiftlet, *C. salangana*.

C. v. amelis (Philippines, on Luzon, Mindoro, Cebu, Bohol and Mindanao) Uniform upperparts. Smaller than *palawanensis*. Wing: 111-127. Tail: (distinctly forked, 7-8 in three BMNH specimens): 46-56 (McGregor 1909, in Dickinson 1989a).

C. v. palawanensis (Palawan, Philippines) Back and rump fairly uniformly blackish-brown. Longer-winged than *amelis*. Wing: 123-134. Tail: 53-56. Tail-fork: 6-9mm (Dickinson 1989a).

C. v. aenigma (C Sulawesi) Dark race, blackish above (darkest on crown and forehead) with blue-green gloss and darker grey-brown below. Undertail-coverts slightly darker than rest of underparts, with slight whitish fringes. Rump in some individuals gives impression of being slightly paler, as a result of extensive white rami. Wing: male (16) 114-120 (116.6); female (11) 112-121 (117.2) (Stresemann 1940). Tail: 47-52 (49.4). Tail-fork: 5-9 (6.4) (Mayr 1937).

C. v. heinrichi (S Sulawesi, from Maros and Makasar) Paler below than *aenigma*, more green-glossed. Wing: (5) 111-115.5 (112.5) (Stresemann 1932).

C. v. moluccarum (Seram, Ambon, Banda, Gorong, Tayandu and Kai Islands) As *aenigma*, but larger eye patch and distinctly pale-fringed underwing-coverts. Wing: (23), 110-118 (114.5) (Stresemann 1932).

C. v. waigeuensis (Morotai and Halmahera in N Moluccas, Misool, Batanta, Waigeo and probably Salawati) Slightly darker than *aenigma*, but throat and upper breast slightly paler. Crown and forehead uniform with upperparts. The only race lacking dark shaft streaks. Small race. Wing: male (5) 109.0-112 (110.3). Tail: (5) 44-47. Tail-fork 5.5-8 (6.9) (Salomonsen 1983).

C. v. steini (Biak and Numfor Islands in Geelvink Bay off New Guinea) Larger than *waigeuensis* and usually darker. Wing: (6) 115-123 (118.2). Tail: (8) 49-55 (52.6). Tail-fork: 5-12 (9) (Salomonsen 1983).

C. v. granti (Throughout lowland New Guinea, Yapen Island and nearby D'Entrecasteaux and Aru Islands) Blacker above than *vanikorensis*, but with similar underparts. Relatively short and shallowly forked tail. Wing: (10) 116-122 (118.6). Tail: (3) 46-50 (48.7). Tail-fork 7.5-8 (7.8) (Salomonsen 1983).

C. v. tagulae (Louisiade Archipelago, Trobriand and Woodlark Islands) Similar to *vanikorensis*, but undertail-coverts paler and greater tendency to have exposed brown bases of the throat as a dark area on chin and central throat. Wing: (2), 118.5-122 (120.3). Wing: (11) 121-129 (123.5). Tail: 54-56. Tail-fork: 9-12 (10.1) (Salomonsen 1983).

C. v. pallens (New Britain, New Ireland, Djaul and New Hanover) Similar to *tagulae*, but darker upperparts (extensive white bases to mantle and nape can produce ill-defined nape band and indistinctly paler rump band), white supraloral spot can be better defined, ear-coverts paler and silvery-grey, almost whitish throat patch contrasts with rest of underparts. Browning (1993) considers the upper back to be paler than other races. Wing: male (44) 113-124 (117.2); female (21) 114-124 (118.9). Tail: (12) 46-50.5 (49.0). Tail-fork: 4.5-8.5 (6.6) (Salomonsen 1983).

C. v. lihirensis (Nuguria and Hibernian Islands; Tabar, Lihir and Feni Islands) Intermediate between *coultasi* and *lugubris*, similar above to *lugubris*, back as crown but rump slightly paler, mantle rami white (not densely distributed, never shows suggestion of nape band),

underparts slightly paler than *lugubris* and therefore much darker than *coultasi*, throat slightly paler than rest of underparts. Wing: male (11) 120-127 (123.4); female (13) 120-128 (122.8). Tail: 54-57. Tail-fork: 6-9 (8.0) (Salomonsen 1983).

C. v. coultasi (St Matthias Group, Bismarck Archipelago and Manus, Rambutyo and Los Negros in Admiralty Islands) Paler above than *pallens* with crown black contrasting with mantle and back, with rump still paler (light grey in some). White bases and rami to upperparts (especially nape). White supraloral patch. Underparts as *pallens*. Measurements (11 Mussau including some immatures): wing 121-129 (124.8); tail 53-57 (54.9); tail-fork 8-11 (9.3). Five males and 11 females from the Admiralty Islands averaged slightly smaller: wing; male 118-124 (121.4), female 118-130 (123), tail; 54-56, tail-fork; 8-11 (9.3) (Salomonsen 1983).

C. v. lugubris (Solomon Islands) Dark form; uniform and blackish upperparts (white rami virtually missing or much reduced), ear-coverts blackish, only concealed bases of supraloral feathers whitish and underparts dark grey. Wing: male (5) 111-116 (113.4); females (9) 111-117 (114.2). Tail: (7) 49-52 (50.9). Tail-fork: 6-8 (7.1) (Salomonsen 1983).

C. v. vanikorensis (Santa Cruz Islands; Duff Group, Swallow Group, New Hebrides and Bank Island) Described above. As *lugubris* but much larger.

VOICE Echolocates within human spectrum. Coates (1985) describes flight call heard from flocks in New Guinea as a descending then rising squeaky song-like trill: *zoo-zu-chee-chee*, repeated three or four times.

HABITS Typically gregarious, often in flocks of hundreds. Frequently associates with other swiftlets. Very mobile, seen feeding in numbers at one site and then rapidly moves on. In areas of open ground (including seashore), areas of open water and over forest canopy will feed very low. More often found at great height flying rapidly over forest. Feeding occurs all day and like many members of the genus still active at nightfall.

BREEDING The nests of all forms are not known, but those that have been described are somewhat rounded, although tending to the characteristic bracket-shape, composed largely of strands of vegetable material bound together with a moderate to sparse application of a firm nest cement. In some cases nests are noted as resting on a supporting surface, rather than being suspended as is the true bracket-shaped type (Medway 1966). Coates (1985) describes the nest in New Guinea as a cup of mossy material with saliva that has a shallow depression for the eggs. Usually sited in total darkness in caves or sinkholes, and attached either to vertical walls or steeply sloping surfaces. Mayr and van Deusen (1956) describe a nest located in the shelter of an overhanging bank of earth and rocks held together by tree roots at the side of a small stream. Colonial. One or two white eggs are laid.

REFERENCES Bregulla (1992), Browning (1993), Coates (1985), Dickinson (1989a), Mayr (1937), Mayr and van Deusen (1956), Medway (1966 and 1975), Medway and Pye (1977), Salomonsen (1983), Sibley and Monroe (1990), Stresemann (1932 and 1940), White and Bruce (1986).

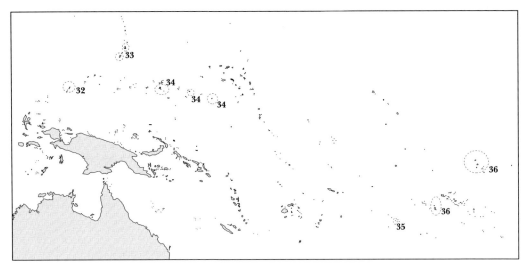

Distribution of Pacific island swiftlets

32   PALAU SWIFTLET
Collocalia pelewensis   **Plate 5**

Other name: *Aerodramus pelewensis*

IDENTIFICATION Length 11cm. This small species with a proportionally short forked tail is the only swiftlet occurring on the island of Palau and as such identification is straightforward. The pale rump distinguish this species from other Micronesian swiftlets.

Wing/tail ratio is different from that of Caroline Swiftlet (which occurs at a similar latitude suggesting that this character is not a function of geographic variation). Differs from Uniform by the lack of great contrast between throat and abdomen or a supraloral spot, pale rump (only *C. v. pallens* shows a slightly paler rump) and shorter tail. Resembles Guam Swiftlet but wing/tail ratio differs (0.41-0.46, mean 0.44, as opposed to 0.47-0.51, mean 0.489 in Guam). Palau Swiftlet has a more deeply forked tail than Caroline or Guam Swiftlets. Marshall (1949) noted that this species differs from Guam Swiftlet in that it has a pale rump patch (indeed this feature is not found in any other Micronesian *Collocalia*), is greyer (i.e. less brown) ventrally, more glossy black on the upperparts and slightly longer-winged. Tarsus naked.

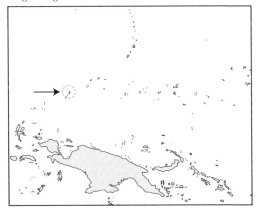

DISTRIBUTION Found only on the Palau Islands, west Pacific where it is present on the larger islands. Noted as absent on Angaur Island (Pratt *et al.* 1987).

Recorded as abundant on the islands by Marshall (1949). Common to abundant from Badeldoab to Peleiu on sizeable islands (Engbring 1988).

MOVEMENTS Resident.

HABITAT Avoids dense forest. Usually encountered in large flocks over ridges or steep canyons. Smaller groups may spend all day feeding in a small area.

DESCRIPTION *C. pelewensis*
Sexes similar. **Head** Dark black-brown with a slight fuscous-green sheen. White restricted to bases of supraloral patch and therefore inconspicuous. Base of throat feathers dark but tips silvery grey so throat appears pale, but not contrastingly. Ear-coverts not notably paler than upper head. **Upperparts** Mantle and back as crown or slightly paler. Much variation in rump colour but always at least slightly paler. Apparently varies in the tip colour which can be rather green-glossed or grey (may be age related). Bases, however, always pale. **Underparts** Darker below throat, dull grey with some inconspicuous shaft streaks, most noticeable on the undertail-coverts. Longest undertail-coverts slightly darker and green-glossed. **Wing** Very uniform on upperwing with only innerwebs slightly paler (but apparently less so than in some other *Collocalia*). Remiges paler below and contrasting with darker median and lesser coverts. **Tail** As remiges.

Measurements Wing: 107-113 (Browning 1993).

GEOGRAPHICAL VARIATION None. Monotypic.

VOICE Typical call of the genus described as consisting of slight chirps and twitterings (Marshall 1949).

HABITS Gregarious, gathering in large feeding flocks at dusk.

BREEDING Brandt (1966) described nests attached on ceilings of caves as very flat structures that had linings of moss and fine grass.

REFERENCES Brandt (1966), Browning (1993), Engbring

129

(1988), Marshall (1949), Mayr (1935), Pratt *et al.* (1987), Salomonsen (1983).

33   GUAM SWIFTLET
Collocalia bartschi   Plate 5

Other names: Guam Cave Swiftlet; Mariana Swiftlet; *Aerodramus bartschi*

IDENTIFICATION Length 11cm. The only swiftlet on the southern Mariana Islands rendering identification straightforward. Separated from its closest relative geographically by its uniformly coloured upperparts, lacking the pale rump of Palau Swiftlet.

Plumage similar to Caroline Swiftlet but slightly paler; uniform upperparts (lacking sheen), underparts lacking dark shaft streaks and supraloral patch much reduced or lacking. Like Uniform Swiftlet has paler throat (not as contrastingly pale as in nominate Uniform). Has shallower tail-fork than Caroline Swiftlet but similar wing/tail ratio (0.47-0.51 mean 0.489 in Guam, as opposed to 0.47-0.53 mean 0.494). Palau is also superficially similar, but it has a deeper tail-fork and a wing/tail ratio similar to Uniform Swiftlet: 42-46, mean 44 (Browning 1993). Marshall (1949) noted that this species differs from Palau Swiftlet in that it is uniform above. Tarsus bare in 7 of 10 individuals studied by Browning (1993), with the remaining three having single feathers on the middle of one tarsus.

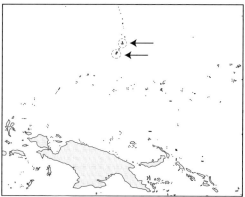

DISTRIBUTION Endemic to the S Mariana Islands; on Saipan, Tinian and Guam Islands. It was introduced onto Oahu in the Hawaiian Islands in 1962 from Guam.

Marshall (1949) considered it abundant on Saipan and Guam. However, more recently the species appears to have declined with Pratt *et al.* (1987) noting that the species was eradicated from Rota and close to extirpation on Guam and uncommon on Saipan, Tinian and Agiguan. In 1989 12 pairs nested in the Halava Valley, Oahu (Browning 1993, quoting Engbring pers. comm.).

MOVEMENTS This species exhibits at least some degree of nomadism or seasonal activity as evening congregations on any one island relocate regularly and at times may be missing from one island or another. For example Marshall recorded the species on Tinian during mid October 1945, but absent by mid November of the same year.

HABITAT Found over both cleared and forested areas (including the interior and coastal mangroves).

DESCRIPTION *C. bartschi*
Sexes similar. **Upperparts** Supraloral spot much reduced or lacking. Ear-coverts slightly paler than upper head. Uniformly warm brown above lacking paler rump band. Plumage lacks any noticeable sheen. **Underparts** Silvery grey-white on throat and upper chest with remainder of underparts darker and greyer. Lacks any dark ventral streaking.

Measurements Wing: 100-108 (Browning 1993).

GEOGRAPHICAL VARIATION None. Monotypic. This form is sometimes considered to be conspecific with Caroline Swiftlet *C. inquieta*.

VOICE Described as weak chirps and twitterings (Marshall 1949).

HABITS Large flocks form in the evening with all the birds from the surrounding area congregating and feeding close to the ground until, when it is dark, they ascend and move away. The feeding time overlaps with that of the islands sheath-tailed bats (*Emballonura*) and Marshall (1949) noted that both operate together in a very similar style.

BREEDING Jenkins (1983) notes that the nest is often cone-shaped and composed of moss tightly held together and firmly secured to cave walls with copious amounts of hardened mucus-like saliva. According to Jenkins the preferred nest site is on cave ceilings. Browning (1993), quoting J. Reichel and J. Engbring, notes that in the Marianas the nests are usually self-supported and contain sparse to copious amounts of nest cement, whereas one nest collected from Hawaii was composed of vegetable matter consisting mainly of a liverwort (*Herbertia* sp.) and sparse nest cement. All of nests observed were supported by niches in the cave wall.

REFERENCES Browning (1993), Jenkins (1983), Marshall (1949), Mearns (1909), Medway (1966), Pratt *et al.* (1987).

34   CAROLINE SWIFTLET
Collocalia inquieta   Plate 5

Other names: Caroline Islands Swiftlet; Micronesian Swiftlet; *Aerodramus inquietus*

IDENTIFICATION Length 11cm. The only member of the genus in the Carolines. A very dark swiftlet, especially below, which can show a very indistinct paler rump.

From Uniform Swiftlet by the less obvious supraloral spot, reduced ventral streaking, reduced sheen to plumage, lack of silvery-white throat (*C. i. ponapensis* can be pale-throated) contrasting with underparts typical of that species and darker ear-coverts. Proportionally longer-tailed than Uniform (nominate form), with a wing-tail ratio of 0.47 - 0.53 (mean 0.49) compared with Uniform 0.45-0.48 (mean 0.46) (Browning 1993). Tarsus usually bare but Browning found one individual with a solitary feather located on the inside of its tarsus.

DISTRIBUTION Endemic to the Caroline Islands where it is found from Yap east to Kosrae (=Kusaie). Several authors include Yap in the range of this species on the strength of a sighting of *Collocalia* sp. on the island from the last century. However, Pyle and Engbring (1985) have doubts over the authenticity of the record.

Noted as abundant on Truk and common on Pohnpei

(=Ponape) by Ralph and Sakai (1979) (Browning 1993) notes population estimates, of 27,900 on Kosrae, 29,800 on Pohnpei and 25.800 on Truk (from unpublished data collated by J Engbring *et al.*).

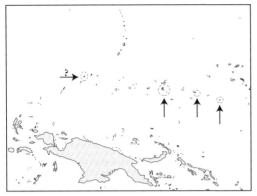

MOVEMENTS Resident.

HABITAT Breeds in caves and is found over open country and forests.

DESCRIPTION *C. i. ponapensis*
Sexes similar. **Head** Crown oily black-brown and slightly green-glossed. Blackish-fringed feathers with slightly greyer bases most visible on rear crown and hind neck. Ear-coverts greyer brown with only a slight contrast with crown. Extensive greyish-white bases to supraloral feathers which are narrowly tipped black. Throat dark grey-brown with little or no contrast with the ear-coverts. **Body** Mantle and upper back black-brown (faintly green-glossed), slightly paler than the crown. Very faint rump band caused by few visible pale grey bases. Uppertail-coverts as mantle but not glossed. Underbody uniform with throat and similarly lacking shaft streaks. Proximal undertail-coverts slightly paler with distal ones a little darker. Undertail-coverts have prominent shaft streaks. Some white bases to underbody feathers. **Upperwing** Very uniform black-brown. Darkest on outer wing and smaller coverts (which show greenish gloss). Browner on innerwebs of remiges, secondaries and greater coverts of innerwing. **Underwing** Paler and greyer on remiges and greater coverts, contrasting with darker black-brown median and lesser coverts. **Tail** As remiges. Similar in colour above to the uppertail-coverts.
Measurements Wing: 96-119 (Browning 1993).

GEOGRAPHICAL VARIATION Three races. Guam Swiftlet is sometimes considered to be a race of this species.
   C. i. ponapensis (Pohnpei, Carolines) Described above.
   C. i. rukensis (Truk islands, Carolines) Larger than *ponapensis* but smaller than *inquieta.* Wing: 103-109 (Browning 1993).
   C. i. inquieta (Kosrae, Carolines) Wing 110-120 (Browning 1993).

VOICE Medway (1966) stated that it was not known whether or not this species could echolocate.

HABITS Presumably similar to other members of the *vanikorensis* superspecies.

BREEDING Medway (1966) reviewed all known details on the nesting of the species. Nests from Pohnpei are distinctly rounded, suppressed bracket-shaped, composed of moss and other soft vegetable matter with nest cement in evidence only at the point of attachment to the cave wall. Nests are usually in clusters and the clutch is a single egg. Nests from Truk are described as well cupped, constructed of grass stems and small tendrils; and occasionally small sticks, fern stems, and more rarely, mosses are employed. A few feathers are generally found embedded in the nest which is glued together with copious secretions of saliva and firmly attached to the cave rocks. No lining is used. Some nests are entirely supported from underneath by stone, whereas others protrude and are fastened only at the back of the nest. According to Brandt (1966) *ponapensis* has a distinctly different nest than the other two races, with a depth of 6-16cm and frequently attached to perpendicular cave or crevasse walls. He considered the nest of *rukensis* to be deeply cupped and only infrequently to be composed of moss, whereas other races had shallow nests composed mainly of moss adhered together by small amounts of saliva. Brandt (1966) notes that *rukensis* has a clutch of two in totally dark deep caves whereas *inquieta* and *ponapensis* have single clutches in less dark caves.

REFERENCES Brandt (1966), Browning (1993), Mayr (1935), Medway (1966), Pyle and Engbring (1985).

35   SAWTELL'S SWIFTLET
Collocalia sawtelli   Plate 5

Other names: Atiu Swiftlet; Cook Islands Swiftlet; *Aerodramus sawtelli*

IDENTIFICATION Length 10cm. This rather dark species is the only swiftlet occurring on Atiu. Its nearest relative, Polynesian Swiftlet, occurs 750 nautical miles away on Tahiti. In plumage the two are similar, but Sawtell's has a greater tendency to show an indistinct pale grey band on the uppertail-coverts and is slightly paler on the underpart. This, coupled with its dark upperparts, creates a greater upperpart/underpart contrast than in Polynesian Swiftlet.

   Both this species and Polynesian Swiftlet (both races, *leucophaeus* and *ocista*) are medium-sized, long-tailed, dull-coloured swiftlets with the pale supraloral patch, present in many associated species, much reduced or lacking. They also typically have a distinctive soft feel to the plumage. Sawtell's Swiftlet can be distinguished from *leucophaeus* by its proportionally shorter tail, smaller size, paler bases to the uppertail-coverts, generally paler underparts, different bill structure (weaker with a rather long hook and abrupt attenuation distally see figure 25) and dark feet and claws.

Figure 25. Heads of Sawtell's Swiftlet (left) and Polynesian Swiftlet of the nominate race (right) showing differences in bill shape.

Ocista has a proportionally longer tail, is larger, averages darker on the bases to the uppertail-coverts, darker underparts, averages a stronger bill structure and slightly darker feet and claws, and is therefore somewhat intermediate between *leucophaeus* and Sawtell's Swiftlet.

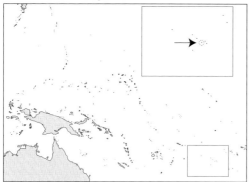

DISTRIBUTION Endemic to Atiu Island in the southern Cook Archipelago.

One of the rarest swifts with only 60 nests recorded when the species was originally discovered in 1973 in Annatake-take Cave. Local inhabitants, however, reported that there were several smaller colonies elsewhere on the island. Fieldwork in 1987-1988 recorded 190 nests in use in two caves and found the primary cause of fledgling mortality to be predation by crabs and starvation after falling from the nest (Tarburton 1990). Not apparently declining in numbers, but it has been noted that tourist disturbance may be a future problem. The species is listed by BirdLife International as Vulnerable (Collar *et al.* 1994).

MOVEMENTS Resident.

HABITAT Reported as feeding over forest and open country (Collar *et al.* 1994).

DESCRIPTION *C. sawtelli*
Sexes similar. **Head** Very dark, with whole of upper head and ear-coverts black with a greenish gloss. Throat grey-brown. **Body** Remainder of upperparts similar to upper head but less glossy and more brownish. The shortest uppertail-coverts are grey-brown with paler bases that are partly exposed. The underparts are very similar to the throat. **Wing** The upperside is very blackish with a green gloss, although the innerwebs are browner and paler. The underwing is paler with the darkest area being the blackish lesser and median coverts. **Tail** Similar above and below to remiges.

Measurements Wing: male (2) 117.5-118 (117.75); female (4), 115.5-118.5 (117.63) Tail: male (2) 53-54.5 (53.75); female (4) 53-56 (54.88) (Holyoak 1974).

GEOGRAPHICAL VARIATION None. Monotypic.

Figure 26. Nest of Sawtell's Swiftlet. (Based on Holyoak and Thibault 1978).

VOICE An echolocating species that utters the typical rattling call.

HABITS Tends to feed by flying slowly through the foliage and canopy of the forest and its edge.

BREEDING Breeds in caves where, according to Thibault and Holyoak (1974), it constructs shallow cups of dry stems and leaves glued together with saliva which never hardens. Medway and Pye (1977) characterise the nest as externally supported; vegetable materials; sparse, sticky nest cement (figure 26).

REFERENCES Bennett (1987), Collar *et al.* (1994), Holyoak (1974), Holyoak and Thibault (1978), Medway and Pye (1977), Tarburton (1990).

36 POLYNESIAN SWIFTLET
Collocalia leucophaeus   Plate 5

Other names: Tahiti Swiftlet (race *leucophaeus*); Marquesan Swiftlet, Marquesas Swiftlet (race *ocista*); South Pacific Swiftlet; *Aerodramus leucophaeus*

IDENTIFICATION Length 11cm. A medium-sized swiftlet with a long tail that is usually quite dark-rumped (most show slight paling on rump, most apparent in the race *ocista* and '*thespesia*' from Tahiti). Sawtell's Swiftlet is the closest relative, both geographically and taxonomically, and is slightly smaller with a proportionally shorter tail, more distinct rump patch and slightly paler underparts.

Although the two races differ in some respects they can both be separated from Sawtell's Swiftlet in that they have a different bill structure, being heavier with a short hook that is not heavily attenuated towards the tip (*ocista* is intermediate with Sawtell's in this respect), the bases to the rump feathers are darker (*ocista* is intermediate), and the legs and claws are paler (intermediate in *ocista*). The upperparts are blackest in Sawtell's and palest in *leucophaeus* (intermediate in *ocista*) and the underparts are somewhat darker. Both forms of Polynesian Swiftlet are larger than Sawtell's. *Ocista* has a proportionally longer tail than Sawtell's Swiftlet.

DISTRIBUTION Endemic to the Polynesian Islands in the Eastern Society Islands of Moorea and Tahiti, and the Marquesas Archipelago: Eiao, Nuku Hiva, Ua Huka, Ua Pou, Hiva Oa and Tahuata.

Listed by BirdLife International as Vulnerable. However, this refers only to *leucophaeus* and not to *ocista* (Collar *et al.* 1994). This latter form has apparently declined this century, possibly due to the introduction of Common Myna *Acridotheres tristis*, which may prey on both the swiftlets' eggs and chicks. Between 1986-1991, the species was searched for in 39 Tahitian valleys, but found only in 6. In 1984 the population was thought to number 200-500 birds. Previously *leucophaeus* occurred on Huahine and it is thought that it may only be a vagrant to Moorea. (Collar *et al.* 1994).

MOVEMENTS Both races are exclusively resident.

HABITAT On Tahiti feeds over rivers and over moist, forested, rocky valleys at greater height over the canopy than either *ocista* or Sawtell's Swiftlet. On the Marquesas feeds slowly through the canopy of the forest and forest edge in much the same manner as the Sawtell's Swiftlet.

DESCRIPTION *C. l. leucophaeus*
Sexes similar. **Head** Dark grey-brown upper head, including ear-coverts. White in the base of the eye patch is very restricted. Throat pale grey-brown. **Body** Upperparts dark grey-brown with only an impression of a paler rump, due to the paler feather bases. Below, the grey-brown feathers are lightly streaked brownish, with little contrast on undertail-coverts. **Wing** Uniformly blackish-brown above with paler innerwebs. The underwing shows the rectrices somewhat paler with the lesser and median coverts appearing darker and browner.

Measurements Tahiti (10). Averages: wing 126.1; tail 57.4 (Holyoak and Thibault 1978).

GEOGRAPHICAL VARIATION Two races which are sometimes considered separate species.
  C. l. leucophaeus (Moorea and Tahiti, East Society Islands) Described above. Peters (1940) included the form *thespesia* which was collected from Tahiti. Browning (*in litt.*) notes that the type of *thespesia* has a contrastingly paler rump than the back (only 1 out of a series of 7 from Marquesas had a rump close to that of *thespesia*) and averages duller and browner than those from the Marquesas. Another specimen in the United States National Museum collected one day apart from the type of *thespesia* on Tahiti is closer to the darker-rumped birds from Marquesas in everything other than size.
  C. l. ocista (Marquesas Islands) The differences between this race and the nominate are discussed above. Holyoak and Thibault studied differences in wing and tail proportions as an index expressed by calculating the tail length as a percentage of the wing-chord. The nominate race has an index of 45.5 whereas the various populations of *ocista* on the different Marquesas Islands ranged between 49.8 and 52.0. Sawtell's Swiftlet has an index of 46.6. Measurements (range of average measurements for various Marquesas Islands). Eiao (6): wing 119.1; tail 61.4. Nuku Hiva (37): wing 122.4, tail 62.6. Ua Huka (12): wing 121.6; tail 60.5. Ua Pou (5): wing 127.4; tail 66.3. Hiva Oa (21): wing 123.1, tail 62.8. Tahuata (4): wing 121.7; tail 62.3 (Holyoak and Thibault 1978).

VOICE Thibault (1975) gives the call of the nominate race as a short trill *trrr* delivered in flight. Observations have shown that *ocista* can echolocate, but the position of *leucophaeus* is uncertain.

HABITS The nominate race, with its smaller population, is seldom seen in the large flocks of up to 100 individuals in which *ocistus* can be found.

BREEDING A variety of sites are used from shallow depressions under overhanging rocks or coastal cliffs, to caves

either deep or shallow. Colonies up to 300 individuals recorded. Medway and Pye (1977) characterised the nest as self-supporting or externally supported; vegetable materials: sparse, sticky nest cement (figure 27).

Figure 27. Nest of Polynesian Swiftlet. (Based on Holyoak and Thibault 1978).

REFERENCES Collar *et al.* (1994), Holyoak and Thibault (1978), Medway and Pye (1977), Pratt *et al.* (1987), Thibault (1975).

37   BLACK-NEST SWIFTLET
Collocalia maxima   Plates 6 & 7

Other names: Low's Swiftlet (race *lowi*); Robinson's Swiftlet (race *maxima*); Indo-Malayan Swiftlet; *Aerodramus maxima*

IDENTIFICATION Length 14cm. A large grey-brown swiftlet with moderately forked tail. Varies throughout range in rump colour. The race *maxima* occurs from the north of the range to W Java and has an indistinct greyish rump band, whilst *lowi* (Sumatra, Nias Islands, W Java, N and W Borneo) has a darker rump and *tichelmani* (SE Borneo) has the palest rump. In the extreme north of its range may overlap with Himalayan Swiftlet (which also winters south to Malaysia and possibly Sumatra) which has a more deeply forked tail. Black-nest Swiftlet is noted as having a heavier build and broader wings than Himalayan Swiftlet, a paler rump and averages paler below. Various races of Edible-nest Swiftlet are sympatric with Black-nest Swiftlet but all are smaller and average deeper tail-forks, and additionally differ in plumage: *C. fuciphaga germani* (S Thailand, Malay Peninsula and coastal N Borneo) has much paler underparts and a well-defined broad whitish rump band; *C. f. amechana* (Malay Peninsula) has a well-defined pale grey rump and averages paler below than *maxima*; *C. f. vestita* (Sumatra and Borneo) is very similar but some pale can be apparent on the rump - lacking in *lowi*, but present in *tichelmani*. The slightly smaller Mossy-nest Swiftlet is sympatric on Sumatra, Java and N Borneo and always has a dark rump, is slightly darker below and shows a squarer tail. This species has occurred as a vagrant to Palawan.
  Averages longer-winged (120-140, mostly over 125) than Mossy-nest Swiftlet (113-128) and the Edible-nest Swiftlet group (106-125). In the hand, from Himalayan Swiftlet by the heavily feathered tarsus (figure 28) and the less deeply forked tail (invariably less than 13% of tail length as opposed to over 15% in Himalyan). Rami of back mainly dark-tipped with only traces of white.

Figure 28. Tarsus feathering of Edible-nest Swiftlet of the *germani* race (left), Himalayan Swiftlet (centre) and Black-nest Swiftlet (right). Note that the tarsal feathering in the first two species can be individually and racially variable (the Himalayan Swiftlet depicted is a heavily feathered individual).

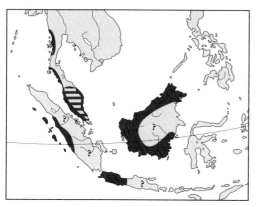

DISTRIBUTION Wide ranging in SE Asia and throughout the Sundas. Occurs in Tenasserim, in Burma, and the west coast of the Thai peninsula from just north of Phuket south into Malaysia. In W Malaysia widely distributed at all elevations. Known colonies occur south to Belitung (Satul) on the west coast and offshore islands and from Redang island (Terengganu) south to the Pahang-Johore archipelago on the east coast. The only inland site is Kuala Lumpur Town Hall where the species has been noted breeding in a colony of Edible-nest Swiftlets. In January-April 1967 a large population was believed to be roosting locally in the vicinity of Gunong Benom (Medway and Wells 1976). On Sumatra, status obscured by the relatively few colonies found almost exclusively on the coast and on islands. A presumed mixed colony found in 1914 with Mossy-nest Swiftlet was in a limestone cave at Buo, Padang Highlands, Barat. Also on Simeulue and Nias Islands. Thought to be one of three species occurring in enormous numbers on Berhala island in the Malacca Straits and on Sipura in the Mentawi Islands and probably also Siberut. Occurs throughout W Java. Smythies does not give exact details of the Bornean distribution, but it is thought to occur in N, E, and SE Borneo, including Labuan and the Anambas Islands. *Lowi* has been collected once from Palawan by Mosely on 6 September 1887, and its status on that island is probably that of a vagrant.

Black-nest Swiftlet has been considered to have a disjunct northern population in E Bhutan and Arunachal Pradesh, based on specimens collected by Ludlow at 2100-3900m (see Ripley 1982 and Ali & Ripley 1970). These specimens were examined by T. P. Inskipp and C. Inskipp and found to be misidentified Himalayan Swiftlets *C. brevirostris* (T. P. Inskipp *in litt.,* in Beaman 1994).

Rather scarce in the Thai and west Malaysian range, elsewhere often abundant.

MOVEMENTS Usually sedentary but its occurrence on Palawan suggests occasional vagrancy.

HABITAT Occurs at all elevations from the coast to the highlands. Feeds over a variety of habitats close to colonies. Regularly in quite dense forest in amongst trees.

DESCRIPTION *C. m. maxima*
Sexes similar. **Head** Very glossy black-brown upper head with slightly paler and less glossy lores, forehead and narrow line of feathers over eye. Ear-coverts are paler and browner than the crown and clearly darker than the pale grey throat. The throat is marked by brown feather shafts. **Body** The upperparts are uniform with the crown. The rump varies in darkness and can appear quite uniform with the rest of the upperparts. However the rump has extensive pale basal barbs so when feathers are disturbed some pale may be seen here. Some individuals have paler grey, indistinct rump bands. The grey underparts are palest on the throat, with the shaft streaking becoming heavy from the chest and the grey tone a little darker from here to the undertail-coverts which are darker grey with pale grey fringes. **Wing** The upperwing is very uniform and glossy-black, darkest and most glossy on the outerwebs and palest and brownest on the innerwebs. This paleness is most apparent on the tertials. The underwing is generally paler with the dark brown lesser and median coverts being clearly darker than the grey greater coverts and remiges. **Tail** This is similar above and below to the remiges.

Measurements Wing: 128-135. Tail: 50-58. Tail-fork: 11-15% (Medway 1961). Wing: (4) 126-133 (130) (Stresemann 1931). Weight: c.28 (Cranbrook 1984).

GEOGRAPHICAL VARIATION Three races.
    C. m. maxima (S Burma, peninsular Thailand, Malay Peninsula and W Java) Described above.
    C. m. lowi (Sumatra, Nias Island, N and W Borneo, Labuan Island, W Java, Palawan) Darker rump than *maxima* and than the smaller *tichelmani*. Medway (1962) states that this race and *maxima* intergrade in Western Java. Wing: 125-140 (Stresemann 1931). Measurements of 4 intergrades from Java at BMNH: 131-133 (132.5).
    C. m. tichelmani (SE Borneo) Small race, rump paler than *lowi*. Wing: 120-129 (Stresemann 1931).

VOICE Similar to Himalayan. Known to echolocate, at least in case of *maxima* and *lowi*.

HABITS Like many other swiftlets, most active at dusk and dawn when it can be seen in large numbers at cave entrances. Highly gregarious.

BREEDING All nests are bracket-shaped, incorporating feathers from all parts of the plumage, and held together by a firm translucent nest cement without the inclusion of any vegetable materials (Medway 1966). Nests characterised by Medway and Pye (1977) as self-supporting; bracket-shaped; feathers and copious firm nest cement.

REFERENCES Ali and Ripley (1970), Cranbrook (1984), Deignan (1955), Dickinson (1989a), Medway (1961, 1962 and 1966), Medway and Pye (1977), Stresemann (1931).

38   EDIBLE-NEST SWIFTLET
Collocalia fuciphaga   Plates 6 & 7

Other names: German's Swiftlet (race *germani*); Brown-rumped Swiftlet (race *vestita*); Thunberg's Swiftlet (race *fuciphaga*); Hume's Swiftlet; *Aerodramus fuciphagus*

IDENTIFICATION Length 12cm. Medium-sized grey-brown swiftlet showing variation in rump colour throughout its range. Where it occurs with dark-rumped races of Black-nest and Mossy-nest Swiftlets it is one of the hardest field identification challenges. However, through-out range it has a more deeply forked tail than either of these two species. The similarly-sized Mossy-nest Swiftlet has a very shallow tail-fork (some racial variation but only 4-7% of tail length in Sumatra, as opposed to 10-19% in Edible-nest Swiftlet) but the larger, bulkier Black-nest Swiftlet has a moderately forked tail (usually less than 13%). The race *vestita* in Sumatra and Borneo has exten-sive white rami in the region of the rump and although the feathers are broadly dark-tipped it is often easy to dis-cern a paler rump which is quite unlike Black-nest or Mossy-nest Swiftlets in this region except the pale-rumped *tichelmani* race of Black-nest Swiftlet in SE Borneo. In Java this species always shows a rump at least slightly greyer than the rest of the upperparts, whereas Mossy-nest is still uniformly-rumped but Black-nest is paler-rumped. Struc-tural differences between the three species remain constant. Throughout the sympatric ranges of Mossy-nest and Edible-nest Swiftlets, Mossy-nest tends to have darker underparts and a less pronounced capped appearance. In the Andamans and Nicobars this species only needs to be separated from the distinctive Glossy Swiftlet. In the northern part of mainland SE Asian range the distinctive race *germani*, with its very pale underparts and broad whit-ish rump, occurs. Further south in peninsular Malaysia (and possibly the extreme south of Thailand) the greyer-rumped *amechana* is found. Himalayan Swiftlet winters within the range of these two races and is larger and bulk-ier with a greyish rump (some individuals have darker rumps) that is typically clearly paler than the rump of *ger-mani*, but similar to that of *amechana*.

Averages shorter-winged (106-125) than Himalayan Swiftlet (116-141) and Black-nest Swiftlet (120-140, in most over 125). From Mossy-nest Swiftlet by rump paler than mantle (in some *vestita* it may be uniform with the man-tle) and (except in some *fuciphaga* and some *vestita*) by white tips to rami (greyish-brown in Mossy-nest). Tarsus lightly feathered or unfeathered, heavily feathered in Black-nest Swiftlet.

DISTRIBUTION Extensive, though rather disjointed, SE Asian and Indonesian range. Found throughout the An-daman and Nicobar Islands. Coastal SE Asia from Dazhou Dao, and SE coast of Hainan Island in the north, Vietnam (including the Con Son Islands) and Cambodia into Thai-land where it occurs locally in Bangkok and then from Khao Sam Roi Yot south on the east coast of the peninsu-la and from close to Khlong Nakha southwards on the west coast, southwards into the Malay Peninsula where it is present as far south as Singapore. It also occurs on many of the peninsula's offshore islands, including Phuket, Phi Phi, Lanta, Phangan, Samui, the Tenggul Group (Tereng-ganu), Penang, the Sembilan group, Tioman and the Pahang-Johore archipelago. It is also present on the Mer-gui Archipelago off S Burma. On Sumatra, found

throughout the mainland and the islands of Belitung, Simeulue, Banyak, Batu and Mentawai. Throughout Java, including Kangean Island. Throughout Borneo in suita-ble habitat. Vast, and well studied, colonies in caves at Niah in NE Sarawak and Gomantong near Sandakan in Sabah. However, probably present in most limestone caves throughout the island. Off Borneo, occurs on islands off the north coast, Satang and Laki Islands near Kuching, Mantanani Island and Berhala Island near Sandakan. Also on Anambas islands. In the Philippines, from Cagayancil-lo, Cagayan Sulu, Calamianes, Cuyo and Palawan. Dickinson (1989a) expresses doubts regarding records from Panay and Ticao. In Wallacea known from Batu (near Tanahjampea), probably Lombok, Sumbawa, Flores, Sum-ba, Sawu and Timor.

Common on the Andamans, breeding in sea caves on many islands. In Sumatra common to locally abundant. It is apparent, as a result of the small amount of observa-tions and specimens, that it is rather local and uncommon in Wallacea.

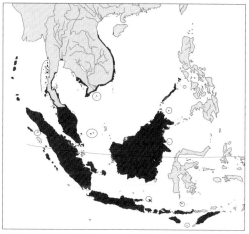

MOVEMENTS Resident throughout range.

HABITAT On the Andamans and Nicobars it behaves as other swiftlets. Hawks all day over mangroves, rubber plan-tations, forest and cultivation (Ali and Ripley 1970). Found feeding over a wide range of habitats from small coastal islets and continental lowlands to 2800m on Sumatra. Can occasionally be seen far out to sea (especially *germani*).

DESCRIPTION *C. f. fuciphaga*
Sexes similar. **Head** Crown rather glossy dark grey-brown (almost blackish) with the forehead marginally paler. The nape and ear-coverts are slightly paler but the whole pat-tern is rather uniform and less capped than in *germani*. The throat is greyish but most examples show considera-ble darker mottling. Pale supraloral patch. **Body** The whole of the upperparts are rather uniform but it in some examples the crown is slightly darker than the mantle. The rump is usually marginally greyer than the mantle but this is variable and contrast is virtually lacking in some cases. In the palest examples the longest uppertail-cov-erts can have slight pale fringing. Underparts follow a similar pattern to *germani* with the throat being palest and greyest and the remainder being slightly browner-grey from chest to vent with a variable amount of brown feath-er streaking, and with the undertail-coverts being clearly blackest with pale grey fringing. **Upperwing** Very uniform

135

and blackish with strong gloss (stronger than in *germani*) that is green when fresh and more purple when worn. Innerwebs are browner than outerwebs and this is most apparent on the tertials. Outerwing may appear blacker than mantle. **Underwing** Typical of the genus with the remiges and greater coverts paler than on upperwing, contrasting with blacker median and lesser coverts. **Tail** Similar to remiges.

Measurements Wing: 116-121. Tail: 54-58. Central tail feathers: 46-50 (Stresemann 1931). Wing: (7 BMNH) 111-115 (113). Tail 48-51 (49.1). Tail-fork: 5-8 (6.8).

GEOGRAPHICAL VARIATION Eight races are recognised here. *Germani* is often considered to be a separate species.

    C. f. fuciphaga (Java, including Kangean Island and Belitung) Described above.
    C. f. inexpectata (Andaman and Nicobar islands) Wing: 113-121. Tail 49-53. Tail-fork 6-8 (Baker and Sims, in Ali and Ripley 1971). Wing: 114-120 (Stresemann 1931).
    C. f. dammermani (Flores, Lesser Sundas) Only known from a single specimen and it is uncertain to what extent it varies from the nominate. Rump only slightly paler than mantle.
    C. f. micans (Sumba, Sawu and Timor, Lesser Sundas) Paler and greyer than *fuciphaga*, less brown below and rump slightly paler than mantle.
    C. f. vestita (Sumatra and Borneo) Similar in general pattern to *fuciphaga* but averages blacker on the upperparts with the rump barely showing as paler than the mantle (although pale rami can cause the effect of a paler rump). However shows considerable variation. One *vestita* at BMNH lacks white rami to upperparts. Wing: 112-120 (Smythies 1981). Wing: (10 BMNH) 112.5-121. Weight: 15-18 (Cranbrook 1984).
    C. f. perplexa (Maratua Archipelago, Indonesia) Rump slightly paler than upperparts. Gloss to remiges and rectrices rather purple. Wing: (9) 119.5-125 (121.8) (Riley 1927).
    C. f. germani (In the north from Hainan Island, China south along the coast of Vietnam and Cambodia, the Burmese Mergui Archipelago and Thailand south into the Malay Peninsula. Also S Philippines) Very pale underparts and broad whitish rump. Wing: (10) 113-121. Tail: 50-53 (Stresemann 1931). Wing: (13 BMNH) 114-123.5 (119.5). Two specimens from Hainan: wing 106 + 108; tail 50 + 55; weight 13 and 14 (Xian and Zhang 1983).
    C. f. amechana (In the Malay peninsula south of *germani*. Lekagul and Round (1991) consider that this race is found north to the extreme south of Thailand) Rump greyer.

VOICE A variety of calls can be heard from *germani* and these may be typical of all races. The most distinctive call is a loud metallic *zwing* not unlike that of the treeswifts. More commonly a series of *chip* notes and a wetter *crip* can be heard. The races *fuciphaga*, *vestita*, *amechana*, *germani* and *perplexa* have been proved to echolocate (Medway and Pye 1977).

HABITS This is a very gregarious species freely associating with other swifts and with hirundines in large flocks. Most frenzied feeding occurs late in evening.

BREEDING Nesting usually occurs in caves, often in vast colonies. Buildings are sometimes also employed and due to the financial rewards associated with this species the disturbances that are sometimes associated with building-nesting species are not an issue. However, over-collecting is another problem. On Belitung over one hundred pairs nest over the kitchen of a Chinese Restaurant! *Germani* and *amechana* commonly breed in urban areas. The only known inland breeding site in west Malaysia is in Kuala Lumpur Town Hall. According to Ali and Ripley (1970) nests are white, opaque and translucent, of clear inspissated saliva with little or no admixture of extraneous matter; thus of the best edible and commercial quality. More or less half-cup shaped, c.6cm across x 1.5cm deep; attached bracket-wise, close to one another, to the rock wall in sea caves, in small colonies of a few nests to large ones of several hundred. Weight of each nest c.14g. Eggs: 2, white, long ovals with little or no gloss (Osmaston 1906). Average size of 48 eggs: 20.2 x 13.6 (Baker). Characterised by Medway and Pye (1977) as self-supporting; bracket-shaped; pure, firm nest cement.

REFERENCES Ali and Ripley (1970), Cheng (1987), Cranbrook (1984), Lekagul and Round (1991), Medway (1966), Osmaston (1906), Riley (1927), Stresemann (1931), White and Bruce (1986), Xian and Zhang (1983).

---

39   PAPUAN SWIFTLET
Collocalia papuensis   Plate 9

Other names: Three-toed Swiftlet; Idenburg River Swiftlet; *Aerodramus papuensis*

IDENTIFICATION Length 14cm. A large and relatively rare species, with a shallowly forked tail and dark rump. Most distinctively, has a silvery-grey throat that contrasts with the remainder of the sooty-grey underparts. Mountain and Uniform Swiftlets are appreciably smaller and do not share this species' underpart markings. The similarly sized (and similarly uncommon) and largely allopatric Bare-legged Swiftlet has wholly sooty-grey underparts and matt black as opposed to slightly glossed, dark fuscous-brown upperparts.

In the hand, identification easily achieved as it is the only known swiftlet to have only three toes (see figure 23). A large species (wing of 119-137), with pale throat and dark streaking on abdomen and undertail-coverts, and whitish bases to nape feathers that contrast strongly with the black tips (cf Bare-legged). Densely feathered along length of tarsus.

DISTRIBUTION Endemic to New Guinea, where it has a very restricted range. Currently known from only three localities with certainty (collected specimens): the Idenburg River and the northern slopes of the Snow Mountains, and from Jayapura, Irian Jaya. Sightings of swiftlets that are probably this species have been made from Lae and Yalu River in the lower Markham Valley, the Watut and Bulolo Valleys and throughout the area between Lae and Wau. South of the watershed believed to occur in lowland areas around Port Moresby.

Considered to be fairly common in the Idenburg River area.

MOVEMENTS Resident.

HABITAT Ranges over a variety of habitats from sea-level to 1800m. If swiftlets observed at Bulolo are this species, it is interesting to note that they are found mainly in the gullies of that area.

DESCRIPTION *C. papuensis*
Upperparts Uniformly dark fuscous-brown, slightly glossed. The crown and back have white concealed barbs contrasting markedly with blackish tips. The neck can appear slightly paler than the crown or mantle. Brownish-grey feather bases in front of the eye create a small supraloral spot. The line of feathers immediately above eye are narrowly white-fringed, forming an indistinct su-percilium. **Underparts** Silvery-grey throat contrasting with the remainder of the grey-brown underparts. Below throat the underparts have dark shaft streaks.

Measurements Wing: male (9) 119.5-136.5 (129.5); female (4) 125.0-136.0 (131.6). Tail: male (9), 48.5-56.5 (52.4); female (4) 49.0-53.0 (51.7). Tail-fork: male (9), 3.0-5.5 (4.2); female (3), 2.5-3.5 (Somadikarta 1967). Wing: male (5) 128-141 (136.9); female (2) 129+133 (Salomonsen 1983).

GEOGRAPHICAL VARIATION None. Monotypic.

VOICE Not known.

HABITS Usually feeds in flocks of 20-30 individuals, often associating with the New Guinea Spinetail and Uniform Swiftlet. Tends to feed high, but sometimes low over marshes.

BREEDING Undescribed.

REFERENCES Coates (1985), Rand (1941), Rand and Gilliard (1967), Salomonsen (1963 and 1983), Somadikarta (1967).

SCHOUTEDENAPUS

This genus, which is endemic to Africa, comprises two small species that are superficially *Apus*-like. Both are apparently resident and one is amongst the rarest of all swifts, only being known from five collected individuals.

Both have very uniform grey-brown or black-brown plumage. The feet are anisodactyl (figure 29), the wings are long and sharply pointed, the tail is deeply forked and the extent of emargination in the outer rectrices is an age-related character.

Figure 29. Foot of *Schoutedenapus* swift (left) and Common Swift (right).

Separation from sympatric genera.

The fine pointed wings and long forked tails of this species ensure that only the *Cypsiurus* and *Apus* species need to be considered when separating these species. The wings are distinctly shaped being comparatively broad, though very straight-sided, on the innerwing, but sharply pointed and tapering on the outerwing with the trailing edge remaining very straight while the leading edge tapers sharply. The tail, like the palm swifts or White-rumped Swift, is usually held tightly closed. Most distinctively, Scarce Swift habitually flocks in large numbers and emits a strikingly unique ticking call with a rather metallic tone.

40 SCARCE SWIFT
Schoutedenapus myoptilus   Plate 19

IDENTIFICATION Length 16.5cm. Small, slim rather featureless swift which nevertheless possesses a distinctive jizz. Initial views recall White-rumped Swift due to long thin tail and wings and the slim body. Tail typically held closed appearing needle-thin, but when opened it is deeply forked. The wings are very thin especially the primaries, which taper sharply from secondaries. This coupled with the rather bulbous head and relatively fat body creates a less streamlined appearance than White-rumped Swift. Flight action less elegant than *Apus*, often portraying the impression of putting in much more effort with many seemingly ineffectual flappy beats. When gliding, wings are angled down slightly and appear very stiff recalling *Collocalia*. The grey-brown plumage appears almost silvery when caught in light especially on wings. Upperparts appear uniform and dark with underparts appearing quite pale from throat to breast and then darker grey. Never gives the impression of showing a defined patch as in a typical *Apus*. When a larger flock calls, identification can be rendered very straightforward.

DISTRIBUTION Sub-Saharan endemic, with rather scattered, mainly East African range. Two populations are considerably more continuous than the others. One stretches from Mount Moroto in E Uganda, east into W and C Kenya, and south into N Tanzania at Oldeani, Mount Kilimanjaro and Arusha. The other is largely west of this from SW Uganda (in the Rwenzoris, as far north as Laropi and Murchison Falls National Park), south in Rwanda and the Lua valley, in Burundi, and into E Zaïre, south to 5°S. South of these populations a number of smaller populations occur: in E Zambia and NW Malawi in the Nyika

Plateau and the Mafinga Mountains, S Malawi at Phalombe and Mount Mulanje, in E Zimbabwe at Inyangani, Stapleford, Chimanimani and Melsetter in the Inyanga Highlands, and Mount Gorongoza in Mozambique. Recent sight record from Johannesburg, S Africa in July 1993 (P. Hayman pers. comm.). Has also been recorded in west Ethiopia although it is uncertain what its status is there. Also recorded in the extreme south of Sudan and S Ethiopia. Away from E and S Africa recorded once from Mount Moco in Angola, and widespread on the island of Bioko (Fernando Po).

Throughout the range shows great variation in abundance. Commonest in moist craggy mountains of eastern Zaïre and Rwanda and described as abundant over Nyungwe forest, in Rwanda and Kitutu, Idjwi Island, Mount Kabobo and in Kivu, in Zaïre. Elsewhere far more local. Around twenty pairs breed around Mapopo Falls in Zimbabwe. On Bioko (Fernando Po) it is regularly seen around Sipopo and Fishtown. The scattered nature of the population and the extralimital records from Angola and Ethiopia, suggest the species may have a wider range than is thought and has been overlooked.

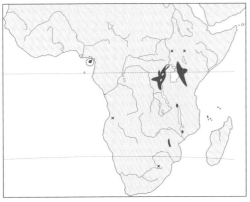

MOVEMENTS Mainly resident with a tendency for some regional dispersal. This is most apparent in eastern Zaïre, at Kamituga and Tubangwa, where it is common from July to September, and uncommon during the rest of the year, indeed it is rare or absent from October to December. The small scale extent of this dispersal becomes apparent when it is realised that species is believed to be resident in nearby mountains.

HABITAT High altitude species preferring wet conditions. This is particularly apparent in the Inyanga Highlands, where it is considerably commoner on the wet slope than the dry. Occurs through an altitudinal range of 500-2450m.

DESCRIPTION *S. m. myoptilus*

Sexes similar. **Head** Dark grey-brown upper head, palest on forehead, lores and narrow line of feathers over the eye which are pale grey-fringed and clearly paler than crown. When fresh grey fringes extend across head but are less distinct than on forehead. Throat shows some variation but is always clearly paler than the upper head, being pale grey-brown, with a rather flared and indistinct border with the darker mid chest. Feather bases darker brown and the effect can be rather mottled when worn. Darker shaft streaking sometimes apparent. The ear-coverts are intermediate in tone between throat and crown and do not grade into either. Black eye-patch. **Body** Upperbody very similar in tone from crown to uppertail-coverts. Un-

derparts clearly darker than throat being mid brown across chest and dark, almost chocolate, brown from breast to undertail-coverts. Some variation occurs in the depth of brown on underparts with some individuals appearing a little paler. Some pale grey fringing can be apparent on the rear belly and flanks when plumage fresh. **Upperwing** Remiges uniformly glossy blackish, and can appear a little darker than wing-coverts or mantle. Outer margins of innerwebs slightly paler and browner. **Underwing** Remiges appear paler than above and slightly paler than greater coverts. Lesser and median coverts notably blacker and can appear paler than the underparts. **Tail** Similar to remiges above and below and certainly will appear darker than uppertail-coverts of some individuals.

Measurements Kenya (4 BMNH): 134.3-141 (137.4). Chimanimani Mountains (1 BMNH): 129.8. Weight: male (2) 28 + 29.5; immature female (1) 30 (Fry *et al.* 1988).

GEOGRAPHICAL VARIATION Three races.

S. m. myoptilus (East Africa, from Ethiopia in the north, southwards to Zimbabwe) Described above.

S. m. chapini (Zaïre, Rwanda, SW Uganda) Blackest subspecies. Wing: male (14) 128-138 (133); female (13) 127-138 (133). Tail: male (14) 65-76 (70.3); female (13) 64-71.5 (68). Depth of fork: male 27-33 (30); female 24-31 (27.4). Weight: female (1) 22 (Fry *et al.* 1988).

S. m. poensis (Bioko (Fernando Po)) Slightly darker than *myoptilus*. Wing: (6) 126-133 (130) (Fry *et al.* 1988). The individual collected in Angola is either *poensis* or *chapini*.

VOICE Highly vocal species with unusual call. Produces a metallic *tic*, preceded by a number of short trills (lasting approximately 125msec), which in turn are followed by weak nasal harmonic twitterings (1.1s), and then, finally, more metallic *tics* (Fry *et al.* 1988). To the European birder individual calls can resemble Greenfinch at times, and especially the chatter of Barn Swallow. When repeated quickly and at greatest frequency calls can sound like the scream of Common Swift, but more often the effect is closer to the echolocating calls of *Collocalia* swiftlets or bats. Calls are mostly heard from large flocks, often at considerable height, rarely from single birds.

HABITS In areas of abundance, such as E Zaïre, occurs in huge (many thousands), mainly monospecific flocks from April to August. Also recorded with Common and African Swifts and Mottled Spinetails. It is important to note that Schouteden's Swift has been collected from swift flocks thought to comprise this species. Preferred habitat is mountainous gorges and cliffs, around waterfalls and over evergreen forests. In W Kenya feeds in large numbers over Kakamega Forest, especially in the afternoon, but apparently roosts at higher altitude in nearby highlands. Often flies low over open areas including gardens, especially in dry weather. This behaviour is thought to be linked with feeding on termites. When large flocks fly close to ground they are easily netted and even killed by children hitting at them with sticks (Fry *et al.* 1988).

BREEDING No nest has ever been found, no doubt due to the extreme inaccessibility of the nest sites on cliff faces, or in rock clefts, high on mountains and close to waterfalls.

REFERENCES Fry *et al.* (1988), Williams (1980).

41   SCHOUTEDEN'S SWIFT
Schoutedenapus schoutedeni   **Plate 19**

Other name: Congo Swift

IDENTIFICATION Length 16.5cm. One of the most enigmatic of all species, it has never been recorded in the
field. Structurally very close to Scarce Swift and it is perhaps safe to assume that behaviour, and even call are also
similar. Although identification of such a rare and possibly endangered species in the field is probably not safe,
observers of Scarce Swift flocks in E Zaïre should be aware
of this species' plumage differences. Schouteden's Swift
differs from Scarce in its darker plumage. All areas of
plumage are slightly blacker, but the throat shows the
greatest difference. Underparts appear very uniform as a
result with the dark brown throat appearing only marginally paler than the remainder of the underparts. The
situation where a Schouteden's Swift might be most apparent amongst a group of Scarce Swifts is when in strong
light the often extensive pale grey area on the throat of
Scarce Swift will contrast strongly with the underparts
whilst this species would remain very uniform. However,
as with all swift identification, shadow effect must be considered and sustained views would be necessary for even
tentative identification to be made.

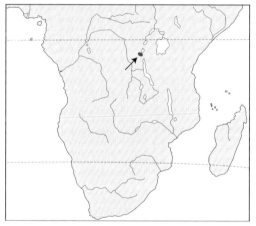

DISTRIBUTION An East African endemic with an extremely restricted range. Only known from E Zaïre where
it has been collected from Mubandakika (28 50°E, 04 12°S,
November 1956), Butokolo (28 16°E, 02 42°S, October
1959), Bionga (28 08°E, 03 21°S) and Kamituga (28 11°E,
03 04°S).
   Believed to be very rare. Only known from 5 specimens
and currently considered a candidate Red Data species.

HABITAT Very poorly known. Most were collected from
flocks of other swifts that were thought to be Scarce.

DESCRIPTION *S. schoutedeni*
Sexes similar. **Head** Upper head black-brown uniform
from forehead to nape. Very slightly paler on lores and
just over eye. Throat slightly paler and ear-coverts similar
to crown becoming slightly paler towards throat. **Body** Upperparts uniformly black-brown with slight gloss from head
to uppertail-coverts. In some there is a suggestion that the
rump is slightly paler. In most individuals mid brown throat
area appears uniform to mid chest and then dark brown

to undertail-coverts. In the individual with the palest throat
there appears to be some form of body moult occurring
on the throat (oldest feathers appear slightly paler) and
there is some contrast between mid throat and lower
throat; otherwise the pattern is as the other 4 individuals.
Upperwing Very uniformly blackish, slightly glossed.
Remiges very slightly paler on outer margins of innerwebs.
Generally very uniform with mantle. **Underwing** Remiges
appear very slightly paler than above but less so than in
vast majority of swift species. All coverts appear very black.
It is likely that the very blackish underwing, especially the
coverts, will appear darker than the browner underparts
in the field. **Tail** Similar to remiges.

Measurements Wing: male (3) 125-131; female (2) 127.
Tail: (3) 60-62 (60.8). Tail-fork: 22-25 (Fry *et al.* 1988).

GEOGRAPHICAL VARIATION None. Monotypic.

VOICE Not recorded.

HABITS Like Scarce Swift, possibly mainly a high altitude
species, as collected from 1470m at Butokolo.

BREEDING Not known.

REFERENCES Collar and Stuart (1985), Fry *et al.* (1988).

MEARNSIA

This small genus has two medium-sized species and is endemic to the eastern Oriental and Papuan regions with one species found on the Philippines and the other on New Guinea.

Both species are highly glossed in plumage, but one has extensively white underparts and blackish upperparts, whilst the other has wholly blackish glossed plumage except for a contrastingly white throat and large white patches on the underwing-coverts. This last feature is unique within the order *Apodiformes*. It has been postulated that the large white patches might act in much the same way as the highly contrasting underwings of some moths if shown during a wing-raising threat display, in the darkness of a nesting or roosting site. Both species have protruding rectrix spines but these are shorter and less substantial than in any other spinetails.

Separation from sympatric genera

Within their ranges both species, with their relatively long butterknife wings and very short tails, are unmistakable. On the Philippines tail shape alone will separate the genus from all others with the exception of the *Hirundapus*, which is far larger with a considerably longer tail and very different plumage, and the obviously smaller more dainty and very differently plumaged swiftlets. On New Guinea only the very different swiftlets have representatives with square tails and the distinctive White-throated Needletail is only a seasonal visitor. The typically rapid flickering flight of this genus coupled with both species' distinctive plumages further facilitate separation.

42 PHILIPPINE SPINETAIL
Mearnsia picina   Plate 12

Other names: Philippine Spine-tailed Swift; Philippine Needletail

IDENTIFICATION Length 14cm. This is the only spinetail occurring on the Philippines and is unlikely to be mistaken for any sympatric swift. Typical spinetail jizz and shape. Wings very long and tail very short and square. Unique amongst all swift species in having large white patches on the underwings caused by white greater and median coverts. Plumage very black with blue-green gloss, relieved only by the conspicuous white patches on the throat and the underwing.

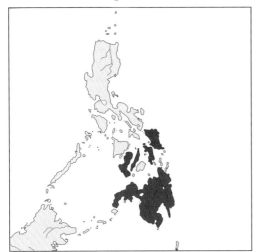

DISTRIBUTION Endemic to the Philippines. Found only on the SE Philippine islands of Leyte, Negros, Cebu, Biliran, Samar and Mindanao.

Little known, though scarcity of both specimens and records suggest this is a rather uncommon species.

MOVEMENTS Believed to be resident.

HABITAT Presumed to be a forest dweller, which could account for its apparent current scarcity.

DESCRIPTION *M. picina*
Adult Sexes similar. **Head** Upper head uniformly blue-black with a purple gloss. Ear-coverts and side of head uniform with upper head, not becoming paler at all towards throat patch. Black eye-patch. Highly contrasting pure white throat patch, extending along the length of the gape and then cutting back sharply to join other side of lower edge at a point on lower throat. Patch is therefore rhombus-shaped and very clearly defined at its edges from blue-black lower throat, although some white feathers at the edges of the patch have brown tips. **Body** Upperbody darkest on mantle and upper back very slightly paler on the nape and head and more obviously paler on rump and uppertail-coverts, which are, however, still blue-black. Gloss across upperparts can appear green. Some paler brown bases can be seen when worn, especially on the nape. The underbody is uniformly black-blue but with a less glossiness and showing more brown bases when worn. **Upperwing** Very uniform with remiges blue-black showing a green gloss when fresh, most apparent on the outerwebs becoming more purple when worn. The innerwebs are slightly paler. The coverts are similarly glossed and probably can appear darker than the remiges in flight. There is no fringing on the coverts or remiges.

Figure 30. Underwing of Philippine Spinetail.

Underwing The remiges appear slightly paler and greyer. The two outer greater primary coverts are uniform with the remiges, otherwise all greater and median coverts are white, except for the black bases of the outerwebs of the outer two median coverts (figure 30). The lesser coverts are black-blue. The leading edge coverts are uniform with the lesser coverts but have some pale grey fringing. **Tail** Glossy black-blue, but invariably cloaked by the tail-coverts and not visible in the field. The small spines are typical of the genus extending only 1.2mm and 1.0mm beyond the web in the central and outer rectrices respectively.

Juvenile Not known.

Measurements Wing: 161 + 164 (only 2 specimens). Tail: 32 (Hachisuka 1934).

GEOGRAPHICAL VARIATION None. Monotypic.

VOICE Not known.

HABITS Typically seen at height, singly or in small groups.

BREEDING No information.

REFERENCES Dickinson *et al.* (1991), Hachisuka (1934).

43 PAPUAN SPINETAIL
Mearnsia novaeguineae   Plate 12

Other names: New Guinea Needletail; Grey-bellied Needletail

IDENTIFICATION 11.5cm. Small spinetail with black-brown upperparts and dark breast and throat contrasting with paler belly. The plumage is very strongly glossed blue-black. Typical spinetail jizz separate it from all sympatric swift species. In plumage it is not unlike sympatric Glossy Swiftlet, having dark upperparts, throat and breast, with the remainder of the underparts white. Differs from all swiftlets in its rather stocky aspect, short square tail, protruding head and typical spinetail wing shape (narrow secondaries, broad inner primaries and rather hooked outer primaries). The much larger White-throated Needletail, which is regular on Papua New Guinea during the northern winter and on migration, has a very different plumage pattern. An additional plumage feature is an indistinct pale band that can be seen in the remiges when viewed from beneath, caused presumably by translucency. Typically, very rapid flight, much faster than swiftlets.

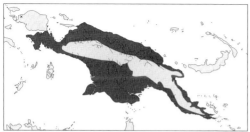

DISTRIBUTION Restricted to New Guinea. Found throughout the lowlands of New Guinea, except the westernmost region of the Vogelkop and the islands. A record of ten birds near Sorong at the western tip of the Vogelkop in August 1992 may represent a range extension or simply local dispersal (N. J. Redman verbally).

Locally common throughout range, though scarce or abundant in some areas.

MOVEMENTS Thought to be wholly resident but may

disperse locally at times.

HABITAT Found only in the lowlands, both in the interior and on the coast. Not recorded above 550m. Occurs primarily over forest including second growth, but also around human habitation. Apparently requires the presence of dead trees which it feeds around.

DESCRIPTION *M. n. novaeguineae*
Sexes similar. **Head** Upper head uniformly blue-black with bright blue-green gloss. Ear-coverts black-brown paler than upper head and darker than throat. Throat dark black-grey with pale fringing. Black eye-patch. **Body** Upperparts uniform from head to uppertail-coverts, perhaps blackest on the mantle. The bases to the feathers are browner and can be seen if the plumage is worn - this can be especially apparent on the nape. The black-grey throat patch is uniform across the throat and then grey across the breast. Below the breast the underparts are off-white/grey-white separated from the breast by a fairly clear cut-off. The feathers although whitish have grey shaft streaks. The undertail-coverts are very black and glossy. **Upperwing** Very uniform and blackish. The remiges are black-brown on the outerwebs and slightly paler on the innerwebs. The outer primaries often appear darkest as do the coverts, especially the median coverts and lesser coverts, which are also highly glossed. **Underwing** The remiges appear paler and greyer than the upperwing with the outer primaries appearing darkest. The greater coverts appear slightly darker than the remiges but the median coverts and lesser coverts are strikingly blacker than the rest of the underwing (and the adjacent grey body). **Tail** Black-brown similar to the remiges above and below. Weak rectrix spines very similar in strength on the outer and inner tail: up to 1mm beyond the web in the central tail and 0.9mm in the outer tail.

Measurements Male: wing (6 BMNH) 128-135 (131.7). Only 2 females measured, both 128.

GEOGRAPHICAL VARIATION Two races.
    M. n. novaeguineae (S and SE New Guinea) Described above.
    M. n. buergersi (N New Guinea, between the Mamberamo and Sepik rivers) Differs from *novaeguineae* in its paler grey throat and breast with the remainder of the underparts (apart from the black undertail-coverts) pure white.

VOICE Unusual series of 3-4 squeaks which have been likened to the noise made by toy rubber animals (Beehler *et al.* 1986). A series of calls transcribed as *zizz* or *sizz* has been heard (Coates 1985). The species is generally rather silent.

HABITS Usually seen in pairs or small groups, although groups of over 40 have been recorded. Associates freely with swiftlets. Typically flies at tree level or lower and most apparent at dawn and dusk. Coates (1985) postulates that this species may occasionally take insects in passing foliage.

BREEDING Breeds in tree hollows and has been seen collecting fibres from withered palms and these are presumed to be used as nesting materials (Coates 1985). Nest and eggs not known.

REFERENCES Beehler *et al.* (1986), Coates (1985).

ZOONAVENA

This genus comprises three rather small spine-tailed swifts. The genus is widely dispersed with one species occurring in the Indian subcontinent, whilst the other two are found on islands off Africa, namely Madagascar and the Comoros, and São Tomé and Príncipe. All species have dark plumage except for pale areas on the undertail-coverts, belly and rump. The pale feathers have dark shaft streaks. The feet are anisodactyl and the tail is square with relatively fine projecting spines.

Separation from sympatric genera.

All three members of this genus are found exclusively allopatrically from other spinetail genera. Of the *Apus* species found in sympatry with two of the species only Little Swift shows a square tail but is easily excluded both on wing shape and plumage details. All other swift species within the range of the genus have forked tails, except in India where the very different swiftlets and *Hirundapus* needletails are easily excluded.

44   MADAGASCAR SPINETAIL
Zoonavena grandidieri   Plate 11

Other names: Madagascar Spine-tailed Swift; Madagascar Needletail; Malagasy Spinetail

IDENTIFICATION Length 12cm. This is the only spinetail occurring on Madagascar and the Comoros and therefore the distinct spinetail jizz allows straightforward identification. Other swift species occurring on Madagascar and the Comoros, Alpine, African and African Palm, have deeply forked tails as opposed to the shallow fork of Madagascar Spinetail. Furthermore, this is the only species with a whitish rump, otherwise the plumage is rather drab with nondescript pale brown underparts. The wings are blackish with slightly paler secondaries. The flight action is typical of the spinetails.

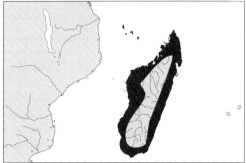

DISTRIBUTION Found throughout Madagascar except the high plateau. Occurs on all larger islands in the Comoros.

Appears to be rather common throughout Madagascar, though particularly in the more humid east as opposed to the west and arid areas. Early authors, such as Lavauden (1937), considered this to be a rather scarce species. This viewpoint is not borne out by more recent studies and there has perhaps been a genuine increase in numbers.

MOVEMENTS Believed to be resident throughout its range.

HABITAT Found in the lowlands, but has been recorded to 1000m. Can be encountered hunting over a variety of habitats (including open country), but especially savanna and forest (both primary and second growth). In certain areas tends to be associated with very large trees which it often feeds around.

DESCRIPTION *Z. g. grandidieri*
Sexes similar. **Head** Black-brown above with a paler grey line of feathers over the eye. Ear-coverts black-brown merging on lower ear-coverts to paler grey-brown, with throat being pale whitish grey-brown with some heavy brown shaft streaking. The contrast between the upper head and the throat, gives the species a somewhat hooded appearance when viewed side-on. **Body** Upperbody uniform from head to rump. Rump is a narrow white band with brown shaft streaking. Uppertail-coverts uniform with the mantle. Underbody fairly uniform from throat with a slightly darker brown breast band. Undertail-coverts dark-brown. **Upperwing** Very uniform black-brown as upperbody with slightly paler innerwebs on the remiges. Underwing only slightly paler than upperwing, and consequently less contrast than in most species. Underwing clearly darker than underbody. **Tail** Black-brown, slightly paler below. Shows little contrast with tail-coverts. The rectrix spines are strongest and longest at the centre where they extend up to 3mm beyond the web in the male and 1.5mm in the female.

Measurements (BMNH). Wing: male (11) 122-131 (127.2); female (13) 121-132 (125.32).

GEOGRAPHICAL VARIATION   Two races.
> *Z. g. grandidieri* (Madagascar) Described above.
> *Z. g. mariae* (Comoros) This race is very similar in plumage to *grandidieri*, but in the small sample measured averaged shorter-winged. Wing (2 BMNH): 122+124.

VOICE Rather quiet swift, which occasionally utters a shrill, abrupt *zree* whilst feeding (Langrand 1990). This call can be uttered as part of a series.

HABITS This species is not particularly gregarious; it is usually encountered in pairs or small groups of up to 15 birds. Habitually hunts low over the canopy.

BREEDING Nesting behaviour has not been closely studied. The nest is a construction of twigs and other plant materials and is located in hollow trees, rock fissures or hollows and wells (Langrand 1991).

REFERENCES Dee (1986), Lavauden (1937), Langrand (1990).

45   SÃO TOMÉ SPINETAIL
Zoonavena thomensis   Plate 11

IDENTIFICATION Length 10cm. Sympatric with only two other swift species, African Palm and Little, from which it can be separated with ease mainly by its highly characteristic jizz and plumage. In the event of any other African white-rumped and white-abdomened spinetails being considered, either within the range of São Tomé Spinetail or vice versa, the following features should be considered. Mottled can be excluded primarily by virtue of the white on the abdomen being exclusive to the area around the central belly, rear flanks and vent, and not from the lower breast to the undertail-coverts as in São Tomé. Sabine's can be excluded as it has wholly white uppertail-coverts usually cloaking the tail. São Tomé can have grey uppertail-coverts but these contrast with the rump and do not give the strikingly white-tailed appearance of Sabine's. Cassin's has a narrow white rump band restricted to the proximal uppertail-coverts. Böhm's has a more clear-cut divide between the breast and the underparts and is a purer white on the rump and underparts than São Tomé. São Tomé has a rather more weak and fluttering flight and is significantly smaller than the other species except for Böhm's which appears a little shorter-tailed.

DISTRIBUTION Confined to the islands of São Tomé and Príncipe in the Gulf of Guinea. Appears to be found throughout both islands.
  Believed to be fairly common on both islands, though no population estimate is available.

MOVEMENTS Resident.

HABITAT Found over forest, both primary, degraded and second growth. Also plantations and over adjacent cultivation. Not found in close proximity to human habitation. Found from coastal lowlands on both islands to the mountains of São Tomé at 1500m and 500m on Príncipe. Usually encountered in small flocks of between 4-8, although 10 have been recorded together. On Príncipe occurs amongst Little Swift flocks, but on São Tomé only unmixed flocks have been recorded.

DESCRIPTION *Z. thomensis*
Sexes similar. Upper head very uniform glossy black with a dark blue gloss (green in some lights). The lores are ashy-grey uniform with the throat. Black eye-patch. Ear-coverts dark ashy-brown and uniform with throat. Throat can appear a little mottled as a result of the slightly paler grey fringes, most apparent when fresh. **Body** Upperbody uniform with head across mantle, scapulars and back. Rump white with heavy shaft streaking, becoming greyer onto proximal uppertail-coverts and then black on distal uppertail-coverts. In some individuals all uppertail-coverts are grey. The breast is uniformly dark ashy-grey with the

throat, but on the lower breast and belly it grades indistinctly into grey-white feathers with heavy shaft streaking. The point of divide is not a straight line. It is narrowest at the centre of the lower breast and then widens, but with the mid flanks still uniform with the upper breast. The feathers become whiter around the vent and undertail-coverts, but the shaft streaks are blacker and heavier on the tail coverts. The white extends onto the rear flanks and is continuous with the rump. **Upperwing** Very uniformly glossy black, with no obvious fringing. In the field there are some birds with the remiges often appearing slightly paler than the coverts. **Underwing** The remiges appear slightly paler and contrast slightly with the greater primary and greater coverts, and more markedly with the very black median and lesser coverts. The leading edge coverts are uniform with the lesser coverts. The darkest underwing-coverts appear obviously darker than the underbody in the field. **Tail** Glossy black, palest on the underside. The shafts form rectrix spines of 1-2.5mm. The uppertail-coverts reach within 15-20 of the tail tips and the undertail-coverts within 9-13.

Juvenile Very similar to the adult but the throat and upper breast are browner-grey and the undertail-coverts are pale brown.

Measurements Wing: male (4) 108-114 (111.5); female (6) 110-116 (113). Tail: male (4), 32-42 (36.7); female (6) 33-39 (37) (Fry *et al.* 1988).

GEOGRAPHICAL VARIATION Monotypic, with no difference of any kind noted between São Tomé and Príncipe birds.

VOICE Call given almost continuously in flight, a high-pitched bat-like squeaking *trirritritri* (Fry *et al.* 1988). De Naurois (1985) described the call as having a high-pitched piercing quality.

HABITS Often seen feeding around very large forest trees in figures of eight or circles, over, or in gaps in, the canopy. Rarely seen feeding over 15m above ground and not recorded feeding close to ground.

BREEDING Solitary nester, breeding in hollow trees or amongst buttress roots. The nest is made of dry twigs fashioned into a tiny open cup 15-25m long and adhered to a vertical surface up to 3m high in a hollow tree or, when amongst buttress roots, 20-40cm above. Two or three eggs are laid (four on one occasion) and average (17) 16.4 x 12.1 (Fry *et al.* 1988).

REFERENCES de Naurois (1985), Fry *et al.* (1988).

---

46   WHITE-RUMPED SPINETAIL
Zoonavena sylvatica   Plate 12

IDENTIFICATION Length 11cm. Within range combination of a white rump and lower underbody with otherwise glossy black plumage is diagnostic. Highly distinctive jizz separates this species when the plumage features are not seen. Separation from the geographically closest spinetail, Silver-rumped of SE Asia, would be easily achieved in the unlikely case of one occurring in the other's range, or in territories between the two ranges, by virtue of Silver-rumped's entirely black underbody and its white uppertail-coverts. White-rumped Spinetail is typical of the other spinetails with its broad 'butter-knife' wings, relatively far-protruding head and short square tail (with spines noticeable under good viewing conditions). The flight action is typical of the spinetails. In active low speed flight it sometimes appears rather fluttery with some rocking motions. High speed flight is perfected with minimal wing beats.

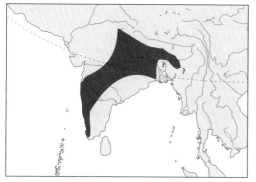

DISTRIBUTION Found in the Indian subcontinent, east of Rajasthan, south of the Himalayas and west of E Burma. Distribution throughout most of range is patchy, though formerly when forest cover was more extensive this would have been far more continuous. Found in the lower Himalayas from Garhwal, Kumaon, through Nepal and into Sikkim. Also the Assam Hills, Manipur, W Bengal, Chota Nagpur, Inani Forest and Sylhet in Bangladesh and into Burma. Range in Burma and Bangladesh uncertain - the record from the Inani Forest was the first sighting in recent years (Johnson 1994). Southwards, the species occurs through Madhya Pradesh (as far west as 79-80°E) and through the Western Ghats from Goa southwards to Kerala. No records from Sri Lanka.

Within range is generally highly localised, though quite abundant in some localities. In Nepal it is local and uncommon, seen with regularity only over lowland forest at Chitwan. Inskipp and Inskipp (1991), believe that this species is only possibly a Nepalese resident.

MOVEMENTS Thought to be largely resident, but dispersive to some extent. It is not known if there is any distinct pattern to these movements.

HABITAT Found primarily in forest areas and plantations and particularly in clearings. It can be found in any forest type across its extensive range. Will visit adjacent areas to feed. Found from lowland plains into highlands of up to 1770m in the Himalayas, although in the peninsula they are generally found at lower altitudes.

DESCRIPTION *Z. sylvatica*
Adult Sexes similar. **Head** Upper head black-blue with a faint blue gloss. Feather bases browner but seldom visible. Lores and forehead unglossed and paler grey-brown. Some pale fringing on forehead when fresh. Line of feathers over eye grey-fringed. Black eye-patch. Ear-coverts clearly paler than upper head, but slightly darker than throat. Grey throat and upper breast, chin and central throat palest, with some paler fringing. Throat can appear mottled when worn, as bases are paler. **Body** Mantle and upper back uniform with head, nape uniform or a little paler, especially when worn, as pale bases most apparent here. Lower back slightly paler and greyer than mantle. Broad pure white rump patch. Some indistinct darker shaft streaking. Bases are darker and in a particu-

larly worn individual these may be apparent and show as mottling. Uppertail-coverts uniform with mantle. Underbody pale grey immediately below grey breast and throat. The belly and undertail-coverts are dirty white. The cut-off is emphasised by a slightly darker border to the lower breast, but then is pale grey before becoming white on the mid lower belly. Mid flanks grey. The white on the rear flanks is continuous with the rump. **Upperwing** Remiges blue-black with a purple sheen most apparent on the outerwebs. Innerwebs grey-brown and paler except at tip. Coverts similarly glossed and slightly darker. **Underwing** Remiges paler and greyer being uniform or slightly paler than the greater coverts. The greater coverts can appear slightly darker towards tips. The median and lesser coverts are dark grey-black and clearly darker than the rest of the underwing and the underbody. The leading edge coverts are not glossed and have pale grey fringes. **Tail** The tail is black-brown and is slightly paler below. The uppertail-coverts do not cloak the tail, unlike the undertail-coverts. The shafts extend beyond the webs to form spines which are shortest in the outer tail, but up to 3.9mm at the centre.

Juvenile Similar to the fresh adult.

Measurements Wing: (107) 112-16. Tail: (100) 34-37. Weight: (1) 13 (Ali and Ripley 1970).

GEOGRAPHICAL VARIATION Monotypic with no marked variation in size or plumage throughout the range.

VOICE A rapidly repeated *swicky-sweezy* (Ali and Ripley 1970) is heard on the wing, also rendered as a twittering *chick-chick* (Nichols, in Ali and Ripley 1970).

HABITS Highly gregarious rarely being seen alone. Usually in small groups but up to 50 birds can be seen together.

BREEDING Probably only a solitary breeder. Two or three nests have been recorded together although this suggests that perhaps the birds make new nests in traditional nest sites. Makes its nest at the base of a hollow tree trunk (overmature trees in humid forest, such as *Vateria indica*) in the form of a depression in the debris. This is lined with dry straw and leaves. Three or four eggs are usual, but five has been recorded. The average size of eggs (100) is 30.7 x 22.2 (Ali and Ripley 1970).

REFERENCES Ali and Ripley (1970), Inskipp and Inskipp (1991), Johnson (1994).

TELACANTHURA

This genus with two species is endemic to Africa. Both are large spinetails. The elaborate patterns of the throat, formed by dark-edged pale-centred feathers, are unique. One species, apart from the throat patch, has essentially blackish plumage while the other is browner in plumage with a white rump patch and a white ventral patch. The squarish tail is long (for a spinetail), with developed shaft spines, and the feet are anisodactyl.

Separation from sympatric genera

Of the African spinetails this is the only genus that can be readily mistaken for other genera. The long wings are not so obviously butterknife-shaped as the other spinetails but the square tails of both species (when closed) exclude all species except for the other spinetails and Little Swift. The latter species can be excluded by its less protruding head, shorter wings and less powerful, more fluttery flight. The *Neafrapus* can be easily excluded by virtue of their very short tails and far more obvious spinetail wing-shape. Sabine's Spinetail has a rather different plumage and appears more graceful with a smaller head, a more streamlined body and a rather truncated tail, unlike the obviously tailed *Telacanthura* species (figure 31).

Figure 31. Mottled Spinetail, Sabine's Spinetail and Böhm's Spinetail (from left to right). Note structural differences between the genera. (Not to scale).

47   MOTTLED SPINETAIL
Telacanthura ussheri   Plate 10

Other names: Ussher's Spinetail; Mottle-throated Spinetail

IDENTIFICATION Length 14cm. A medium-sized spinetail which is closest in jizz to the *Apus* swifts than any other spinetail. This is largely a result of the proportionally long, prominent tail and the restricted nature of the white on the underparts, when compared with most other African spinetails. This combination of features ensures that Little Swift is the main confusion species. Under good and preferably sustained views the conclusive observation of either a narrow white band across the vent joining the prominent white rump or a small white belly patch will confirm identification as Mottled Spinetail. Although diagnostic and sometimes very well pronounced these markings are somewhat variable in extent and hard to see in the field. In this case the following features will exclude Little Swift: Mottled Spinetail has an indistinct pale throat caused by mottled brown feathers and not a clear cut white throat surrounded by black as in Little Swift; Little Swift has a pale grey not black forehead. Jizz features further help identification: Mottled shows the typical spinetail wing shape with narrow secondaries and bulging primaries, and a relatively long, protruding head in comparison with the scythe-winged, short-headed Little Swift. The wings also appear longer than in Little and when gliding or shearing are held in an elegant arc as opposed to the stiff downturned shape of Little. Although the tail spines are very hard to see in the field the tail appears rather spiky and indeed, when fully spread, reminiscent of a sweep's-brush. It is important to remember that the Mot-

tled Spinetails occurring in southern Africa are essentially brownish in plumage and those around the equator and further north-west are blackish, and those of east Africa are relatively short-tailed. The flight action can be very rapid with powerful though rather laboured wing beats, less fluttery than the smaller spinetails and very graceful. The species can regularly be seen shearing at great speed or gliding in wide circles at height. When flying low down close to nest sites or feeding amongst baobabs the speed of this species can be phenomenal.

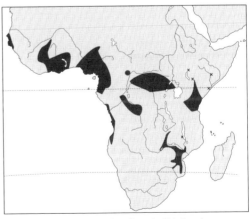

DISTRIBUTION Extensive though disjointed west, central and east African range. In the north-west of the range it occurs mainly in the coastal strip of Senegal, Gambia and Guinea Bissau as far north as 14-15°N. Further east the range is continuous from S Mali and SW Guinea east and south through N and SE Ivory Coast, all of Ghana except the extreme north-west and Togo except the coastal strip. It does not occur in Dahomey or W Nigeria, but the range is then continuous throughout S, C and NC Nigeria, the western half of Cameroon, Equatorial Guinea and NW Gabon. The species occurs along the coastal strip of Angola and Zaïre around the mouth of the Congo. A population is found in SC Zaïre in a broad arc along the Kasai river system, from the Congo River in the west. It occurs along the whole of the northern border of Zaïre, in SW Central African Republic and into W and S Uganda as far as Lake Victoria and throughout the coastal lowlands of S Kenya and N Tanzania. In S Africa it is found in S Zambia along the Luangwa and Zambezi valleys, into S Mozambique and in adjacent areas in N and E Zimbabwe. The species is also found in the north of the Kruger national park in South Africa.

Throughout its range the species is very local and could not be described as abundant in any location, although it is reported as common in parts of the Ivory Coast and East Africa. Throughout the West African range it is uncommon to locally common. The Angolan coastal population is locally common with up to 100 being recorded at one roost. The species is perhaps commonest in its central and East African ranges. In S Africa it is locally fairly common but rare in its limited South African range.

MOVEMENTS Believed to be resident; infrequent records beyond known breeding range are possibly the result of the species being overlooked.

HABITAT Due to wide range the species has adapted to a variety of climate and habitat requirements. Throughout much of its range it is found mainly in low-lying areas, but in East Africa in can be found as high as 1800m, and in Zambia it has been encountered up to 900m. Preferred habitat is dry deciduous woodland with a predominance of baobab trees (used as roosting and nesting sites), around which the species is frequently observed feeding. In the Ivory Coast preferred habitat is *Borassus* palm savanna. Often seen around gallery forest or in the larger clearings of evergreen forests. In the course of feeding can be encountered over many habitats. Will nest in abandoned buildings, but not normally associating with human habitation.

DESCRIPTION *T. u. ussheri*
Adult Sexes similar. **Head** Upper head uniformly dark grey-brown, palest on the forehead, lores and line over eye. Ear-coverts paler grey-brown, slightly paler towards throat, leading to a slightly capped impression. Black eye-patch. Mottled throat palest on the chin and central throat. White feathers are delicately marked: dark brown bases, fringes and shaft streaks, though not to the distal quarter of the feather, create a three pronged mark. When these feathers lay one upon another the dark base cannot be seen and the effect is of white lines across a mottled surface. This is particularly apparent on the lower throat, upper breast and around the gape. This patch can appear to extend high onto the sides of the head towards the rear nape. **Body** Upperbody from head to lower back dark grey-brown though with some paler bases showing through especially around the nape. White rump patch becoming slightly mottled towards lower edge. Uppertail-coverts blacker than mantle. Underbody is lightly mottled brown, becoming blacker on undertail-coverts. Small round white belly patch in front of vent. The extent of this white patch

is variable and can join with the white rear flanks and rump in some cases. When at its smallest it can be very hard to discern in the field. **Upperwing** Black-brown upperwing rather uniform, though with more contrast than on other spinetails. Black-brown remiges with paler innerwebs and tips. Coverts can appear blacker towards the lesser coverts and are more glossed than the remiges; very uniform with body. **Underwing** Remiges appear greyer than on upperwing and are uniform with the white-fringed greater and greater primary coverts. The median coverts are also white-fringed and with the lesser coverts, are distinctly blacker-brown than the remiges. Lesser coverts are black-brown with grey fringes. **Tail** Black and more distinctly green-glossed than the remiges. The undertail is greyer. The tail has shaft spines that protrude 2-4mm beyond the webs, which are grey-brown fringed. The outer rectrix has a white spot midway along the fringe of the outerweb.

Juvenile Differs in that it shows pale fringes when fresh on the head and upperparts (especially the back and uppertail-coverts). The rectrices and all the wing coverts above and below, are narrowly pale-fringed, especially on the underwing.

Measurements Wing: (33) 140-151. Tail: (16) 33-37 (male average 35, female average 33.5) (Fry *et al.* 1988).

GEOGRAPHICAL VARIATION Four races.

T. u. ussheri (W Africa from Senegambia to Nigeria) Described above.

T. u. sharpei (Cameroon to Gabon, S and N Zaïre to Uganda. Blackest in plumage of the different races, with a browner tint to the throat patch. Wing measurements similar to *ussheri* but tail longer. Wing: (69) 140-152. Tail: (7) 34-41. Weight: (6) 27-37 (32) (Fry *et al.* 1988).

T. u. stictilaema (E Africa) Plumage blacker than in *ussheri*, throat patch browner, wings shorter and tail longer. Wing: male (7) 139-146 (143); female (12), 132-147 (142). Tail: (4) 44-49 (45.5). Weight: male (3) 30-32 (31); female (5) 28.5-35 (31.5) (Fry *et al.* 1988).

T. u. benguellensis (SW Angola and SE Africa) Upperparts browner than all other races, throat patch browner than *ussheri*, and belly greyer. It lacks the white abdomen spot having instead a white line through the vent. In Angola this race occurs in dry deciduous woodland in the south-west, whereas *sharpei* occurs in the north-west in evergreen forest (Brooke 1993 pers. comm.). Wing: male (8) 141-151 (146); female (10) 142-150 (147). Tail: (2) 44. Weight: (7) 31-36 (33.3) (Fry *et al.* 1988).

VOICE Usually a silent species but four distinct flight calls have been recorded: a rasping far-carrying twitter *kak-k-k-k* (Chapin 1939), a soft liquid *tt-rrit, tt-rrit, tt-rrit*, a creaking chatter and a seldom recorded disyllabic *zi-zik* (Mackworth-Praed and Grant 1970).

HABITS Usually seen in pairs or in small groups of up to 12 individuals, although exceptionally a group of 100 was recorded in Malawi. Forms groups with other swift species and swallows. Away from the breeding site usually observed only fleetingly.

BREEDING Mottled Spinetail breeds in hollow trees, with a preference for baobabs, or on vertical walls or under beams within buildings, including chimney stacks. The nest is rather bracket-shaped and has been likened to a half-saucer. It is a sturdy construction consisting of thin dry twigs (usually around 2cm long, although one measured 5.5cm) bound together by saliva and incorporating other materials such as leaf-ribs or fragments of banana leaves. What has been described as a 'beard of twigs' hangs below the saucer. The nest is 7.5-8cm wide where attached to wall. Four eggs have been recorded in a clutch and measured 20.6-21 x 13.8-14.3 (Fry 1988).

REFERENCES Chapin (1939), Fry *et al.* (1988), Serle and Morel (1977), Williams (1980).

---

48   BLACK SPINETAIL
Telacanthura melanopygia   Plate 10

Other names: Chapin's Spinetail; Ituri Spinetail; Ituri Mottle-throated Spinetail

IDENTIFICATION Length 15cm. A large dark spinetail, unique in that it lacks any white on the rump or upper-tail-coverts and in having pale markings on the underbody restricted to pale brown mottling on the throat. Its jizz, however, is still typical of the spinetails and it is this that will separate it from sympatric dark *Tachymarptis* and *Apus* species. The tail is square and the wings are typically narrow in the secondaries and broad in the inner primaries before shortening and then lengthening again in the outer primaries. The tail is comparatively long for a spinetail, but the head is typically protruding. The flight is rapid and powerful, soaring and gliding at speed with few beats. A powerful active flight is also performed, when wing beats appear rapid and shimmering.

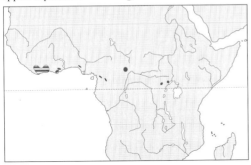

DISTRIBUTION Central and western Africa. Disjointed distribution points to the species being widely overlooked. In W Africa found in Mount Nimba in Liberia, Ivory Coast (from 5 localities), S Ghana and from Benin, Nigeria, Cameroon and Gabon. The species has also been recorded from NE Zaïre on two occasions, from SW Central African Republic and from Angola once at the Mussengue River, Dundo.

Locally common only in S Ghana, elsewhere locally uncommon to rare.

MOVEMENTS Presumed to be resident, with those records from Zaïre, Central African Republic and Angola probably a result of infrequent recording as opposed to vagrancy.

HABITAT Restricted to the rainforest zone, though not only to primary forest, being found also over secondary forest, plantations and around stony bluffs in forest.

DESCRIPTION *T. melanopygia*
Adult Sexes similar. **Head** Upper head uniformly deep black-brown, faintly glossy. Feathers above eye indistinctly grey-fringed. Black eye-patch. Upper ear-coverts and side of head uniform with crown, becoming more mottled towards intricately mottled throat. Throat mottling is formed by feather pattern: chocolate white with a dark shaft streak, except for c. 1mm before the tip, and with a sooty brown fringe. The appearance in the field is of white-spotting on an otherwise brown underbody. The patch extends to the lower throat. The throat is palest on the chin. **Body** From head to uppertail-coverts great uniformity. On the underbody the throat pattern becomes markedly darker from the lower throat to the breast. The pattern is not apparent on the belly and the underparts become progressively darker beneath the belly, reaching their darkest point on the vent and undertail-coverts. **Upperwing** Very uniform wing appearing a little blacker than the mantle. Remiges paler on the outer vanes of the innerwebs with very narrow pale fringes to the outerwebs. A purplish gloss is present on the wings, most marked on the outerwebs. The greater primary and greater coverts have indistinct paler fringes. The coverts are very uniform. **Underwing** The remiges are paler grey than above and are uniform with the greater primary and greater coverts. The smaller coverts appear much darker. The degree of contrast between the upperwing and underwing is markedly less than in most species. The leading edge coverts are paler brown than the lesser coverts. **Tail** The blackish uppertail is more glossed than the wing, and the undertail is slightly paler. The uppertail-coverts come within c. 20mm of the tail tip, and the uppertail-coverts within about 15mm. The rectrix spines extend up to 5.8mm beyond the webs in the central tail and are shorter and weaker in the outer tail.

Juvenile Unknown.

Measurements Wing: (7) 162-176. Tail: (7) 49-53.5. weight (1 female) 52 (Fry *et al.* 1988).

GEOGRAPHICAL VARIATION None. Monotypic.

VOICE Loud harsh call with three main variants: a metallic *crr tchi*, an often repeated *creeou* and a tuneless clicking trill (Fry 1988).

HABITS Occurs singly or in flocks of up to 12. Freely mixes with other forest swifts and also Square-tailed Roughwing *Psalidoprocne nitens*. Typically feeds at low level.

BREEDING The breeding biology of Black Spinetail remains largely unknown. Heim de Balsac and Brosset (1964) believed that this species was nesting in crags rising above the Gabonese forest.

REFERENCES Colston and Curry-Lindahl (1986), Fry *et al.* (1988), Heim de Balsac and Brosset (1964), Lockwood *et al.* (1980).

RHAPHIDURA

This small genus comprises two small spine-tailed swifts: one in South-east Asia, the other in equatorial Africa. Both species have very glossy black plumage relieved only by a white rump and uppertail-coverts in one and a white rump, uppertail-coverts, belly and undertail-coverts in the other. Anatomically they are unique in having the tail-coverts entirely cloaking the tail. The tail spines are protruding, although rather fine.

Separation from sympatric genera

Both species are typically located flying over the forest canopy often at great speed and with rapid flickering wing beats coupled with very graceful high-speed gliding. Both have very typical spinetail structure with butterknife wings and relatively short, truncated tails. In South-east Asia Silver-rumped Spinetail has no similar sympatric species. Within its range only the much larger and very differently plumaged needletails and the much smaller and very different swiftlets have similar tail shapes. In Africa greater care is required. The *Neafrapus* spinetails are far less graceful in flight and both have much broader wings and very short tails. The *Telacanthura* spinetails are longer-tailed and thinner-winged and in consequence appear more *Apus*-like than the other spinetails, with the tail in particular being very evident and appearing rather full, rounded and spiky when spread.

49 SILVER-RUMPED SPINETAIL
Rhaphidura leucopygialis Plates 10 & 12

Other names: White-rumped Spine-tailed Swift; Silver-rumped Swift

IDENTIFICATION Length 11cm. Within range unmistakable, being the only swift species with wholly white rump and uppertail-coverts, and otherwise very dark, blue-black plumage. Other swifts with white rumps in range, House and Pacific, both have, to a lesser or greater extent, forked tails and white throats, and typically slender *Apus*-jizz. Pale-rumped swiftlets in range have generally mid-brown to pale brown plumage and are quite different in jizz. When seen from below distinctive butterknife wing shape, broad truncated body, with relatively far-protruding head, and square tail (often with white uppertail-coverts showing beyond tail tip or 'hanging' over outermost rectrix), create very distinctive jizz. Dark plumage makes rump and uppertail-coverts appear very white, although, when viewed in the hand, the rump and uppertail-coverts are grey-white with a heavy black shaft streak. This can be seen in the field, under optimum conditions, as can the rectrix spines. Rapid flight, fast shimmering wing beats, with some rocking from side to side, though at great speed, and not like that of *Chaetura* or *Collocalia*. Also can fly in a manner like *Hirundapus* swifts, with broad high-speed circling and shearing.

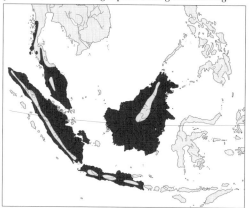

DISTRIBUTION S South-east Asian and Sundanese range. Found from S Burma, throughout peninsular Thailand and the Malay peninsula. Present in Borneo, Sumatra (including Bangka island) and Java.

Locally common in suitable habitat throughout most of range, uncommon on Java. No population estimates.

MOVEMENTS Resident with no extralimital records.

HABITAT Found over forests, second growth and plantations, especially near water. Found both inland and at coastal sites. Rarely seen over human habitation. In Sumatra occurs to 600m, on Java to 1500m and on the mainland peninsula to 1250m.

DESCRIPTION *R. leucopygialis*
Adult Sexes similar. **Head** Upper head very uniform, deep blue-black with green-blue sheen. No fringing to feathers and brown bases not visible. Ear-coverts and throat marginally paler, being slightly less black. Black eye-patch. **Body** Mantle can appear very slightly darker than head but with the same feather pattern as head. Rump and uppertail-coverts grey-white with heavy black shaft streaks. Grey-white does not extend onto rear flanks. Uppertail-coverts often longer than tail and can therefore be viewed from beneath as white tips overhanging the outermost rectrix. Underparts quite uniform although throat not as deeply black as belly. In some individuals longest undertail-coverts are white-tipped and have a more extensive brown base. Otherwise feather pattern very similar to upperparts. When worn some individuals can appear slightly brown-mottled. **Wing** Very uniformly black-blue, remiges less glossed than coverts which are similar to body. Little contrast between outer primaries and inner primaries, slightly more so between remiges and coverts. Remiges black-blue with paler and less glossed innerwebs and tips. Worn feathers appear less glossed, browner and a little paler. Upperwing-coverts very uniform, though often seeming darker towards leading edge in field. Underwing slightly paler (less contrast than in any *Apus*), coverts appear obviously darker than remiges (greaters marginally so, medians and lessers markedly so), and outer primaries appear a little darker, contrasting more with inner remiges than on upperwing. Greater coverts have darker tips. Darkest underwing-coverts appear uniform with body. **Tail** Uppertail: innerweb grey-brown, outerweb blue-black

with gloss. Rectrices have slightly paler fringes. Tail paler than mantle. Under tail paler than above, and both webs quite uniform. Rectrix spines up to 4mm, and longest in central tail. As mentioned above, tail cloaked by tail-coverts and hard to discern.

Juvenile Similar to fresh adult, though less glossy and longest uppertail-coverts shorter, not reaching end of tail.

Measurements Wing: male (15 BMNH) 114-124 (119.4); female (12 BMNH) 119-126 (122.1).

GEOGRAPHICAL VARIATION Monotypic. The small sample of birds measured from Borneo have (5 individuals: 3 females, 1 male and 1 unsexed) an average overall wing length of 119.8 as opposed to an average of 120.6 for 27 individuals (15 male and 12 female) from the mainland range.

VOICE High pitched call in flight *tirrr-tirrr* (MacKinnon 1988). Also rapid chatter not unlike House Swift.

HABITS Usually encountered in small flocks of up to 5 or 6 birds, or individually throughout most of range. In Borneo, where it can be very common, often seen in large groups. Tends not to associate, more than very loosely, with other species. Particularly fond of feeding in large clearings or around rock outcrops in forests. Very evident in late evening feeding groups.

BREEDING Little is known of this species breeding behaviour. Smith (1977), in Borneo, recorded many birds nesting in a hollow tree (Smythies 1981).

REFERENCES King and Dickinson (1975), MacKinnon (1988), Lekagul and Round (1991), Medway and Wells (1976), van Marle and Voous (1988), Smythies (1981).

---

50 SABINE'S SPINETAIL
Rhaphidura sabini   Plate 10

Other name: Sabine's Spine-tailed Swift

IDENTIFICATION Length 11.5cm. Within Africa only species with wholly white uppertail-coverts. Distinctive spinetail jizz, with broad butterknife-shaped wings (narrow in the secondaries, bulging in the inner primaries and then appearing hooked as the outer primaries are considerably longer), short, square tail and a broad, rather truncated body, make for easy separation from those *Apus* species with white rumps. Within range confusion only possible with sympatric white-rumped spinetails, Cassin's and Mottled Spinetails. The rather larger and longer-winged, though proportionally shorter-tailed, Cassin's is best separated by its black rump and distal uppertail-coverts, with only a narrow band of white across the proximal uppertail-coverts. The slightly longer, longer-winged and proportionally longer-tailed Mottled has white restricted to the rump and only a narrow white band across the hind belly; the upper and undertail-coverts are black, and the throat is more obviously pale. Böhm's and São Tomé Spinetails share, with Sabine's, white undertail-coverts, but have black uppertail-coverts, and are geographically separated. Adult Sabine's, with its long tail-coverts cloaking the tail, would present no problem, but juveniles, with their shorter tail-coverts would present a pattern closer to the other two species. However, it would still show less black in the uppertail. Furthermore, the tiny São Tomé Spinetail has

a quite different jizz, and a less clear-cut distinction between the dark breast and white belly; Böhm's Spinetail is also smaller and has a very short tail, and accordingly a quite different jizz: rapid flight action on rather fluttering wing beats, often seen circling at great height and speed when wings held still.

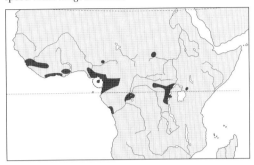

DISTRIBUTION Endemic to Africa, with a predominantly West African range, and two isolated East African populations. Found throughout the Fouta Djalon in a continuous range from extreme S Guinea (from the coast to the Ivory Coast border) the northern half of Sierra Leone, from Freetown in the west, N Liberia (Mount Nimba) and into NW Ivory Coast. Along the length of the Ivory Coast coast from San Pedro in the west to Ayame. Found from S Nigeria through W Cameroon and Equatorial Guinea and into NW Gabon. Found in the Congo basin at Cabinda and then again in E Congo and NW Zaïre. Also recorded in S Central African Republic. In E Africa, two populations exist: one in W Uganda and extreme E Zaïre and the other around Kakamega Forest and Mount Elgon (one record) in W Kenya.

Fairly common to common throughout range, though no detailed population studies or figures. It is probably at its least common in the easternmost part of its range in W Kenya's Kakamega forest, although it has been recorded as not uncommon there in the past.

MOVEMENTS Presumably sedentary.

HABITAT Restricted to rainforest zones, from the lowlands (normally above 700m) to 1700m. Feeds either over forest or in clearings and forest edge, but can be found over adjacent grasslands and mangrove habitats. Often seen over areas of open water. Normally close to the ground but can be seen circling over forest at very great height.

DESCRIPTION *R. sabini*
Adult Sexes similar. **Head** Upper head uniformly blue-black from forehead and lores to nape. Feathers glossed blue-green with warm brown bases. Ear-coverts become progressively paler towards uniformly mid grey or mid grey-brown throat and chin. When fresh, throat feathers pale-fringed, especially chin which can be broadly white-fringed. Black eye-patch. **Body** Head uniform with blue-black mantle and back. Rump and uppertail-coverts white, with indistinct black shaft streaks. Uppertail-coverts cloak tail with the longest ones being as long as the tail, or up to 4mm longer. However when the tail is fanned, or even on occasions when closed, can be seen as black corners beyond the uppertail-coverts. Underparts become darker away from the mid throat being uniformly grey-brown from the lower throat to the grey-brown breast,

149

though there is only marginal contrast with the mid throat. The extreme flanks are black-blue. The belly to the undertail-coverts is white, sharply demarcated from the dark breast. The black shaft streaks are more marked than on the uppertail-coverts, and the bases are grey. The undertail-coverts can be longer than the tail. **Wing** Very uniformly blue-black, with little contrast between tracts. Innerwebs of blue-black remiges green-glossed and slightly paler. In sunlight wing-coverts appear darker than remiges. On underwing, remiges look slightly paler. The greater and greater primary coverts are slightly darker than remiges and are distinctly darker-tipped. The median and lesser coverts are clearly darker than the greater coverts. The leading edge coverts are indistinctly paler-fringed, otherwise there is no marked fringing in the wing. When worn the gloss is reduced and feathers appear very slightly browner and paler, which can increase contrast. **Tail** Blue-black tail similar on both webs and highly glossed as body. Rectrix spines extend up to 4mm, with the central ones being the longest.

Juvenile Plumage generally less glossy, and throat appearing browner. Longest uppertail-coverts 10mm less than tail tip.

Measurements Liberia. Wing: (5) male average 123, female (5) (127). Nigeria to Gabon. Wing: male (19) 118-125 (122); female (15) 116-126 (121.5). Zaïre to Uganda.

Wing: male (22), 113-124 (average 118); female (18) 116-126 (120.5). Tail: male (5) average 45; female (5) average 45.7. Weight. Liberia: male (5) average 19.9. Uganda: male average 17; female (6) 15.5-18 (16.6). Kenya male (4) average 16; female (5) 15-17 (16) (Fry *et al.* 1988).

GEOGRAPHICAL VARIATION Monotypic. Variation noted only in size and weight, with the western populations tending to be longer-winged and heavier than the eastern populations.

VOICE Likened to the weak, high-pitched call of African Palm Swift (Williams 1980).

HABITS Usually encountered singularly or in small flocks, though up to 50 have been recorded together. These flocks are often very tightly formed, and commonly involve other spinetails and rough-winged swallows.

BREEDING Sabine's Spinetail nests either solitarily or with two pairs together, inside hollow trees or in cavities amongst roots of old trees. It seems likely that on occasions breeding in human habitations or abandoned pits occurs. The nest is a small half-cup of twigs adhered to a vertical surface. Two or three eggs are laid averaging (5) 17.4 x 12.25 (Fry 1988).

REFERENCES Colston and Curry-Lindahl (1986), Fry *et al.* (1985), Serle and Morel (1977), Williams (1960 and 1980).

NEAFRAPUS

This genus is endemic to Africa and has two sedentary members, one small, the other medium-sized. Both have similar plumage with blackish upperparts with narrow white bands on the rump or uppertail-coverts, and greyish throat and upper breast with the remainder of the underparts white. Both have diagnostically short tails which are compensated for by bulging secondaries. The rectrices have projecting spines and the feet are anisodactyl.

Separation from sympatric genera

The unusual structure of both of these species renders generic separation straightforward. No other African swifts have such exceedingly short tails and such broad secondaries, coupled with rather plump bodies and heads. This leads to a distinctive jizz personified by a very unsteady flight in which the wings have to compensate for the short tail by acting as stabilisers with constant adjustments that give both species the impression of airborne detritus when not in fasting flickering active flight.

51   CASSIN'S SPINETAIL
Neafrapus cassini   Plate 10

IDENTIFICATION Length 15cm. Medium-sized spinetail, with very short tail and long wings. The narrow white rump band positioned across proximal uppertail-coverts is diagnostic. Like smaller spinetails it has extensive white from the lower breast to the undertail-coverts (which excludes Mottled Spinetail and Little Swift). As well as the rump band the smaller species differ in plumage as follows: Sabine's has white uppertail-coverts, São Tomé has more extensive grey on the breast and is not as clear-cut on the area of divide, Böhm's is purer white on the underparts, lacking the heavy shaft streaking of Cassin's. A subtle difference can be seen in the black rear flanks of the two species. The black is broader on the rear flanks of Cassin's and the connection between the white underparts and the rump is very narrow (figure 32).

Figure 32. Cassin's Spinetail (upper) and Böhm's Spinetail (lower) showing plumage and size differences.

150

Cassin's has the typical broad wing shape of the spine-tails, but especially because of its long wings the tail looks incredibly short and appears a little ungainly. Has a powerful flight action rather like Mottled and can often be seen circling high in the sky, as well as having a lower more active flight with rather loose-jointed beats.

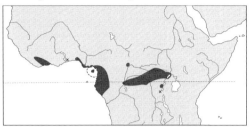

DISTRIBUTION Found in the forest belt of W and C Africa. In West Africa from Mount Nimba in Liberia eastwards from EC Ivory Coast to the Ghanian border through S Ivory Coast. Scattered distribution in the forests of SW Nigeria and the Niger River delta. Further east and south, in SE Nigeria, Cameroon, Equatorial Guinea and throughout Gabon and into Cabinda. In the Congo basin occurs from E Congo through N Zaïre to the Budongo forest, W Uganda. Isolated records from Bioko (Fernando Po), SW Central African Republic and E Zaïre.
Believed to be uncommon throughout range.

MOVEMENTS Presumably sedentary.

HABITAT Found exclusively in and around rainforest, occasionally visiting neighbouring habitats, such as second growth and plantations. Often seen around forest lakes. This may be the optimum habitat, as indicated by its relative abundance in this habitat.

DESCRIPTION *N. cassini*
Adult. Head Upper head, including lores, black-brown, slightly green gloss. Ear-coverts and side of neck a little paler, though still darker than lower throat. Upper throat and chin white with heavy shaft streaking. Feathers around gape and lower throat to mid breast uniformly grey, with some slightly paler mottling caused by paler tips to feathers when fresh. **Body** Upperbody, from nape to lower rump black-brown, uniform with head, though mantle and back slightly darker due to very deep blue gloss (feather tips blue-glossed, bases greener). Distinct white band across proximal uppertail-coverts (distal uppertail-coverts black) very narrow; far narrower comparatively than other African spinetails or Little Swift, but still very conspicuous. The proximal uppertail-covert feathers have black bases and shaft streaks. On the underbody the grey upper breast is distinctly cut-off from the strikingly white lower breast and belly, through to the undertail-coverts. These white feathers have dark shaft streaks and bases, which are least defined on the undertail-coverts. Rear flanks black becoming paler on mid flanks. **Upperwing** Very black-brown, uniform with mantle and similarly glossed blue. Gloss most noticeable on outerwebs of the primaries. All feather tracts very uniform though in flight the outerwing may appear darkest. There is no fringing on the upperwing other than the tips and fringes to the innerwebs of the remiges which are faintly pale brown. **Underwing** The remiges are very dark but appear paler than upperwing. They are fairly uniform with the greater and greater primary coverts, which are white-tipped, but contrast greatly with the black median and lesser coverts. The black leading-edge cov-

erts are paler-fringed. **Tail** The tail is black-brown, paler below and with slightly paler tips. The undertail-coverts reach the end of the tail but the uppertail-coverts fall short of the upper tail. The rectrix spines are strongest and longest at the centre where they extend up to 6mm beyond the webs in the male, and 4.5mm in the female. This apparent sexual size difference in the spine length is based on a small sample of specimens.

Figure 33. Uppertails of Cassin's Spinetail (left) and Böhm's Spinetail (right).

Sexual Differences Sexes very similar, but the female appears to have a greater tendency to be mottled on the breast.

Measurements Wing: male (34) 146-165 (156); female (23) 148-164 (158). Tail: male (4) 22-27; female (3) 25-27. Weight: (3) 37-41.5 (39.5) (Fry *et al.* 1988).

GEOGRAPHICAL VARIATION None. Monotypic.

VOICE Not recorded.

HABITS Often seen very high over forest, tending to feed low only at dawn and dusk. Usually in small groups, though as many as 40 have been recorded together. Regularly with Sabine's Spinetail, and less often with Mottled and Black Spinetails.

BREEDING Thought to nest in hollow trees but a nest has never been found. It is reputed to build a clump of mud nests (Bannerman 1953), but this has not been verified.

REFERENCES Bannerman (1953); Colston and Curry-Lindahl (1986), Fry *et al.* (1988), Serle and Morel (1977).

52   BÖHM'S SPINETAIL
Neafrapus boehmi   Plate 10

Other name: Bat-like Spinetail

IDENTIFICATION Length 10cm. A tiny, short-tailed spinetail, with unique jizz making it the most distinctive of all African swifts. The very small, short, square tail protrudes just beyond the very broad secondaries, but it is the wings that appear most unusual. They are typical of a spinetail with hooked outer primaries, shorter mid primaries and massively bulging inner primaries and secondaries, except for the innermost secondaries which cut in sharply before the tail, but it is their width relative to the size of the bird that appears so striking. Flight appears very uncertain with all manoeuvres made with movements of both wings, usually in unison, with a correction up or down in one wing compensated for by the opposite movement in the other wing. On occasions the secondaries appear to be moved sharply upwards and the leading edge angled downwards to change altitude and, even more comically, direction changes are sometimes achieved by holding one wing still and flapping the other vigorously. The whole action appears very erratic with the species seldom flying far before veering off in another direction. Although the wing beats can appear fluttery and the flight rather slow on occasions, especially when flying high, progress can be very rapid when the wings are held stiffly angled down and then the bird descends rapidly by rocking from side to side in a manner that can only be likened to a jet-fighter going into a tumble! Plumage further distinguishes other spinetails: Cassin's has a broader rump band, Sabine's has black uppertail-coverts, Mottled has more extensive white on the underparts. São Tomé is the closest in plumage, and although they are unlikely to ever be encountered together Böhm's has a more clear-cut border between the dark breast and white upperbody and São Tomé is dirtier white with more extensive grey on the breast.

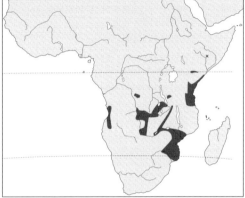

DISTRIBUTION Widely scattered sub-Saharan range in SW, E and SE Africa. Occurs in W Angola south of the Luanda River in a narrow strip along the escarpment of the Bie Plateau often close to the coast, south to Vila Arriaga. In Kenya mainly in the far south-east including the coastal strip, and also in the central valleys of the rivers Athi and Tana. The range in S Kenya continues into N Tanzania south to the Malandu river. Isolated populations occur in W Tanzania from the Nyamanzi river to Kakoma, and in Kasia in S Zaïre. A complicated but continuous range occurs from extreme E Angola through NW Zambia and in a narrow strip into S Zambia south to the Zambezi, through extreme S Zaïre and into NE Zambia. Occurs in a narrow strip along the Luangwa River in E Zambia, and into the Zambezi valley extending west along the north of Zimbabwe and S Zambia but not joining with the W Zambezi population. The range in Mozambique is extensive south of the Zambezi south to the Limpopo valley, but also north a little into Malawi. Also in E and SE Zimbabwe and NE South Africa in the N Kruger National Park. An isolated population exists in NE Mozambique. There are two records from S Somalia.

Generally the rarest of the white-rumped spinetails occurring in Africa. Highly localised and nowhere considered more than locally common; often rare.

MOVEMENTS Resident, though apparently commoner in Zambia in the dry season.

HABITAT Occurs over a variety of primary savanna woodlands mainly in lowlands below 600m. Often in fairly open arid deciduous forest with baobab, miombo, *Cryptosepalum*, but also dense evergreen woodland, both riverine forest

and more extensive lowland forest. In these thicker forests tends to occur on edges and in clearings. Found to 900m in Zimbabwe and 1300m in Tanzania.

DESCRIPTION *N. b. boehmi*
Adult Sexes similar. **Head** Brown forehead and lores. Feathers immediately over eye pale grey-fringed. Rest of upper head black-brown with bronze gloss. Ear-coverts uniform with crown, though paler on lower ear-coverts. Throat and upper breast mottled grey, palest on central throat and chin and slightly darker lower on breast and on sides of breast. **Body** Upperparts uniformly black-brown with a bronze sheen from crown to lower back. Some paler bases evident when worn. Broad white rump band highly contrasting, with black-brown back and uppertail-coverts. Rump has no heavy shaft streaking unlike most other spinetails. Underparts white from lower breast contrasting sharply with grey upper breast. Grey sides of breast merge with black rear flanks, making line of division between white lower breast and grey upper breast a gentle U-curve - not a straight line. Rest of underparts white, to undertail-coverts with dark shaft streaking less evident away from breast. **Upperwing** Uniformly black-brown and slightly bronze-sheened. Some upperwing-coverts have a slightly blue gloss. Innerwebs of remiges paler brown. **Underwing** Remiges paler grey, uniform with white-fringed greater coverts. Smaller coverts contrastingly black. **Tail** Black-brown though paler-tipped and outerweb of outer rectrix white-fringed. Tail greyer beneath and can be seen protruding beyond the undertail-coverts (though not as far as beyond the uppertail-coverts from above). The uppertail-coverts come within 4mm of tail tip. The tail has shaft spines from each rectrix projecting 2.5-3.5mm beyond webs.

Juvenile As fresh adult.

Measurements W Angola. Wing: male (22) 109-128 (122); female (20) 120-131 (126). Tail: (2) 18.5 + 20. Wing measurements of central African birds average 2.4 longer. Weight: male (7) 13-16 (14.3); female (5) 13-16 (14.6) (Fry *et al.* 1988).

GEOGRAPHICAL VARIATION Two races.
  N. b. boehmi (W and E Angola, Zaïre, W Tanzania, and N Zambia) Within range some difference in size noted between west Angola and central African birds.
  N. b. sheppardi (E Africa, E and S Zambia to Mozambique, Zimbabwe and S Africa) *Sheppardi* intergrades with *boehmi* in upper Zambezi valley, south Zambia. Differs from *boehmi* in that the upper breast is uniformly grey, lacking the darkening on the lower edge and hence the degree of contrast is somewhat reduced. Also shaft streaks are more evident on the underparts and the rump has dark shaft streaks which are missing in *boehmi*. In fresh plumage the primaries are white-fringed on the outerwebs. Slightly smaller than *boehmi*. Wing: (53) 109-124 (117). Weight: male (5) 12-15.5 (13.5); female (7) 12-15.5 (13.6) (Fry *et al.* 1988).

VOICE Seldom heard call is a silvery rippling quadrisyllabic twitter *Tir-it-treetree*, or *Ti-tititeieie*, with particular emphasis on the initial note which can also be heard in isolation as *Ti* or *Tit*.

HABITS Usually feeds closely around the crowns and canopy of large trees. Highly gregarious forming flocks of up to 50 birds, though usually far fewer and occasionally singly. Associates with Mottled Spinetail and African Palm

Swift and a wide variety of roughwings (*Psalidoprocne* spp.) and swallows (*Hirundo* spp.). Brooke (1993 pers. comm.) notes that as well as nesting in hollow trees this is the only old world swift known to breed in underground sites such as wells, exploration pits and so on. It is therefore the only swift likely to be seen disappearing into or below the ground.

BREEDING Böhm's Spinetail is a solitary nester. All nests that have been discovered are in subterranean pits, shafts or wells, usually amongst *Brachystegia* woodland. Howev-er, they do associate with baobab trees and have been ob-served leaving hollows in these trees, but they might only roost in them. The nest is a U-shaped construction of thin twigs (2-3cm long) with one or two feathers and is 6-8cm wide and 2.5-6.5cm deep. It is usually stuck on an area of slight overhang 3-9m down the shaft or well. The clutch is usually 3 eggs (although only two have been recorded) averaging (9) 19.1 x 12.5 (another series of eleven eggs averaged 17 x 12 (Fry *et al.* 1988).

REFERENCES Brooke (1966), Fry *et al.* (1988).

HIRUNDAPUS

This is a group of large swifts which are largely Oriental in distribution. Two species are migratory with one of these being a long distance migrant that is regularly recorded well outside its normal range. The other two species are partially migratory or sedentary.

Separation from sympatric genera

This genus has several distinctive features. The plumage is highly glossy. Three of the four are unique in having contrastingly pale mantles, two have distinctive white spots on the innerwebs of the tertials and all possess the unique horseshoe pattern at the rear of the underbody caused by white flank stripes extend-ing from white undertail-coverts. There is no sexual dimorphism in plumage, but it is possible to distin-guish adults from at least some immatures in all species.

The genus is as distinctive in structure as it is in plumage. The broad wings are rather triangular with very full secondaries, the head is large and protruding and the broad, rather square or slightly rounded tails are relatively short.

The size of the rectrix spines is rather variable with the two smaller species (with the squarest tails) having rather slight spines, whereas the more rounded-tailed large species have very prominent broad spines especially those on the central rectrices (see figure 34). Furthermore, the smaller species show less differ-ence in spine length between the central and outer rectrix spines than the larger species.

Figure 34. Tails of the *Hirundapus* species. White-throated Needletail, Silver-backed Needletail, Brown-backed Needletail and Purple Needletail (left to right). Note more rounded tail shapes of the two larger species compared with the more square tail shapes of the two smaller species. Also note gradation of the differential in spine length between central and lateral rectrix spines from White-throated Needletail (least difference) to Brown-backed Needletail (greatest differ-ence).

53 WHITE-THROATED NEEDLETAIL
Hirundapus caudacutus   Plate 13

Other names: Needle-tailed Swift; Spine-tailed Swift; Northern Needletail

IDENTIFICATION 19-20cm. Highly distinctive jizz creat-ed by combination of plumage features and structure. Dark plumage with highly contrasting pale saddle, white horseshoe mark, broad wings, bulky body with relatively large head and short tail; all serve to immediately identify this species as a needletail. Distinctive flight action: rapid powerful wing beats propel the species at the high speeds the genus is renowned for. In level direct flight (on mi-gration or when feeding) equally rapid but wing beats interspersed with periods of gliding, most often watched, however, at much lower speed, gliding for long periods on slightly bowed wings with few wing beats. Most easily distinguished from other needletails by highly contrast-ing broad white throat patch. All other needletails (excluding Purple) often show paler throat patches and these are enhanced by the darker ear-coverts and sides of head. Silver-backed in particular can show an off-white patch, but it is never as striking or as clear-cut as White-throated Needletail (see figure 35). Furthermore, this is the only needletail showing white on the lores and across the forehead, although juveniles lack this feature as does *nudipes*, and variation within adult *caudacutus* is consider-able with some birds showing no white or simply white lores and dark foreheads. The pale saddle is clearly paler than in Brown-backed or Purple but similar to Silver-backed Needletail. Juvenile *caudacutus* and *nudipes* show a slightly darker saddle. White on the innerwebs of the tertials can be seen in good views. However, it is greyer in *nudipes* and in fact very similar to Silver-backed which also shows this feature under excellent viewing conditions. The plumage of White-throated is generally more glossy than the other species, although in dull light this is hard to see. Structurally differs from Brown-backed and Purple in having thin short tail-spines and a rather square tail shape. Tail shape in the needletails is affected by how closed or open the tail is held. When fully spread tail shapes appear rounded and when tightly closed they appear very pointed, with variations between these two extremes.

Figure 35. Heads of adult White-throated Needletail (nominate race, left) and Silver-backed Needletail (right). Note well-defined white throat patch of White-throated and ill-defined patch of Silver-backed.

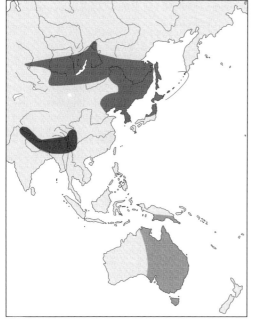

DISTRIBUTION Northern populations breed from C Siberia to Japan, wintering in Australia. Southern population in Himalayas. In Siberia breeds from the Vasyugan River basin and then east through N China (and into SE Mongolia), N Manchuria and Korea in a broad band to Sakhalin, Kuril Islands and Japan (Hokkaido and N Honshu). Winters primarily in E Australia, and in lesser numbers in S New Guinea, New Zealand and possibly on Lombok. Possibly winters as far north as Thailand (Khao Yai); if this is the case it is possible that the species has been overlooked as a winter visitor throughout SE Asia. Passage migrants recorded throughout E China, SE Asia, Greater Sundas and Wallacea. Breeds throughout the Himalayas from Hazara in Pakistan, east through Nepal,

Sikkim and Bhutan to Arunachal Pradesh and W Yunnan, and north to Sichuan. A single bird recorded from the Chin Hills, W Burma in April 1995 (C. Robson verbally), may have been a migrant but could possibly represent an undiscovered disjunct population.

Uncommon and local throughout Himalayan range. Within Siberian range, rarer in west becoming commoner further east.

MOVEMENTS Northern populations migrate south through China, E Thailand and Indochina, crossing Wallacea and New Guinea to reach Australia. The lack of records from the Indian subcontinent indicates that central Siberian birds migrate south to south-east in Siberia, north of the C Asian mountain ranges before moving southwards. The relative scarcity of passage migrants in Wallacea and SE Asia is indicative of the rapid high-altitude migration and the difficulty in identifying fast-moving needletails. Breeding birds leave Siberia from late August and through September, with late records from W Siberia in late September. Migration through SE Asia continues from September to November, with arrival in Australia from December. Departure from Australia starts in March, with arrival in Siberia from mid May. The departure is not synchronised, and some movement still occurs through SE Asia in May. Spring records from Hong Kong 17 March - 8 May (P Kennerley pers. comm.). In SE Asia most migration occurs through Indochina, the species being surprisingly uncommon in Malaysia and Thailand. The Himalayan race *nudipes* is probably resident.

Has occurred as a vagrant in western Europe on 12 occasions in the period April-July, with one record in November. Seven of these records are from Britain. In North America has been recorded from the Pribilofs. From the southern Hemisphere vagrants have been recorded from Seychelles (twice), Fiji, New Zealand and Macquarie Island (54°S 159°E).

HABITAT In the Himalayas occurs between 1250-4000m, feeding over upland grasslands and river valleys. In Siberia over wooded lowlands and hills with open spaces. Extensive daily movements in search of food take the species over a variety of habitats, with weather conditions dictating feeding altitude. Birds can be found from lowlands to the snow line. In Australian wintering grounds occurs in both mountainous and coastal areas.

DESCRIPTION *H. c. caudacutus*
Adult Sexes similar. **Head** Broad white throat patch, extending from the gape, along lower ear-coverts and down to lower throat, contrasting highly with surrounding dark plumage. White band across lores and forehead (broadest on lores and just over black eye-patch). Some individuals browner on centre of forehead. Rest of head and nape dark olive-brown. In fresh plumage top of head and nape glossed greenish-blue, though this is removed with wear, when some paler brown feather bases can be seen. Nape feather bases white though, even with heavy wear, they are not easily seen in the field. Head uniform, slightly paler towards throat patch. Side of neck and ear-coverts appear darker than underparts enhancing the white throat and dark upper head. **Body** Underparts dark olive-brown from breast and upper flanks to vent. When plumage fresh, brown gloss can be apparent. Lower flanks and undertail-coverts clear white, forming highly distinctive horseshoe mark. Saddle pale-brown, palest at centre of lower mantle and back, becoming progressively darker

towards nape, olive-brown scapulars, central rump and clearly even darker olive-brown sides of rump and upper-tail-coverts. Saddle palest when plumage worn. **Upperwing** Remiges black, innerwebs of primaries paler brown though not to tips. All coverts black. Blue gloss to remiges and primary coverts. Other coverts with blue gloss to outerwebs and tips of innerwebs, green gloss on remainder of innerwebs. White on innerwebs of tertials can be seen in the field with good views. Gloss hard to see unless in good light, otherwise appears black. Gloss lost through wear. Inner/outer wing contrasts less than in *Apus*. **Underwing** Remiges appear paler beneath and fairly uniform with greater primary and greater coverts. Median and lesser coverts darker black-brown, uniform with axillaries. **Tail** Black with green gloss. Central rectrix spines up to 6mm, outer rectrix spines to 2.5mm. **Body/Wing Contrast** Upperwing obviously darker than pale saddle. Underwing coverts appear slightly darker or uniform with underbody.

Juvenile Differs in grey-brown forehead with pale grey-brown lores. Throat slightly less contrasting. Black areas of upperparts less obviously glossed green than in adult. Pale saddle less obvious or as pale as in adult. Underparts differ most markedly on the lower flanks with some black streaking and spotting on the white feathers. Feathers of undertail-coverts have black fringes to tips. Upperwing and tail black, less gloss than adult. Innerwebs of tertials more broadly bordered black (shortest tertial only narrowly tipped). Note that juveniles show great variation and can be closer to adult in plumage with regard to all above named differences, with the black fringes to the tips of the longest undertail-coverts being the most consistent feature. Prior to the post-juvenile moult on the winter quarters, some paler grey bases (white bases on the nape and rump) to the body feathers can be seen giving the body an untidy mottled effect.

First-year After post-juvenile moult, body similar to adult, but with greater tendency to show grey-brown forehead, and typically has black fringe on tip of shortest tertial and some black fringes to tips of the undertail-coverts. Juvenile remiges and larger wing-coverts retained to post-breeding moult and therefore clearly worn.

Measurements Siberia. Wing: male (7) 200-213 (208); female (10) 198-211 (206); juvenile male 203-215 (209); juvenile female 195-215 (206). Tail (including rectrix spine): male (11) 47-53 (49.7); female (7) 47-51 (48.8); juvenile male (9) 47-55 (50.3); juvenile female (10) 48-53 (50.6) (Cramp 1985). Weights (Siberia): male (11) 109-140 (122); female (5) 101-125 (114) (Collins and Brooke 1976).

GEOGRAPHICAL VARIATION Two races.

H. c. caudacutus (range of species except Himalayas) Described above. No significant variation in size or plumage throughout extensive range.

H. c. nudipes (Himalayas) Primary difference from *caudacutus* is in uniformly black forehead and lores. Saddle not quite as strikingly pale, being browner and less silvery-grey, the underparts are slightly darker. Those areas of plumage that are glossed green are darker and bluer, except the underparts that are slightly green in gloss. Measurements. Wing: (6) 194-214 (206) (Cramp 1985). Weight (type): 127g (Hodgson, in Collins and Brooke 1976). Central rectrix spines to 5mm, outer rectrix spines to 2.5mm.

VOICE A rapid insect-like chattering much softer than *Apus* screams *trp-trp-trp-trp-trp-trp......* Calls of varying duration and speed can be heard.

HABITS More solitary in behaviour than most swifts on the breeding grounds. On migration and in winter can form large flocks, though on migration still just as likely to be seen singly. Mixed species flocks usually involve other needletails rather than other genera, although often seen migrating with Pacific Swift.

BREEDING Uses a shallow depression at the base of a hollow in a tree or scrape such a depression in the accumulated debris. Eggs, of the nominate race, average (28) 32.3 x 22.3 (Collins and Brooke 1976). Clutch size varies between two and seven eggs.

REFERENCES Ali and Ripley (1970), Chantler (1993), Cheng (1987), Collins and Brooke (1976), Cramp (1985).

54   SILVER-BACKED NEEDLETAIL
Hirundapus cochinchinensis   Plate 13

Other names: White-vented Needletail; White-vented Spine-tailed Swift; Grey-throated Needletail; Rupchand's Needletail (race *rupchandi*)

IDENTIFICATION Length 20cm. Easily identified as a needletail by virtue of distinctive plumage and jizz. Similar in shape, plumage and behaviour to White-throated from which it can be excluded with care by close examination of throat: at its palest, an indistinct pale grey patch which merges with the lower throat in Silver-backed and often hardly paler than rest of the underparts. In White-throated always a broad white highly contrasting patch. The relative darkness of the upper head, ear-coverts and sides of head can in the palest examples produce a notable but mottled patch but further examination of its lower edge will show it to be highly indistinct. Many adults of the migratory race of White-throated *caudacutus* show white lores and forehead, which is missing in Silver-backed. Caution must be exercised when using the pale inner tertial webs as a feature for Silver-backed as the outerweb is contrastingly pale grey-white and under excellent viewing conditions can be seen. Brown-backed and Purple Needletails can be excluded by examination of the saddle, which appears as an extensive silvery-grey patch but in Brown-backed it is dull brown (but nevertheless prominent, especially when seen in bright light). Purple is the darkest-saddled *Hirundapus*, with the upperparts appearing uniformly dark. The distinctly pale saddle of Silver-backed also creates the impression of a darker head and tail, as in White-throated. Furthermore, Brown-backed averages slightly darker in underpart coloration and the widespread race *indicus* has a distinctive white supraloral spot, as does Purple. Tail shape as White-throated, though spines slightly stronger.

DISTRIBUTION Range imperfectly known. Occurs in SE Asia and E Himalayas to C Nepal, E India south of the Brahmaputra in Assam, Mizoram, Manipur, Nagaland and Bangladesh, and then east into Burma. The Thai range is uncertain. It is known with regularity only from Khao Yai and the mountains of Khao Soi Dao, and is probably only a winter visitor. Scattered records throughout the country may refer to passage migrants only. It occurs on Hainan and possibly southwards throughout Indochina. First recorded in N Vietnam at Fan Si Pan in NW Tonkin, May 1995 (C. Robson verbally). Migrant in Hong Kong. An isolated population exists in the Himalayan foothills of C Nepal and another on Taiwan. Both *rupchandi* and *cochinchinensis* winter in W Java, Sumatra and Peninsular Malaysia (and perhaps more widely in SE Asia). The exact wintering grounds of the indistinct *formosanus* are not known.

Few indications of status throughout range, but believed to be common in the Assam hills.

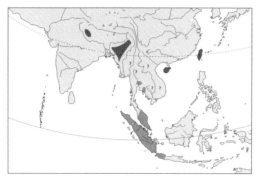

MOVEMENTS Little known. Known as a passage migrant from Sumatra and possibly Thailand, although no doubt much overlooked. Timing of migration and arrival on the wintering grounds poorly understood, although it may be that migration through SE Asia is performed at a similar time to White-throated, during April and May, and September to October, although the presence of wintering birds in SE Asia somewhat confuses the picture. In peninsular Malaysia extreme dates for this species are 28 September and 10 April. Nine migrants were collected at night from Fraser's Hill, over 4 autumns in the period 28 October to 3 December. Regular spring migrant Hong Kong between 25 March - 11 May with maximum count of 100 on 8 April 1988 (P Kennerley pers. comm.). It must be presumed that Hong Kong migrants are going to Taiwan as only known north of Hong Kong on that island.

HABITAT In Nepal from 600m in the central terai (lowlands) where it can be seen over tropical forest at Chitwan. In this respect differs from *nudipes*, the Himalayan race of White-throated Needletail which is found from 1500-4000m. In SE Asia and Sumatra occupies wide range of habitats due to utilisation of all altitudes from lowlands to mountains up to 3350m.

DESCRIPTION *H. c. cochinchinensis*
Adult Sexes similar. **Head** Throat grey-brown to pale grey-brown. Often barely paler than rest of underparts, but in some clearly paler especially central throat and chin. This effect is enhanced by the dark ear-coverts and side of head, which appear clearly darker than the underparts and form a dark frame for the paler throat. Rest of head and nape dark black-brown or olive-brown. In fresh plumage, top of head and nape glossed greenish-blue, though this is removed with wear, when some paler brown feather bases can be seen. Nape feather bases pale grey though even with heavy wear it is unlikely that they are ever easily seen in the field. Upper head uniform. **Body** Underparts dark olive-brown from breast and upper flanks to vent. When plumage fresh, brown gloss can be apparent. Lower flanks and undertail-coverts clear white, forming highly distinctive horseshoe mark. Saddle pale brown, palest at centre

of lower mantle and back, progressively darker towards nape, olive-brown scapulars, central rump and clearly darker olive-brown sides of rump and uppertail-coverts. Saddle palest when plumage worn. **Upperwing** Remiges black, innerwebs of primaries paler brown though not to tips. All coverts black. Remiges and coverts glossed bluish though this wears from greenish-blue to dark blue and then appears dull black. Gloss hard to see unless in good light, otherwise appears black. Tertials have pale grey innerwebs visible in very good field views. The longest tertial has a dark grey tip to the innerweb. **Underwing** Remiges appear paler beneath and fairly uniform with greater coverts. Median and lesser coverts darker black-brown, uniform with axillaries. **Tail** Black with green gloss. Central rectrix spines to 6.25mm, outer rectrix spines to 2.8mm. **Body/Wing Contrast** Upperwing obviously darker than saddle. Underwing coverts appear slightly darker or uniform with underbody.

Juvenile Plumage resembles that of juvenile White-throated but slightly less obvious saddle and some brown crescents on flanks, vent and undertail-coverts.

Measurements Wing: 183-184. Tail: 48-49 (Ali and Ripley 1970). Weight: (2) 76.1 + 85.5 (Collins and Brooke 1976).

GEOGRAPHICAL VARIATION Three races.
    H. c. cochinchinensis (E Himalayas through SE Asia) Described above. Within range some individual variation in relative paleness of throat and saddle, although appearance of these features varies according to moult and wear.
    H. c. rupchandi (C Nepal) Darker-throated, with slightly paler underparts and less glossy upperparts than *cochinchinensis*. Wing: 180-192. Tail: 46-49 (Ali and Ripley 1970).
    H. c. formosanus (Taiwan) Recognised by Collins and Brooke (1976) only by virtue of its geographical isolation from other populations.

VOICE As White-throated Needletail, a soft rippling trill *trp-trp-trp-trp-trp*.

HABITS Highly active, only staying at one site for any length of time if drinking or feeding. Often seen in small groups feeding over forest, open areas or especially rivers and lakes. In Malaysia wintering flocks of 200 have been seen in winter and spring. Often seen in association with Brown-backed Needletail. Where a sudden abundance of food occurs (such as a hatching of insects) will congregate in a highly active feeding group, often with the last species, making repeated high speed passes through the insects.

BREEDING No authenticated nest has been described for this species. One egg from the oviduct of a collected specimen measured 28.1 x 21.0 (Baker 1927).

REFERENCES Ali and Ripley (1970), Baker (1927), Chantler (1993), Collins and Brooke (1976), Lekagul and Round (1991).

55   BROWN-BACKED NEEDLETAIL
Hirundapus giganteus   Plate 13

Other names: Brown Needletail; Giant Needletail; Brown Spine-tailed Swift

IDENTIFICATION Length 25cm. Typical needletail jizz and plumage make confusion possible only with other needletails. Large size and powerful build, even for a needletail, add to its highly distinctive appearance. Brown-backed has a dull pale brown saddle that is far less contrasting than those of either Silver-backed or White-throated, but it is always conspicuous especially in bright light. White-throated can furthermore be excluded by its highly contrasting white throat patch, by the white forehead in the case of many adult *caudacutus* and by the highly contrasting white inner tertial webs. The throat of Brown-backed often appears barely paler than the rest of the underparts but, as in Silver-backed, there is some individual variation. Identification of the widespread *indicus* race is further facilitated by the highly contrasting white supraloral spot. This is especially prominent when viewed head-on. The closely related Purple Needletail is best identified by its very dark plumage. The saddle patch is not present and the upperparts appear uniform. The throat patch is the least marked of the *Hirundapus*, appearing very uniform with the underparts. Furthermore it is highly purple-glossed. The flight action of Brown-backed is similar to the smaller needletails, but when seen in low-speed cruising it can appear more lumbering and slower in turning. Allegedly the fastest bird species. The tail, like that of Purple, differs from the two other species in having longer central feathers, producing a more rounded appearance when open and more pointed when tightly closed. This appearance is enhanced by the tail spines which are the longest and most robust of any *Hirundapus*, and very noticeable in the field (especially those on the central rectrices).

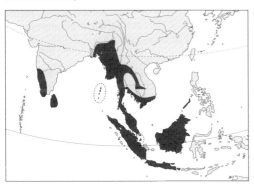

DISTRIBUTION Extensive S and SE Asian range. In SW India in the Western Ghats from Goa and N Kanara southwards through Karnataka and W Tamil Nadu to Kerala. Occurs on Sri Lanka and the Andamans, Bangladesh north to Assam, through the mountains of Mizoram, Manipur and Nagaland. Found throughout Burma, W Thailand from the far north, south through the western highlands and throughout the peninsula. Extends in Thailand from the north-central plains southwards through the Dong Phaya Yen range, and then eastwards into Cambodia along the Phanom Dongrak range, and from the mountains of Khao Soi Dao in SE Thailand to the Cambodian Carda-

mom mountains. From Cambodia the range extends north to NE Laos and east into S Vietnam. From peninsular Thailand south throughout Malaysia and into Sumatra, Java and Bali. Also on the Sumatran archipelagos of Riau and Lingga, throughout Borneo (including the N Natuna islands) and the Philippine islands of Calamian and Palawan.

Generally the most abundant needletail throughout range, commonly encountered in suitable habitat.

MOVEMENTS *Indicus* is a partial migrant to the Malay Peninsula, probably Borneo and possibly Sumatra, during September and November returning during May. South of *indicus* thought to be resident with some dispersive behaviour.

HABITAT Like Silver-backed Needletail occurs in a wide range of habitats, from lowlands to high altitude. Preferred habitat is forest, either primary or secondary growth. Extensive daily foraging brings species into contact with many habitat types. In Thailand occurs to 1800m.

DESCRIPTION *H. g. indicus*
Adult Sexes similar. **Head** Pale brown to brown throat, palest in centre and chin, often hardly paler than rest of underparts, though in some individuals fairly distinct especially as ear-coverts appear dark emphasising throat patch. Upper head and nape dark olive-brown. In fresh plumage top of head and nape glossed metallic black, though this is removed with wear, when paler brown feather bases can be seen. **Body** Underparts dark brown from breast and upper flanks to vent. When plumage fresh, brown gloss can be apparent. Lower flanks and undertail-coverts clear white, forming highly distinctive horseshoe mark. Dark shaft streaks on the undertail-coverts are apparent in some specimens. Saddle brown, clearly darker than White-throated or Silver-backed. However, still paler than head and tail, palest at centre of lower mantle and back, becoming progressively darker towards nape, olive-brown scapulars, central rump and even darker olive-brown sides of rump and uppertail-coverts. Saddle palest when plumage worn. **Upperwing** Remiges black, innerwebs of primaries paler brown though not to tips. All coverts black, with indistinct gloss, in field coverts appear slightly darker than remiges. Tertials paler browner on the innerwebs - not visible in field. Inner/outer wing contrast less apparent than in *Apus*. **Underwing** Remiges paler beneath appearing fairly uniform with greater coverts. Median and lesser coverts darker black-brown, uniform with axillaries. **Tail** Black-brown similarly glossed to head. Tail spines extend to 11mm in central rectrices and 1.8mm in outer rectrices. **Body/Wing Contrast** Upperwing darker than saddle. Underwing coverts appear slightly darker or uniform with underbody.

Juvenile Similar to the adult but has less extensive white supraloral spots, some very narrow dark crescentic fringes on the white rear flanks and undertail-coverts, and very narrow pale fringing to the remiges.

Measurements Wing: 188-200. Tail: 54-60 (Ali and Ripley 1970). Wing : (12 BMNH) 181.5-206 (194.5). Weight: (4) 123-167 (137.13) (Collins and Brooke 1976).

GEOGRAPHICAL VARIATION Two races.
   H. g. indicus (Range of species north and west of *giganteus*) Described above.
   H. g. giganteus (S Thai Peninsula, through Malaysia, Greater Sundas, and Palawan) Plumage closely similar to *indicus* but lacks white supraloral spot. Wing:

(21 BMNH) 191-209 (199.5). Central rectrix spines to 11.5mm and outer rectrix spines to 1.9mm.

VOICE Similar to Silver-backed and White-throated Needletails but a little slower, an insect-like rippling trill. Also, a brittle squeaked *cirrwiet*, repeated 2 or 3 times and a thin squeak *chiek*.

HABITS Gregarious species, seldom seen singly, often in pairs or small groups. Up to 60 birds recorded together. In areas of sympatry often with Silver-backed, and, like other needletails, tends not to form flocks with other swift genera. Most often encountered cruising low over forest or circling in small flocks at high altitude. Frequently comes to water in forest, throughout day but especially in the evening where small groups will circle and then make repeated visits to the lake surface. Large size is apparent as a significant wake is left when surface is skimmed by huge gape, and birds often splash into water in a rather clumsy fashion. The large legs and feet can be clearly seen as they are lowered when water is approached. During the rapid descent to the water the species has been reported to make a drumming sound. It is possible that this noise is generated by tail movement as has been recorded for Mottled Swift.

BREEDING This species is a solitary breeder. Up to three nests have been found in a single hollow tree, but it is unlikely that more than one pair ever occupy the same tree. No actual nest is used, simply a depression is used or a shallow scrape is made in debris at the base of a tree hollow. Average egg size (100) of *indicus* is 29.6 x 22.2 (Baker 1927).

REFERENCES Ali and Ripley (1970), Baker (1927), Chantler (1993), Collins and Brooke (1976), Lekagul and Round (1991).

---

56   PURPLE NEEDLETAIL
Hirundapus celebensis   Plate 13

Other names: Celebes Needletail; Sulawesi Spine-tailed Swift

IDENTIFICATION Length 25cm.This most enigmatic member of *Hirundapus* has the most restricted range and is thought to be purely resident. It is not sympatric with other needletails but White-throated could occur on migration anywhere in the Philippines or Sulawesi. The jizz is typical of the family and it also shares the white horseshoe mark on the underparts. The plumage of Purple is, however, darker both above and below than its congeners and is very glossy. Both the throat, which is uniform with the blackish underparts (far more so than in Brown-backed), and the saddle, which is likewise uniform with the rump and head, are darker than the other needletails. These rather different characters ensure that with good views identification should be straightforward. A further difference from White-throated Needletail is lack of white outerwebs to the tertials. Shares with nominate Brown-backed pale loral markings, though this is often less well-defined in juveniles (see debate in geographical variation). Tail structure similar to Brown-backed, with longer central tail feathers producing a more rounded shape and larger, longer and stronger tail spines.

DISTRIBUTION Occurs in NE Sulawesi (possibly C Sulawesi) and on Philippine islands of Luzon, Mindoro,

Marinduque, Catanduanes, Calayan, Panay, Negros, Cebu, Leyte, Biliran, Mindanao and Basilan.

No certain population details are known and it is possible that this is a rather scarce species, purely on the basis of the scarcity of museum specimens. In central Sulawesi large flocks of needletails have been observed, presumably of this species.

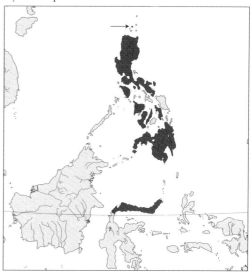

MOVEMENTS Only known from the Philippines (excluding Palawan) and Sulawesi, where it is presumed resident.

HABITAT Over forest and open country in lowlands and hills. Little is known of the habits but it seems likely that owing to its close relationship to Brown-backed Needletail its general behaviour will also be very similar.

DESCRIPTION *H. celebensis*
Adult Sexes similar. **Head** Blackish throat patch not contrastingly paler than rest of underparts. Throat slightly browner than upper head, though probably not distinct in field. Upper head uniformly blackish with blue-black gloss, except for pure white loral spot in front of the black eye-patch. It is likely that gloss is reduced through wear and that in particularly worn birds some paler feather bases can be seen. **Body** Underparts typical of *Hirundapus* but blacker, being black-brown (with distinct blue-black gloss when fresh) from breast and upper flanks to vent, contrasting with clear white lower flanks and undertail-coverts, forming horseshoe mark. Upperparts atypical of the genus, lacking paler saddle. Upperbody from head to uppertail-coverts blackish and blue-black glossed. **Upperwing** Black remiges (slightly glossed most markedly on outerwebs) paler on innerwebs especially outer vane (though not the tip). All wing-coverts similarly blackish, though they can appear a little blacker than the remiges, and are more obviously glossed to a degree similar to body feathers. **Underwing** Remiges paler beneath and fairly uniform with greater primary and greater coverts. Median and lesser coverts appear blacker. The greater covert fringes are whitish. **Tail** Black with bluish gloss. Tail spines (type specimen) project up to 4.5mm in the central tail and 1mm in the outer tail (Dekker pers. comm.). **Body/Wing Contrast** The upperwing appears fairly uniform with body, though across the remiges the wing can appear a little paler. Underbody appears darker than underwing.

Juvenile Differs from adult in having dull brown loral patch and having dull brown fringes to the greater under-wing-coverts.

Measurements Sulawesi. Wing: (2) 203 + 208 (205.5). Negros. Wing: (11) 211-218 (214.9). Luzon. Wing: (2) 222-226 (224). Mindoro. Wing: (5) 214-234 (223.8). Mindanao and Basilan. Wing: (4) 211-217 (213.75) (Collins and Brooke 1976). Weight: (22) 170-203 (179.6) (Morse and Laigo 1969).

GEOGRAPHICAL VARIATION There is debate regarding the differences in plumage between birds from the Philippines and those from Sulawesi. The race *dubius* has been recognised in the Philippines. The plumage is said to be generally browner than *celebensis* apart from the head and wings, which have a more violet-blue gloss. The loral spot is said to be whiter in the Philippines although Collins and Brooke believe this is an age-related feature. Collins and Brooke consider that Purple is monotypic species and that the features were based on too small a series of specimens. Furthermore they believed that supposed differences in wing length were not outside the range one could expect within a species of swift. A further subspecies has been recognised by some authors, *H. c. manobo* from Mindanao, Basilan and Negros.

VOICE Not known. Presumably similar to other members of genus.

HABITS Morse and Laigo (1969) reported that the species was suffering in the Philippines as a result of its unfortunate habit of taking bees by flying low over fields close to hives. Between January and March 1968 it was believed that as many as 450 needletails were captured as a result of this behaviour.

BREEDING It has been suggested that this species is a cave nester but there is no evidence for this. (Collins and Brooke 1976).

REFERENCES Collins and Brooke (1976), Morse and Laigo (1969), White and Bruce (1986).

CHAETURA

A genus of nine small species, seven of which are restricted to the Neotropics, whilst two species breed in North America (one exclusively) and migrate to the Neotropics in winter.

The plumage is quite similar throughout the group with uniform upperparts and paler rumps (or paler rumps and uppertail-coverts). The underparts are pale usually on the throat and breast only, though in two species also on the undertail-coverts.

Most members of the genus have adjusted to nest primarily in man-made nest sites especially chimneys and subterranean manholes. These sites clearly present similar conditions to those found in natural nest sites such as hollow trees. All nests that have been recorded for this genus are very similar being bracket-shaped, stuck to a vertical surface and comprising fine twigs bound together with saliva.

Separation from sympatric genera

All the members of this genus are rather small and have the distinctive and typical spinetail wing shape comprising rather hooked outer primaries and bulging inner primaries and secondaries. The tail is short and square and the head protrudes almost as far as the tail producing the classic 'cigar with wings' impression. White-tipped Swift and some of the shorter, square-tailed members of the *Cypseloides* can be excluded by examining the wing shape. Flight is rapid and fluttering, often rather bat-like with rocking movements made on rather stiff, down-tilted wings (figure 36).

Figure 36. Band-rumped Swift (left), White-collared Swift (centre) and White-chinned Swift (right). Note structural differences between the genera.

57 BAND-RUMPED SWIFT
Chaetura spinicauda   Plate 16

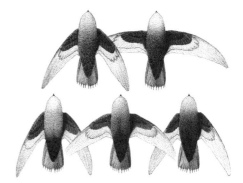

Other name: Smoky Swift (race *fumosa*)

IDENTIFICATION Length 10.7cm. Typical *Chaetura* shape and jizz. Easily identified by narrow whitish rump contrasting markedly with remainder of blackish upperparts. Some other species exhibit great contrast between saddle and rump, but in Band-rumped Swift the highly restricted rump band contrasts strongly with the uppertail-coverts and the lower back. The races *fumosa*, and to a lesser extent *aethalea*, have more extensive and slightly greyer rumps. However, they are very black on the upperparts thus appearing very white-rumped and always appear whiter-rumped than sympatric subspecies of Grey-rumped Swift and clearly darker below with more contrasting pale throat patches (figure 37).

Figure 37. Underparts of Band-rumped Swifts of nominate race (top left) and race *fumosa* (top right), and Grey-rumped Swifts of nominate race (bottom left), race *guianensis* (bottom centre) and race *lawrenci* (bottom right).

160

The upperparts lack the bluish gloss of Grey-rumped Swift and the bronze gloss of Pale-rumped Swift, appearing rather matt by comparison. Generally one of the darkest *Chaeturas* species below. Geographically isolated from the similar Pale-rumped Swift, which has a similar white rump, extending onto the lower back. Like Grey-rumped and Lesser Antillean Swifts, a little longer tailed and rather more slender than other *Chaetura* species. Robert Clay (pers. comm.) considers that Band-rumped is perhaps more rakish than Grey-rumped.

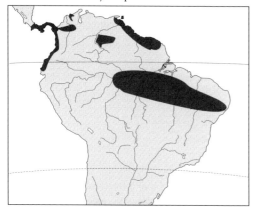

DISTRIBUTION Range extends through both S Central America and N South America, from the most northern point of range in the Golfo Dulce region of Costa Rica's southern Pacific slope, and then south to W Panama. Throughout much of N Colombia, south to Valle on the Pacific coast and across the northern lowlands to the lower Cauca valley in the south and to Santa Marta and Guajira in the north and east. In Colombia east of the Andes occurs in W Caqueta, W Putumayo and Guainia. In Ecuador occurs in the west extending south to north Manabi and western Chimborazo. Further east found in southern and eastern Venezuela at Jobure in Delta Amacuro, throughout Amazonas and in Bolivar along the lower Rio Cauca, lower Rio Paragua and the Cerro Paurai-Tepui. North of the Orinoco at the base of the Paria Peninsula in Sucre, and on nearby Trinidad. Throughout the Guianas and north-east Amazonian Brazil from Purus eastwards to eastern Para and Alagoas.

One of the commonest *Chaetura* swifts, frequently outnumbering all other species where they occur sympatrically. In Ecuador most abundant between 450-900m (Marin 1993).

MOVEMENTS Resident throughout range.

HABITAT From sea-level to 1500m in both forested and cleared areas. In west Andes occurs perhaps exclusively in mountainous regions from 300-1500m. Populations north of Orinoco in Venezuela only at sea-level over open terrain. In Ecuador Marin (1993) noted that this species had the widest altitudinal range (300-1500m) of the *Chaetura.*

DESCRIPTION *C. s. spinicauda*
Adult Sexes similar. **Head** Upper head black-brown and glossed, with indistinct fringes, especially on the forehead and in narrow line over eye (these are reduced by wear). Black eye-patch. Ear-coverts paler grey-brown becoming paler towards throat. Pale grey throat - some mottling created by darker bases especially when worn. Appears dark-capped. **Body** Glossy, black-brown mantle and back

uniform with nape and head. Narrow, highly contrasting pale grey rump (boundary not entirely clear-cut). Uppertail-coverts black-brown uniform with mantle. Some proximal uppertail-coverts can be greyish. Lower throat/upper breast darker grey-brown than throat and then uniformly grey-brown from breast to vent and blacker on undertail-coverts. **Upperwing** Typically uniform, little contrast between tracts. Remiges almost black, darkest on outer primaries and browner on secondaries. Outerwebs blackest with innerwebs paler grey. Greater coverts fairly uniform with respective remiges with median coverts, lesser coverts and alula a little blacker, uniform with outer primaries. Grey-brown leading edge coverts with slightly paler fringing. Tertials grey-brown with paler innerwebs. **Underwing** Remiges appear paler than on upperwing though basic pattern similar. However the contrast between the grey-brown remiges and greater coverts and the blacker-brown smaller coverts is far more obvious. Some indistinct underwing-covert fringing (very hard to discern in the field). **Tail** Dark black-brown, paler beneath. Uppertail-coverts not entirely cloaking tail, though they are very similarly coloured, adding to rump contrast. Rectrix spines strongest in central tail, extending up to 5mm beyond the web, as opposed to up to 2.5mm in outer tail. **Body/Wing Contrast** Dark upperwing can appear darker than body, especially between the secondaries and the rump, with median and lesser coverts more uniform with saddle. Underwing-coverts appear darker than adjoining areas of body.

Juvenile Similar to fresh adult, distinct but narrow white tips to inner primaries, secondaries and tertials.

Measurements Cayenne. Male (4): wing 102-107 (104.7); tail 39.5-41 (40.4). Female (2): wing 102 + 104; tail 38-40 (39). Britiah Guiana. Male: (5) wing 94-106 (98); tail 39-40 (39.4). Trinidad. Male (5): wing 101-105 (103); tail 39-42 (40.2) (Zimmer and Phelps 1952). Wing: (11 BMNH) 101-108 (104.5). Weight: male 14-14.3, female 15-15.8 (Haverschmidt 1968).

GEOGRAPHICAL VARIATION Five races.

C. s. spinicauda (E Venezuela (see comments on distribution of *latirostris*), the Guianas and N Brazil) Described above.

C. s. fumosa (W Costa Rica, W Panama and N Colombia) Darker below and above than *spinicauda* with more pronounced throat patch. Rump patch larger and greyer than *spinicauda*, and a little more so than *aethalea*. Like that race distinguished from Grey-rumped by darker underparts with pale throat and whiter rump patch. Wing: male (13 Costa Rica and Chiriqui) 107.7-115.4 (109.8); female (6 Costa Rica) 107.5-113.4 (109.8). Tail: male (13) 37.7-42.2 (39.4); female (9) 36.8-41.5 (39.8) (Ridgely 1976). Wing: (3 BMNH) 112-115 (113.6).

C. s. aetherodroma (E Panama southwards to S Ecuador) Separable from *fumosa* by smaller size. Wing: male (20) 100.2-107 (103.5); female (13) 100.3-105.8 (103.1). Tail: male (20) 36.3-41.8 (38.8); female (13) 34.6-41.3 (39.3) (Ridgely 1976).

C. s. latirostris (E Venezuela from lower Orinoco at Caicara to Delta Amacuro and south through the Paragua valley to Mount Pauraitepui on the Brazilian border, and apparently in the northern state of Sucre. Some authors give a far more limited range restricting it only to Delta Amacuro) Most similar to *spinicauda* differing in duskier, darker underparts and blacker less brown upperparts. The underparts are paler than in *fumosa* or *aethalea*. Differs structurally from all other races as a result of the larger, broader bill with less sharply decurved culmen. Wing: (17) 100-105. Tail: (17) 38-43 (Zimmer and Phelps 1952).

C. s. aethalea (C Brazil) Somewhat larger than *spinicauda*, darker above and below with more pronounced white throat patch as a result. Rump patch more extensive than in *spinicauda*, and a little darker, with greyer uppertail-coverts and paler lower back. In this respect the race is closer to Grey-rumped but the rump is clearly whiter, and the underparts are darker with a more contrasting pale throat patch. Wing: (4 BMNH) 107-111 (109). Type (Para): wing 109; tail 40 (Todd 1937).

VOICE A variety of different calls have been described: high-pitched squeaking notes, low-pitched chattering and soft twitterings. Said to be quieter and sharper than Vaux's. Call high-pitched ripple *tsoo-si-si-si* faster than Grey-rumped Swift. Sick (1993) transcribes this as *sri-sri-sri*, resembling Grey-rumped more than Ashy-tailed Swift.

HABITS Regularly feeds over open stretches of water particularly in early morning and evening. Gregarious species occurring in small single species flocks (mainly at higher altitude, Marin *et al.* 1992) of up to 10 birds or in mixed species flocks often comprising Grey-rumped Swift, but also Lesser Swallow-tailed and Chestnut-collared Swifts. In mixed species flocks tends to occur in the lower strata (Marin 1993).

BREEDING Band-rumped Swift has been recorded nesting in hollow trees on Trinidad (Snow 1962).

REFERENCES Chantler (1995), Chapman (1917), Haverschmidt (1968), Hilty and Brown (1986), Marin (1993), Marin *et al.* (1992), Ridgely (1976), Meyer de Schauensee and Phelps (1978), Ridgway (1911), Todd (1937), Zimmer and Phelps (1952).

---

58   LESSER ANTILLEAN SWIFT
Chaetura martinica   Plate 15

IDENTIFICATION Length 10.7cm. Dark *Chaetura* with rather limited range. Only sympatric with one other member of the genus: Short-tailed Swift on St. Vincent. Easily separated from that species by longer tail, less exaggerated wing shape and by having a narrow rump band paler than the uppertail-coverts as opposed to uniform with these coverts in Short-tailed. It is possible that Chimney Swift may occasionally occur on the Lesser Antilles. Lesser Anuillean Swift is considerably darker than Chimney Swift in underpart coloration, especially the rather dark indistinct throat patch. Also separable by its upperpart pattern as Chimney has the rump and uppertail-coverts uniform, but only slightly paler than the saddle. In Short-tailed Swift the pattern is the same but the contrast is far greater. This upperpart pattern (amongst other features discussed under each species heading) also excludes Vaux's, Chapman's and Ashy-tailed Swifts. Band-rumped has a similar, although whiter and narrower, rump band, and is paler on the throat. Grey-rumped has a more extensive pale area extending further onto the uppertail-coverts and lower back, and is clearly paler be-low (except in *sclateri* and *phaeopygos*). Pale-rumped Swift, although perhaps the least likely species to occur near to Lesser Antillean, has a whiter rump and is larger. Lesser Antillean Swift, like the closely related Grey-rumped Swift, is rather longer-tailed than the other *Chaetura* and has a quicker more active flight than some of the larger species. Black Swift also occurs on the Lesser Antilles but is considerably larger with a longer forked tail, longer thinner wings, generally darker plumage and the rump uniform with rest of the upperparts.

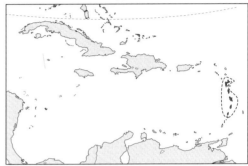

DISTRIBUTION Lesser Antillean endemic. Found only in the C Lesser Antilles on Guadeloupe, Dominica, Martinique, St Lucia and St Vincent. Apparently restricted to the north of Martinique.

Fairly common, except on Martinique. No detailed population counts.

MOVEMENTS Resident.

HABITAT Primarily a forest species. Found over primary forest and secondary growth or plantations. Also occurs over a wider variety of open country habitats and different altitudes within the range.

DESCRIPTION *C. martinica*
Adult Sexes similar. **Head** Lores, forehead, crown and ear-coverts dark black-brown. Indistinct pale grey fringes on head especially on forehead and in narrow line over eye in fresh plumage. Black eye-patch. Grey-brown ear-coverts paler towards throat. Throat pale or mid-grey, with darker grey mottling due to darker feather bases showing (especially apparent when worn). Dark-capped appearance typical of the genus. **Body** Black-brown mantle and back uniform with nape and upper head, but rump contrastingly pale grey. Upperparts then darken again to dark grey-brown uppertail-coverts. Throat patch not strikingly paler than the underparts. Differs from other *Chaetura* as the pale area of the throat is restricted to the throat and does not extend onto the lower throat which is darker grey-brown. The dark grey-brown lower throat/upper breast is darkest distally on the vent and then black-brown on the undertail-coverts. **Upperwing** Typically uniform. Outerwebs of remiges virtually black, darkest on the outer primaries and slightly browner on the secondaries with paler grey innerwebs. Greater coverts are fairly uniform with remiges, the median coverts, lesser coverts and alula being blacker. Leading edge coverts black-brown with slight paler fringing. Tertials palest (black-brown) feathers in wing, with pale grey-brown innerwebs. **Underwing** Remiges paler than on upperwing, appearing fairly uniform with the greater coverts. Considerable contrast with the darker median and lesser coverts. Underwing coverts show some pale fringing when fresh - rarely visible in field.

162

Tail Coloration similar to outer primaries, paler on undertail. Uppertail-coverts do not entirely cloak tail. Weak spines, strongest in central tail where they extend up to 5mm beyond the web, weaker in the outer tail extending up to 2mm. **Body/Wing Contrast** The dark upperwing can appear slightly darker than body especially at rump. Median and lesser coverts more uniform with saddle. Underwing coverts can appear a little darker than adjoining areas of body.

Juvenile Very similar to fresh adult, but distinct narrow white tips to inner primaries, secondaries and tertials.

Measurements Wing: male 106.5-111 (108.7); female 111-113 (112). Tail: male 37.5-39.5 (38.5); female 37.5-39.5 (38.5) (Ridgway 1911). Wing: (4 BMNH) 106-114 (112.25).

GEOGRAPHICAL VARIATION None. Monotypic.

VOICE Utters a soft twittering similar to other members of the genus.

HABITS Highly gregarious, usually encountered in groups of 20-40, sometimes in association with hirundines.

BREEDING Few details have been published regarding this species. Presumably breeding behaviour is typical of the genus and the nest form is said to be similar to that of Short-tailed Swift. One nest has been recorded in a disused oven! A clutch of 3 eggs was recorded (Bond 1985).

REFERENCES Bond (1985), Ridgway (1911).

59   GREY-RUMPED SWIFT
Chaetura cinereiventris   Plate 16

Other name: Sclater's Swift; Ash-rumped Swift (race *sclateri*)

IDENTIFICATION Length 10.7cm. Typical, but rather dainty and long-tailed *Chaetura*, lacking brown tones in plumage. Differs from Vaux's, Chimney, Chapman's, Ashy-tailed and Short-tailed in that the paler rump and proximal uppertail-coverts contrast with both the black lateral and distal uppertail-coverts and tail (figure 38).

Figure 38. Upperparts of Grey-rumped Swift (top) and Chimney Swift (bottom). Note uniformity of rump and uppertail-coverts in Chimney Swift and the contrast between the rump and the distal uppertail-coverts in Grey-rumped Swift.

However the subspecies of Grey-rumped differ in the extent and darkness of the rump patch. The contrast is more difficult to discern than is the case for Pale-rumped and Band-rumped Swifts which have narrower and whiter rump

bands, but can be seen if looked for carefully. Chapman's and Ashy-tailed can be further distinguished from most subspecies of Grey-rumped by virtue of the clearly darker underparts. *Occidentalis, sclateri* and *phaeopygos* are very dark below and rather similar in tone to Ashy-tailed and Chapman's (they lack the contrastingly white throat patch of Ashy-tailed). *Cinereiventris, lawrenci* and *guianensis* are grey on the underparts with great contrast with the deep black undertail-coverts. From Vaux's, as well as the upperpart pattern, the plumage tone is important. The saddle is black with a bluish tone as opposed to grey-brown and the rump in particular is cold grey, as opposed to the warmer brown of Vaux's. The degree of rump contrast is also greater in Grey-rumped, although this varies subspecifically. The underparts are colder-toned in Grey-rumped. The distinctive Band-rumped of the subspecies *spinicauda* has a strikingly narrow white rump band, with the palest area being far more restricted than in most Grey-rumped. Although some Band-rumped subspecies have more extensive rump patches which overlap in size with those Grey-rumped with more restricted rump patches (especially *cinereiventris*) the rump patches of Grey-rumped Swifts are always darker and greyer. The closely related Pale-rumped Swift has a whiter rump patch of similar size and shape, a dark belly, uniform with the undertail-coverts, is rather larger and has a bronzy gloss to the upperparts. In addition, looks noticeably shorter-winged in direct comparison.

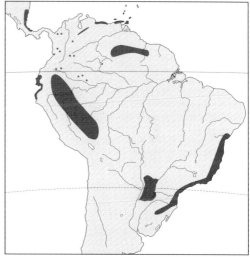

DISTRIBUTION Extensive range in S Caribbean, Central and South America. Found in Central America from eastern Nicaragua through the Caribbean slopes of Costa Rica to W Panama in the foothills and lowlands of Bocas del Toro. In the S Caribbean occurs in Grenada, Margarita Island, Trinidad and Tobago. In Colombia has a disjunct distribution in the lowlands of all three Andean ranges and both the pacific slope and lowlands and also the E Amazonas region. West of the Andes from Colombia south through W Ecuador into NW Peru. The distribution in Venezuela is similarly patchy being found in S Tachira and Merida, the northern serrianas and cordilleras in Yaracuy, Aragua, Miranda, Animate and Sucre. South of the Orinoco in Venezuela in Bolivar, across the cerros of the Gran Sabana and Cerro Guaiquinima, Rio Paragua, and also in SW Amazonas. Occurs in W Guyana. Through E Ecuador and E Peru south into N Bolivia and

W Amazonian Brazil as far west as the Ituxi river, E Acre. First recorded from Paraguay in July 1977 (R. S. Ridgely *in litt.* to Robert Clay 1994). A specimen was collected in November 1987 near to Estancia Primavera, Departmento Concepcion (Robert Clay *in litt.* 1994). Subsequently found to be widely distributed in the Oriental region of Paraguay in 1991 (Brooks *et al.* 1992) and breeding has been proved to occur around the Itaipu Dam (F. E. Hayes *in litt.* to T.M. Brooks 1994). An isolated population occurs in SE Brazil, from Bahia to Rio de Janeiro, Santa Catarina and Rio Grande do Sul, and in NE Argentina in Misiones.

Generally one of the commonest *Chaetura* swifts throughout its large range. Interestingly, although common throughout Ecuador, increases in abundance at the upper limit of its altitudinal range on the eastern slope whilst increasing in abundance towards the lower limit on the western slope (Marin 1993). Outnumbers Pale-rumped Swift where the two occur in sympatry.

MOVEMENTS Believed to be resident throughout range, but in Paraguay may be an austral migrant that winters in the country (Brooks *et al.* 1992).

HABITAT Primarily a species of forested or second growth slopes. However ventures to lowlands and sea coasts. Recorded from sea-level to 1800m. In Ecuador recorded by Marin (1993) from 200-1000m on both the western and eastern Andean slopes.

DESCRIPTION *C. c. guianensis*
Adult Sexes similar. **Head** Black-brown upper head, lightly blue-glossed and unfringed even in fresh plumage. Black eye-patch. Pale grey ear-coverts and neck sides, with some grey fringing to feathers on side of nape. Ear-coverts similar to throat though less pale fringing. Throat pale grey with paler fringes and many dark bases creating a mottled appearance, restricted to chin in some individuals. Very black upper head and pale throat give a very marked capped appearance. **Body** Mantle and upper back uniform with nape and head, contrasting greatly with strikingly pale, cold grey lower back, rump and uppertail-coverts. The feathers of the rump and uppertail-coverts have dark grey shaft streaks and pale grey tips. Lateral, and some distal, uppertail-coverts are darker grey, especially on the outerwebs (figure 39).

Figure 39. Rump and tail of Grey-rumped Swift of the race *guianensis*, showing extent of rump patch and feather pattern.

Throat generally uniform becoming darker grey on upper breast, darkening progressively to the vent. Black-grey undertail-coverts. **Upperwing** Wing typically uniform, but perhaps more contrast between the inner and outer remige webs than other members of genus. Primary outerwebs black-brown (blue-glossed) with pale grey innerwebs, outer primaries appearing slightly darker. Secondaries a little browner than primaries. Greater primary and greater coverts uniform with respective remiges. Me-

dian primary and median coverts somewhat blacker. Alula and lesser coverts black-brown uniform with outer primaries. Leading edge coverts black-brown, slightly paler-fringed. Tertials palest on innerwebs. **Underwing** Remiges paler and greyer than above, though still uniform with respective greater coverts. Greater degree of contrast than on upperwing with black-brown axillaries, median and lesser coverts contrasting with greyer remiges and greater coverts. Median coverts distinctly fringed grey-white. **Tail** Coloration as outer primaries and similarly glossed (paler undertail). Uppertail-coverts do not cloak the tail to the same extent as other *Chaetura*, thus some contrast can always be seen between tail and uppertail-coverts. Strong central rectrix spines extend up to 7mm beyond webs, with outer rectrix spines appearing shorter (up to 2mm) and weaker. **Body/Wing Contrast** Very black upperwing strikingly darker than rump and back, but uniform with saddle. On the underwing the darkest underwing-coverts often appear darker than the adjacent body.

Juvenile Indistinct pale fringes to the tertials, secondaries, and greater secondary coverts.

Measurements Wing: (3 BMNH) 102-106 (104.3). Weight (unspecified race): 18.6 (Marin 1993).

GEOGRAPHICAL VARIATION Seven races.
   C. c. guianensis (E Venezuela and W Guyana) Variation occurs mainly in the darkness of the rump and the underparts. The darkest forms tend to show a more graded underpart pattern, lacking the strong contrast between the black undertail-coverts and mid-grey bellies of the paler forms, and they tend to show less pale fringing than the paler forms.
   C. c. phaeopygos (C America from E Nicaragua to Panama) Slightly paler than *sclateri*, and clearly darker than the paler forms, especially underparts. However, the undertail-coverts are not as black as other subspecies. Rump not as dark as *occidentalis* or *sclateri*. Wing: male (7 Bocas del Toro) 108.3-112.6 (111.1); female (4, Bocas del Toro) 109.0-113.3 (110.5). Tail: male (7) 36.5-40.3 (38.6); female (4) 34.8-39.2 (37.5) (Ridgely 1976). Wing: (2 BMNH) 108-111 (109.5).
   C. c. lawrenci (N Venezuela, Margarita Island, Grenada, Trinidad and Tobago) Close to *guianensis* in general colour, though averages a little darker on underparts. Wing: (6 BMNH) 103-110 (106.5). Two measured by Ridgway (1911): wing 104 + 106.5; tail, 40 + 42.5.
   C. c. schistacea (E Colombia, and west to Tachira and Merida in W Venezuela) Similar to *lawrenci*, but generally darker with steel-blue gloss on upperparts and underparts being a deep neutral grey with the undertail-coverts darker and more slaty. Wing: 113. Tail: 36 (Todd 1937).
   C. c. occidentalis (W Colombia and W Ecuador) Darkest-rumped form, being dark grey though still showing some contrast. Although the rump is usually expansive and quite rectangular some individuals have more square bands reminiscent of *cinereiventris* (Robert Clay pers. comm.). Underparts uniformly grey. May appear longer-tailed than *sclateri* in the field (Robert Clay pers. comm.). Wing: 114-117. Tail: 40-42 (Chapman 1917).
   C. c. sclateri (Upper Amazonia in N Brazil, S Venezuela and S Colombia) Darkest form excepting *occidentalis*, from which it can be separated by greyer

belly. Differs from paler forms on underparts which are blackish-grey from upper breast, as opposed to just undertail-coverts. Wing: (4 BMNH) 106-112 (108.5). One measured by Chapman (1917) from Colombia: wing 107; tail 39.

C. c. cinereiventris (E Brazil) Palest-bodied form, and probably the palest-throated. Light grey underparts. Dark on undertail-coverts, becoming paler on breast. Rump whiter than other forms and uppertail-coverts appear somewhat blacker, emphasising the contrast. Wing: (4 BMNH) 109-115 (112). Two Ridgway (1911): wing 110; tail 42.

VOICE Typical of genus: rather high-pitched and insect-like twittering, *che-che-che-cheee*, with emphasis on 4th note. Sick (1993) gives the call as a harsh *chree-chree-chree*.

HABITS Highly gregarious, usually encountered in groups of 20-40, often in association with sympatric swift species (especially other *Chaetura* swifts but also *Aeronautes* and even *Streptoprocne*), and occasionally hirundines. Recorded primarily in the lower stratifications when with other swift species in Ecuador (Marin 1993).

BREEDING Grey-rumped Swift is typically found nesting in chimneys, although other inaccessible vertical surfaces are also used. The nest is typical of the genus and the clutch size has been recorded as 2-4 (Bond 1985 and Sick 1959).

REFERENCES Bond (1985), Brooks *et al.* (1992), Chantler (1995), Chapman (1917), Hilty and Brown (1986), Marin (1993), Marin *et al.* (1992), Ridgely (1976), Meyer de Schauensee and Phelps (1978), Sick (1959 and 1993), Todd (1937), Zimmer (1953).

---

60   PALE-RUMPED SWIFT
Chaetura egregia   Plate 16

IDENTIFICATION Length 10.7cm. Closely related to Grey-rumped Swift and in common with that species it differs from Vaux's, Chimney, Chapman's, Ashy-tailed and Short-tailed Swifts by the paler rump and proximal uppertail-coverts contrasting with the black distal and lateral uppertail-coverts and tail. However the pattern is far more contrasting, and the rump is grey-white to whitish as opposed to grey. Furthermore the shorter wing length is noticeable in the field (when seen in direct comparison with Grey-rumped) and the underparts appear darker from the belly. The gloss on the plumage is rather bronzy instead of bluish. Band-rumped has a much narrower and whiter rump band. The geographically separated Lesser Antillean is smaller than Pale-rumped and, although the rump pattern is similar, Lesser Antillean has a narrower rump and much darker plumage.

DISTRIBUTION Restricted WC South American range. East of the Andes in E Peru, in Loreto, San Martin, Ucayali and Madre de Dios, and W Brazil as far as the Ituxi river in E Acre, and N Bolivia in Pando and Santa Cruz. A specimen was collected in 1987 and others seen at Tayuntza, Province Morona-Santiago in Ecuador and since this date three more specimens have been collected from the E Napo at Bellavista, Valle de Nangaritza, Province Zamora-Chinchipe, as well as several more sightings throughout the Ecuadorian oriente: from Pachicutza, 35km from Zamora in Province Zamora-Chinchipe and

in the Zamora-Nangaritza valleys.

As with other *Chaetura* no detailed population counts or estimates have been made though it is likely that this is one of the rarer species as there are few museum specimens. Outnumbered by Grey-rumped in mixed flocks recorded by Parker and Remsen in Bolivia (1987), Marin *et al.* (1992) and by Günter de Smet (pers. comm.) in Ecuador. Noted as abundant in the higher limit of the species' altitudinal range on the eastern slope of the Ecuadorian Andes.

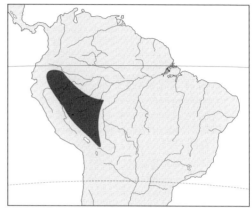

MOVEMENTS Presumed resident throughout range.

HABITAT Occurs over both humid forest and open country in lowlands and foothills.

DESCRIPTION *C. egregia*
Adult Sexes similar. **Head** Blackish upper head, with rather bronzy gloss. Black eye-patch. Ear-coverts paler towards throat. Pale grey throat exhibits some mottling when worn (darker feather bases showing). Rather dark-capped appearance. **Body** Above black from crown to upper back, with bronze gloss, though this is reduced through wear. Lower back, rump and proximal uppertail-coverts very pale grey (to whitish), with distal and some lateral uppertail-coverts appearing darker grey. Lower throat and upper breast similar to upper throat, becoming darker through lower breast, with belly brown then uniform to vent, with undertail-coverts black. **Upperwing** Similar to Grey-rumped with glossy black upperwing and little contrast between coverts and remiges, or between inner and outerwing. Like Grey-rumped marked contrast between inner and outer remige webs. Tertials (especially innerwebs) palest (grey-brown) feathers in wing. **Underwing** Remiges appear paler and greyer than above, and are uniform with respective greater coverts. Other coverts markedly darker than the rest of the underwing. Some very indistinct fringing to underwing-coverts can be seen, but probably not in field conditions. **Tail** Similar to remiges. Adds to the rump contrast as can be seen beyond tail-coverts. Rectrix spines not measured. **Body/Wing Contrast** Strong upperpart uniformity, darkest underwing-coverts appear darker than underbody.

Juvenile Similar to fresh adult, but with greater tendency towards fringing.

MEASUREMENTS Wing: 120. Tail: 40. Todd (1916). Weight: 22.6 (Marin 1993).

GEOGRAPHICAL VARIATION None. Monotypic.

VOICE Not known. Presumably close to Grey-rumped Swift.

HABITS A gregarious species that has been recorded with Grey-rumped Swifts and in mixed flocks with Chestnut-collared, Short-tailed and Lesser Swallow-tailed Swifts. Those collected in Ecuador ranged in altitude between 850-1000m, but has been recorded from 200-1000m (Marin *et al.* 1992, Marin 1993). Noted flying primarily in the low stratifications when present with other swift species in Ecuador (Marin 1993).

BREEDING Pale-rumped Swift has not been recorded at the nest.

REFERENCES Chantler (in press), Marin *et al.* (1992), Marin (1993), Parker and Remsen (1987), Meyer de Schauensee (1982), Todd (1916).

---

61   CHIMNEY SWIFT
Chaetura pelagica   Plate 14

IDENTIFICATION Length 12-14cm. To many observers the most familiar, and typical, *Chaetura*. This is the most uniform of the genus on the upperparts - only slight contrast between dark grey-brown saddle and the grey-brown rump and uppertail-coverts. Across most of North American breeding range it is the only swift likely to be encountered, but as a result of its highly migratory nature it could be encountered in at least part of the range of any other *Chaetura* species. The closely related Vaux's Swift is the main confusion species and is best distinguished from Chimney Swift by its more highly contrasting rump and uppertail-coverts, generally paler grey-brown plumage, and best of all by the underpart pattern: large whitish throat patch extending to upper breast, where it becomes pale grey-brown and then grey-brown on the undertail-coverts. In contrast the Chimney Swift has a more restricted and rather indistinct grey throat patch, bordered on the lower throat by mid grey, becoming progressively black and more sooty grey from the mid breast. The darker southern races of Vaux's are closer in underpart coloration to Chimney, but still average slightly paler and have whiter throat patches (but of a similar size and shape to Chimney Swift). Other more subjective differences are the paler innerwing, and more capped appearance caused by the paler throat. Jon Curson (pers. comm.) notes that Vaux's has a rather quicker more agile flight. Both species show some variation particularly in the degree of contrast on the upperparts and, especially in Chimney, relative darkness of throat, and should only be identified after prolonged close views preferably with both species present. Uniformity in shade between the rump and uppertail-coverts exclude Band-rumped, Grey-rumped, Pale-rumped and Lesser Antillean Swifts. Chapman's Swift has darker underparts than Chimney Swift with a very restricted throat patch, much greater contrast on the upperparts, and rather more glossy plumage. The shorter-tailed Ashy-tailed has similar upperparts to Vaux's but differs on the underparts from Chimney as a result of its (restricted) throat patch contrasting strongly against the blackish underparts (with the exception of the paler grey-brown undertail-coverts). Noted to appear longer-winged than Grey-rumped Swift (Robert Clay pers. comm.). Short-tailed Swift differs in jizz, underpart and upperpart pattern.

DISTRIBUTION Extensive E North American range. Win-

ters in W South America. Throughout Canada and USA east of Rockies from EC Saskatchewan and S Manitoba south (in the east) to SC Florida, and westwards to the SC Gulf coast in Texas to E New Mexico. Breeds occasionally in S California and possibly Arizona. Wintering range is Amazonian Peru, and also the Pacific coast of Peru with records from Libertad to Lima. One sight record from NW Brazil, other records from Arica, Chile and several records from Ecuador (where it is thought to be a rare passage bird all over the country though possibly a winter visitor in the eastern lowlands). This suggests the wintering range may be more extensive. Since 1987 groups of between 16-22 have been recorded during December and January in the Calama Valley, Chile (Demetrio 1993). Occurs as a passage migrant in the West Indies and Central America. Has occurred as a vagrant in Europe on four occasions (including two birds together), all in the British Isles, between 21 October and 4 November. These records are somewhat later than the main migration times in north America as is the case with other instances of transatlantic vagrancy to the British Isles by other species.

Common and widespread breeder throughout range, although rather rare, but regular, in California and perhaps in Arizona.

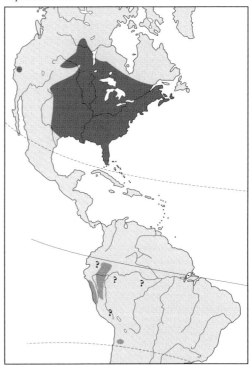

MOVEMENTS Long distance migrant, moving to South American wintering grounds through Central America and across the Gulf of Mexico. Autumn migration from mid August, with the majority departing during late August and September. The peak autumn count from New York state is of 1500 on 30 August, although 450 have been recorded as late as 4 October. However, generally rare in the north of range from early October, but has been recorded as late as 4 November from New York state. Migration through West Indies has been recorded from 22 August - 23 October. The few autumn records from

Panama are all in October and November. Between 12 February and 1 March 1994 6-150 individuals were seen daily at Puerto Quito, Pichincha Province in W Ecuador (Robert Clay pers. comm.). These are the first records for western Ecuador (R. S. Ridgely *in litt.* 1994) and Robert Clay has postulated that the coastal cordillera of W Ecuador could be a migration stopover for northbound Chimney Swifts. Spring migration in the West Indies is noted from 13 April - 18 May and in Panama between March and May. Numbers in southern California peak in late May and early June. Recorded as early as 3 April in New York state, but generally rare before the end of April, numbers peaking in mid May when up to 3000 have been noted at roosts. Also a regular, but uncommon vagrant to Bermuda from 30 March to late June and from 6 August to December 31. Usually singly or in twos or threes, although larger numbers noted in spring. Occasionally noted on passage in the northern Bahamas and on two occasions in the southern Bahamas.

Very few records of migrants from northern South American. Only one from Venezuela, and very few from Colombia (all in spring). Presumably occurs commonly in these countries where it is liable to be overlooked amongst resident *Chaetura* species.

HABITAT Found in a great variety of habitats throughout the wide breeding range. Most often associated with human settlements, but a wide variety of more natural habitats attract the species both for breeding and feeding. Recorded from 300-1000m on the Ecuadorian eastern Andean slope and from c.2500m in the Andean Valle de Quito (Marin 1993).

DESCRIPTION *C. pelagica*
Adult Sexes similar. **Head** Lores, forehead, crown and ear-coverts dark grey-brown. Pale grey fringes on head, especially forehead and narrow line over eye (reduced when worn). Black eye-patch similar in size and shape to *Apus*. Ear-coverts paler towards throat. Throat pale or mid grey. Some darker grey mottling when worn (darker feather bases showing). Head appears dark-capped. **Body** Dark grey-brown mantle uniform with crown and nape. Lower back across rump to uppertail-coverts slightly paler grey-brown (degree of contrast with dark saddle is variable). Lower throat and upper breast usually similar to throat though becoming steadily darker towards breast, where it becomes dark grey-brown (very similar to mantle tone) through to undertail-coverts. **Upperwing** More uniform than *Apus*. Primaries black-brown with paler innerwebs, outer primaries appearing marginally darker (though less so than *Apus*). Secondaries appear somewhat browner than primaries. Greater primary and greater coverts uniform with respective remiges, with median primary and median coverts somewhat blacker. Alula and lesser coverts black-brown. Leading edge coverts grey-brown with slightly paler fringing. Tertials (especially innerwebs) palest (grey-brown) feathers in wing. Broad grey-brown fringes to innerwebs of remiges. **Underwing** Remiges appear paler than above and uniform with respective greater coverts. Greater degree of contrast than upperwing with darker grey-brown axillaries, median and lesser coverts clearly darker than remiges and greater coverts. Some indistinct fringing to underwing-coverts, probably not visible in field. **Tail** Dark grey-brown, little shows beyond long tail coverts except when spread. Little or no contrast between tail and coverts. Long tail spines strongest at centre where they extend up to 7.5mm beyond the webs. In the outer

tail they extend up to 4.8mm. **Body/Wing Contrast** Very dark upperwing can appear darker than body especially between secondaries and rump, with the median and lesser coverts more uniform with saddle. Underwing-coverts appear darker than adjoining areas of body.

Juvenile Similar to fresh adult, but with distinct but narrow white tips to inner primaries, secondaries and tertials.

Measurements Wing: male 126-133 (129.2); female 122.5-133.5 (129.7). Tail: male 39.5-44 (42.3); female 40-45.5 (42) (Ridgway 1911). Wing: (3 BMNH) 127-131 (130.67). Weight: 23.6 (Marin 1993).

GEOGRAPHICAL VARIATION None. Monotypic.

VOICE A variety of different twitterings of varying duration can be heard, the most common being a series of soft, though rather loud, accelerating and decelerating chippings.

HABITS Highly gregarious. Often in huge numbers when roosting. Roosting occurs particularly in spring, when large industrial chimneys may be occupied by several thousand individuals. Considerable fatalities occur if the chimney is used. Marin (1993) notes that this species primarily occupies the low strata in mixed-species flocks.

BREEDING Chimney Swift nests primarily in chimneys although it may utilise other man-made strucures. The nest is typical of the genus and Fischer's (1958) study recorded the following dimensions: width 75 - 113, depth 25 - 31 and from front to back 50 - 75. A semicircular support of saliva is adhered above the nest. Clutches of between two and seven eggs have been recorded and average clutch sizes of between 4.0 - 5.3 have been quoted (Fischer 1958).

Figure 40. Nest of Chimney Swift.

REFERENCES Alström *et al.* (1991), Amos (1991), Bond (1985), Buden (1987), Bull (1976), Chantler (1993), Demetrio (1993), Fischer (1958), Garnt and Dunn (1981), Hilty and Brown (1986), Marin (1993), Ridgely (1976), Ridgway (1911), Meyer de Schauensee and Phelps (1978), Wetmore (1957), Williams (1986), Zimmer (1953).

167

Other names: Dusky-backed Swift; Richmond's Swift (race *richmondi*); Yucatan Swift (race *gaumeri*)

IDENTIFICATION Length 12cm. Western North American counterpart of Chimney Swift. A small *Chaetura*, with the most extensive breeding range. Within range several subspecies are recognised differing primarily in the darkness of plumage, particularly underpart colour, palest in northern and arid areas of range, and largest in size. Vaux's Swift is best separated from Chimney by greater contrast between saddle and rump, generally slightly paler plumage and most importantly the underparts: whitish throat patch extending onto mid breast, then pale grey becoming darker on vent and grey-brown on undertail-coverts. In Chimney the throat patch is more restricted and the underparts darker, appearing sooty-grey from upper breast. Underparts of darker southern races closer to Chimney Swift, differing in their whiter throats. Band-rumped, Pale-rumped and Grey-rumped distinguished primarily by their rumps contrasting not only with the saddle but to a lesser or greater extent with the uppertail-coverts (however, when tail spread contrast can be seen between Vaux's tail and the uppertail-coverts - figure 41).

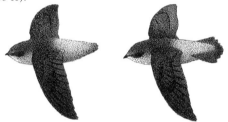

Figure 41. Vaux's Swift showing contrast between uppertail-coverts and tail when spread (right).

Ashy-tailed Swift is closest in upperpart coloration to Vaux's and is best excluded by the rather different underparts which have a highly contrasting though small whitish throat, the remainder being dark blackish-brown except for grey-brown undertail-coverts. Ashy-tailed is also larger with a proportionally shorter tail. Chapman's is much darker on the underparts which are dark grey-brown with throat patch barely paler. It also shows considerably more rump contrast (the lower back is paler and hence the area of pale on the upperparts appears more extensive) and is generally much more glossy and black in plumage, as well as being considerably larger. Short-tailed Swift with its distinctive jizz and very dark underparts is easily excluded.

DISTRIBUTION Extensive American range from Alaska in the north to N South America. In North America from SE Alaska south through NW and S British Columbia, N Idaho and W Montana keeping west of the Cascades and the Sierra Nevada to C California. From C Mexico southwards from Sinaloa and S Tamaulipas. This northern population is migratory, wintering from Mexico to Honduras. In Central America resident from Oaxaca, Veracruz, and Chiapas in S Mexico south to Panama in most suitable habitats. Separate and distinct form in SE Mexico on the Yucatan peninsula and Cozumel Island. The most southerly population is in N Venezuela, in serranias from

Lara and Yaracuy east to Sucre and Monagas. Common throughout much of its range.

MOVEMENTS The northern *vauxi* winters in Mexico and Guatemala. Occasionally winters in S California, with a maximum of 208 in Oceanside on 22 December 1979. Also recorded from Nevada in the winter: 200 with large numbers of White-throated Swifts on 4 February 1978 at Davis Dam. During migration recorded as far east as Louisiana and Florida. Like other migrant swift species, Vaux's tends to pass through in large waves as opposed to a more continuous migration. Departs British Columbia from late August, numbers peaking in the first half of September with few remaining by the end of that month. Migration through southern California peaks in the first week of September and generally involves fewer birds than in spring. Autumn passage through Nevada has been recorded from 12 August - 17 October. Spring migration through southern California mainly mid April to early May, with stragglers recorded to the end of May. Most migration through Nevada occurs in early May. In the far north of the species' range in British Columbia has been recorded as early as late March on the coast and from mid April in the interior. However, most migration starts in May and continues all month.

HABITAT Recorded from sea-level to 1800m over a wide variety of habitats. Due to widely varying ranges of different subspecies the habitats occupied by this species also vary considerably. The N Venezuela form is frequently encountered at high altitude over the cloud forests of the N Cordilleras. Those from Central America can be found in arid habitats. *Vauxi* occupies both temperate and warmer climates. Regularly recorded around human habitation, but less frequently than Chimney Swift.

DESCRIPTION *C. v. vauxi*
Adult Sexes similar. **Head** Lores, forehead, crown and ear-coverts grey-brown. Head fringed pale grey especially on forehead and in narrow line over eye, though these are reduced by wear. Black eye-patch. Grey upper ear-coverts becoming paler towards throat. Throat pale-grey, with some darker grey mottling when worn (darker feather bases showing), and some paler fringes when fresh. Appears rather dark-capped, emphasised by the pale throat. **Body** Grey-brown mantle appears uniform with crown and nape. Lower back across rump to uppertail-coverts distinctly paler grey-brown. Underparts progressively darker distally, with a marked contrast between pale grey (though slightly darker than throat) lower breast and grey-brown upper belly, the darkest point being the undertail-coverts. **Upperwing** Primaries black-brown with paler innerwebs (grey-brown) and tips to outerwebs, outer primaries appearing marginally darker. Secondaries appear somewhat browner and paler than primaries. Greater primary and greater coverts uniform with respective remiges, with median primary and median coverts darker. Alula and lesser coverts black-brown, almost uniform with outer primaries. Leading edge coverts grey-brown with distinctly paler fringing. Tertials (especially innerwebs) are the palest (grey-brown) feathers on wings. **Underwing** Remiges appear paler on underwing, uniform with respective greater coverts. Degree of contrast more marked than upperwing (though less marked than in some species) with grey-brown axillaries, and median and lesser coverts clearly darker than remiges and greater coverts. Pale grey/off-white fringing on underwing-coverts. **Tail** Dark grey-brown tail, short projection beyond tail coverts except if spread, darker than the uppertail-coverts and equal to the undertail-coverts. Outer webs of rectrices have dark fringes though pale at tips and on innerwebs. Tail spines weak and show little variation between central and outer rectrices. Spines extend up to 8mm (usually considerably less) and 3mm beyond web on inner and outer tail respectively. **Body/Wing Contrast** Upperwing palest on the innerwing, though this is still a little darker than rump. The coverts are virtually uniform with mantle and upper back though the wings, especially the darker outerwing, can appear darker than body. Underwing coverts appear darker than adjoining areas of body.

Juvenile Similar to fresh adult, but with distinct but narrow white tips to inner primaries, secondaries and tertials, and more marked fringing to body feathers.

Measurements Wing: male 107-115 (112.8); female 107-117 (111.9). Tail: male 34-37.5 (36.2); female 36-39.5 (37.1) (Ridgway 1911). Wing: (20 BMNH) 112-122 (117.05).

GEOGRAPHICAL VARIATION Six races.
C. v. vauxi (W North America, from S Alaska to NE Mexico, migrating to Mexico and Guatemala) Described above.
C. v. tamaulipensis (E Mexico) Darker overall than *vauxi*, with upperparts being blacker with a strong greenish gloss. The belly and undertail-coverts are clearly darker as are the throat feather bases. Line of feathers over eye not clearly pale-fringed as in *vauxi*. Differs from *richmondi* in its slightly paler underparts and the upperparts are less black and green-glossed as opposed to blue or blue-green glossed. *Ochropygia* has a paler rump. Wing: male (3) 111-113.5 (111.8); female (5) 109-115 (112.4). Tail: male (3) 35-38

(36.6); female (5) 32-40 (36.2) (Sutton 1941).
C. v. richmondi (S Mexico to Costa Rica and W Chiriqui) Similar size to *vauxi*. Rump and uppertail-coverts darker, greyish-brown. Mantle sooty-black, as opposed to olive-brown. Upperparts greenish glossed when fresh. Underparts considerably darker than *vauxi*, closer to Chimney Swift although the throat is clearly whiter and more contrasting as a result of the darker underparts. Black bristles in front of eye reach bill, rather than being restricted to just in front of eye as *vauxi*. Tail spines stronger than *vauxi*, showing greater difference between inner and outer tail. Inner tail spines extend 6mm beyond the webs as opposed to 3mm in outer tail. Measurements (8 males and 11 females, Guatemala, Nicaragua, Costa Rica and Chiriqui). Wing: male 108.7-113.8 (111.4); female 106.0-114.8 (111.0). Tail: male 33.4-38.1 (35.0); female 34.5-37.9 (36.6) (Ridgely 1976). Wing: (9 BMNH) 108-120 (114.11).
C. v. ochropygia (E Panama) Blacker upperparts than *vauxi*, rump appearing slightly paler as result. Measurements (10 males and 10 females from Veraguas, Isla Coiba and Isla San Jose). Wing: male 107.0-111.0 (109.2); female 108.8-114.4 (110.6). Tail: male 33.4-36.5 (34.8); female 32.8-36.4 (34.6) (Ridgely 1976).
C. v. gaumeri (Yucatan peninsula and Cozumel Island) Noticeably smaller than *vauxi* or *richmondi* in the field (Robert Clay pers. comm.). Slightly paler than *richmondi*. Rectrix spines much reduced or absent when worn (through nesting on the side of limestone wells). When fresh can extend 2mm beyond webs in both central and outer tail. Usually slightly stronger and longer spines in central tail. Tail spines appear to be weaker than *richmondi*. More extensive supraloral spot as in *richmondi*. Measurements. Wing: male 105-106 (105.5); female 99-111 (105.2). Tail: male 26-29.5 (27.7); female 28-31.5 (30.1) (Ridgway 1911). Wing: (28 BMNH) 104-117 (109.64).
C. v. aphanes (N Venezuela) Dark race, most similar to *richmondi*. Wing: males (13) 107-117.5 (112.6); females (9) 108-117 (113). Average both sexes 112.8 (Sutton and Phelps 1948).

VOICE Typical call rather softer than Chimney and usually shorter often disyllabic *chee-wee* or, *chee-chee wee*. Believed to have a greater variety of calls than other *Chaetura*, with variety of buzzing notes, thin sharp chippings, high-pitched rippling chatters and sibilant squeaking recorded.

HABITS Typically gregarious, often in large flocks. Associates with other swift species and also hirundines.

BREEDING Vaux's Swift is a solitary or colonial breeder. Of thirteen British Columbian nests eight were in disused chimneys (mainly vacant houses), one was under the roof of a railway water tank and two were in hollow maples. The nests are typical half-cup brackets constructed from grass and small twigs, glued together with saliva and adhered to a vertical wall. One nest measured by Fischer (1958) had a width of 10.0, depth of 4.0 and from front to back was 6.0. In British Columbia clutches vary between 1 and 7 eggs (Campbell *et al.* 1990 and Fischer 1958).

REFERENCES Alcorn (1988), Campbell *et al.* (1990), Chantler (1993), Fischer (1958), Garnt and Dunn (1981), Howell and Webb (1995), Ridgely (1976), Ridgway (1911), Meyer de Schauensee and Phelps (1978), Sutton (1941), Sutton and Phelps (1948), Wetmore (1957).

63 CHAPMAN'S SWIFT
Chaetura chapmani   Plate 14

Other name: Dark-breasted Swift

IDENTIFICATION Length 13-14cm. Typical, though rather large, *Chaetura* which is darker and more heavily glossed in plumage than most. The rather dark underparts give the species a uniformity of overall plumage lacking in most sympatric species and reduce the capped impression also found in many of these species. Belongs to the group of *Chaetura* that are uniformly dark across the upperparts except for contrastingly pale lower back, rump and uppertail-coverts. This feature excludes Grey-rumped, Pale-rumped and Band-rumped. The tail, which can be seen most easily when tail is fully spread, is largely cloaked by the uppertail-coverts. Across much of the species' range the smaller Vaux's is the main confusion species. Chapman's exhibits greater contrast between rump and saddle and considerably darker underparts with the throat only slightly paler than the remainder, which appears very dark grey or blackish. In comparison, the whitish-throated Vaux's appears grey below the breast, being darkest grey-brown on undertail-coverts. Ashy-tailed has very similar upperparts to Vaux's but its underparts are closer to Chapman's appearing rather blackish, although the pale throat is clearly paler than the breast, and although more difficult to discern, the undertail-coverts are paler grey-brown. Furthermore Ashy-tailed, although similar in size, has a notably shorter tail. Short-tailed is the only species that is darker than Chapman's being sooty-black below and showing no contrast on the throat. However, the undertail-coverts are clearly paler than the rest of the underparts and the unique jizz make separation straightforward. Chimney Swift, although closer in underpart coloration to Chapman's, has a larger and clearly contrasting throat and shows far less rump contrast.

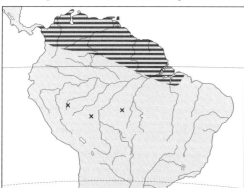

DISTRIBUTION Endemic to N South America where its range is poorly known. Believed to have two distinct populations, with *viridipennis* occurring in central Brazil in the Mato Grosso and around the Rio Ituxi in Acre, and in eastern Peru. This subspecies is thought to be migratory as it has occurred as far north as Antioquia in Colombia. In the north of the range *chapmani* occurs from C Panama to Venezuela: in NW Zulia, Aragua, Sucre and N Amazonas, east to the Guianas and south to NE Brazil in E Para and the Amapari river in Amapa. It is possible that the range is continuous from Panama through Colombia to Venezuela although there is insufficient evidence to

confirm this.
 The paucity of records indicate that this is a rather uncommon species. No population estimates have been published and on Trinidad, where most studies have been conducted, only one nest has been found.

MOVEMENTS *Viridipennis* is considered to be migratory and has twice occurred far north of its presumed breeding range in the department of Antioquia in Colombia, during March-April.

HABITAT Recorded over a variety of habitats from sea-level to 1600m in Colombia, 600m in Venezuela north of the Orinoco and to 200m south of it. In Trinidad, although recorded throughout the island, it forages mainly over wooded and hilly terrain, with an especial preference for higher forested areas. The only recorded nest site was in an area of brushy savanna at Waller Field at the foot of Trinidad's Northern Range.

DESCRIPTION *C. c. chapmani*
Adult Sexes similar. **Head** Upper head uniform black-brown with oily gloss, less fringed than either Chimney or Vaux's Swift. Black eye-patch. Brown ear-coverts a little paler towards throat. Pale brown and rather mottled throat patch, less obviously paler than rest of underparts than either of the latter species. Head appears rather dark-capped, though this is less apparent than in some other *Chaetura* because of the comparatively darker throat. **Body** Glossy black-brown mantle uniform with crown and nape, and strikingly darker than grey-brown back and pale grey-brown rump and uppertail-coverts. Lower throat and upper breast slightly darker brown than throat and from mid breast to undertail-coverts dark sooty-brown. **Upperwing** Very uniform, and generally black-brown in plumage similar in shade and gloss to head and mantle. Primaries black-brown with paler innerwebs, outer primaries appearing marginally darker. Secondaries appear somewhat browner than primaries. Greater primary and greater coverts uniform with respective remiges, with median primary and median coverts somewhat blacker. Alula and lesser coverts black-brown. Leading edge coverts dark-brown with slight paler fringing. Tertials (especially innerwebs) palest (grey-brown) feathers in wing. Broad grey-brown fringes to innerwebs of remiges. **Underwing** Remiges slightly paler than above, but still uniform with respective greater coverts. Greater degree of contrast than on upperwing with darker grey-brown axillaries, median and lesser coverts clearly darker than remiges and greater coverts. Very indistinct fringing to underwing-coverts. **Tail** Dark grey-brown (paler than the remiges), short projection beyond tail coverts except when tail spread. Spread tail visibly darker than the uppertail-coverts and uniform with undertail-coverts. Tail spines rather weak, though strongest and longest at centre, where they extend up to 6mm beyond the webs, as opposed to up to 4mm in the outer tail. **Body/Wing Contrast** Very dark upperwing appears darker than body especially between dark-brown secondaries and pale grey-brown rump, with median and lesser coverts more uniform with the saddle. Underwing coverts appear darker than adjoining areas of the body, though less so than in Chimney or Vaux's Swift.

Juvenile Plumage similar to fresh adult, but with distinct but narrow white tips to inner primaries, secondaries and tertials.

Measurements Wing: male (17) 116-121.5 (118.8); female (16) 116-123.5 (119.9). Tail: male (17) 39-45.5 (41.7); female (16) 40-44 (42.2). Weight: adult 21.75-28 (24.7); juvenile (6) 20.75-24.25 (22.7) (Collins 1968a).

GEOGRAPHICAL VARIATION Two races.

C. c. chapmani (Across N South America from C Panama, probably across N Colombia, N Venezuela, Trinidad, the Guianas and into NE Brazil, in Amapa and Para states) Described above.

C. c. viridipennis (South of *chapmani* in C and W Brazil and E Peru. Records from Colombia regarded as migrants) Larger than *chapmani*. Wing: male (5) 127-135 (130.5); female (5) 127-134.5 (130.2). Tail: male (5) 41-45 (42.6); female (5) 40.5-44 (42.2) (Collins 1968a).

VOICE Not known.

HABITS Usually encountered alone, or in groups of Short-tailed or Grey-rumped Swifts.

BREEDING Chapman's Swift is a rare species and only one nest has been found to date, on Trinidad just 8 inches from the top of a cement manhole in brushy savanna. The nest was typical of the genus and measured: 69 wide, 59 from back to front and was 24 deep. Only two eggs were laid in this nest but there had been disturbance and this was either a replacement clutch or an earlier egg may have been lost (Collins 1968a).

REFERENCES Collins (1968a and 1968b), Hilty and Brown (1986), Ridgely (1976), Meyer de Schauensee and Phelps (1978).

---

64 SHORT-TAILED SWIFT
Chaetura brachyura   Plate 15

IDENTIFICATION Length 10cm. The strikingly distinctive shape renders this the most easily identified *Chaetura*. The stubby tail appears markedly shorter than all other species, including Ashy-tailed, and as a result it appears very long-winged. The wing shape is equally unmistakable being a rather exaggerated version of the classic spinetail butterknife-shape: very short inner secondaries and markedly bulging inner primaries (often appearing broader than the length of the tail), tapering to the 9th primary with the 10th appearing clearly longer, giving the wing a rather hooked tip. Typical flight very flappy and rather unstable with much typical *Chaetura* rocking and often performed at low speeds. Generally rather bat-like. However, can fly with great speed when wing beats appear more rigid and flicking. The plumage is also rather striking. The glossy blackish plumage of the upper head and saddle contrasts strikingly with the rump and uppertail-coverts (these cover the rather pale tail). Of those *Chaetura* that have uniform rump and uppertail-coverts only Chapman's shows the same degree of contrast. The underbody is unique amongst the *Chaetura* by virtue of the great contrast between the pale straw undertail-coverts and the sooty-black remainder. This can be hard to discern and even when seen well the contrast is not as great as between the saddle and the uppertail-coverts. The very dark throat is not noticeably paler than the rest of the underparts. The black wings are not remarkable and are typical of the genus in being very uniform above with the usual underwing contrast.

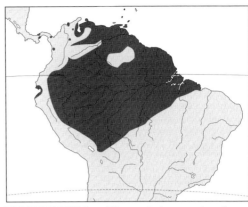

DISTRIBUTION Broad Neotropical range primarily east of Andes, from the S Lesser Antilles and Panama in the north to C Brazil in the south. In the S Caribbean found on the Lesser Antillean islands of St. Vincent, Grenada, Tobago and Trinidad. At its most north-westerly point in the Canal Zone of Panama (both Pacific and Caribbean sides) and then from the lower Turien river in E Darien. Only recorded from Panama as recently as 1960 although this is probably the result of the species having been overlooked rather than any expansion in range and it is probably widespread south of the Canal. In N Colombia has been recorded at Quibdo on the Pacific coast and from scattered locations in the Caribbean lowlands. Throughout Colombia east of the Andes, into E Ecuador and E Peru through N and E Bolivia in the states of Pando, Beni and Santa Cruz and throughout Brazil north of Mato Grosso and Para. An isolated population occurs in SW Ecuador south from Guayas through El Oro to Western Loja and into NW Peru. Throughout Venezuela north of the Orinoco, and south of it in N Bolivar from the upper Rio Cuyuni to Caicara and N Amazonas at San Fernando de Atabapo and around Cerro Duida. East of Venezuela occurs through the Guianas.

Fairly common over much of range and often encountered in suitable habitat. Commonest swift in Colombia east of Andes. First discovered in Panama in 1960 but subsequently found to be common in suitable open lowland habitat. Noted to be the commonest *Chaetura* in the lower parts of the altitudinal range, 200-500m (Marin 1993).

MOVEMENTS Resident throughout range.

HABITAT More often encountered at lower altitude than other members of genus. Not frequently encountered over 800-900m, even in Andean parts of the range, although has been recorded at 1900m in Ecuador. Frequently in arid open areas and llanos. Often near the coast, especially over mangroves. In Surinam large landward movements in the early morning away from mangrove roosting sites (Haverschmidt 1968). Occurs over primary forest or plantations. Occasionally over human habitations. Recorded by Marin (1993) from 200-1100m on the eastern slope of the Andes and this was slightly broader than the altitudinal range on the western slope.

DESCRIPTION *C. b. brachyura*
Adult Sexes similar. **Head** Upper head uniformly deep black-brown and purple-glossed, with little fringing even when fresh. Black eye-patch. Ear-coverts quite uniform with crown, marginally paler at lower edge. Chocolate-brown and mottled throat patch, becoming darker on

171

lower throat and upper breast. The patch is very indistinct and not notably paler than the underparts. In some dark individuals the throat appears uniform with breast. Head appears rather uniformly dark, lacking capped appearance of many *Chaetura*. **Body** Mantle and scapulars uniformly black-brown as head, and similarly glossed. Whole of back distinctly paler grey-brown, rump and uppertail-coverts pale grey-brown. On the underparts breast to vent uniformly deep sooty black-brown only slightly lighter than the mantle. Undertail-coverts distinctly paler grey-brown and continuous with rump and uppertail-coverts creating a very pale-tailed appearance. **Upperwing** Very uniform pattern. Primaries black (purple-glossed) with paler innerwebs (lightly green-glossed), outer primaries appearing marginally darker. Secondaries appear somewhat browner than primaries. All coverts deep black-brown (slightly darker towards leading edge) with purple-glossed outerwebs, innerwebs greener and slightly paler. Leading edge coverts black-brown with slight paler fringing. Tertials like other *Chaetura* with paler greyish innerwebs, but in addition even paler fringes to innerwebs. **Underwing** Remiges only slightly paler than above, and still uniform with respective greater coverts. Greater degree of contrast than on upperwing with black-brown axillaries, median and lesser coverts clearly darker than black-grey remiges and greater coverts. Some indistinct fringing to underwing-coverts can be seen, but probably not in field. **Tail** Pale straw-grey, fairly uniform above and below, appears fairly uniform with very long tail-coverts which entirely cloak tail. Very weak rectrix spines, largely obscured by the uppertail-coverts, extending beyond web to 5mm in central tail and to 2.5mm in outer tail. **Body/Wing Contrast** Very black upperwing uniform with the saddle, but strikingly darker than back, rump and tail. Dark underwing and body appear fairly uniform.

Juvenile Plumage closely similar to fresh adult, but with distinct but narrow white tips to inner primaries, secondaries and tertials.

Measurements Wing: (29) 113-124 (118.8). Tail (29) 27-30 (28.8) (Zimmer 1953). Wing: (15 BMNH) 118-125 (121.67). Weight: male 19-20; female 19-20 (Haverschmidt 1968). Weight: 15.5-22.0 (18.3) (Collins 1967).

GEOGRAPHICAL VARIATION Four races.

C. b. brachyura (N South America from Panama to the Guianas and Trinidad and south to C Brazil and Bolivia) Described above.
C. b. praevelox (Lesser Antilles on Grenada, St Vincent and Tobago) Similar to *brachyura* though slightly paler-throated, and a little browner in body plumage. Wing: (6 BMNH) 115-125 (121.2).
C. b. ocypetes (SW Ecuador and NW Peru) Distinct from other forms by paler grey chin and throat, brown forehead paler than crown and noticeable pale superciliary line. Longer-winged and longer-tailed than other forms. Wing: male (3) 124-127.5 (125.8); female (1) 124. Tail: male (3) 31-32 (31.8); female (1) 32 (Zimmer 1953).
C. b. cinereocauda (E Brazil) Very similar to *brachyura*, averaging blacker on the throat although some *brachyura* are identical. Averages greener-glossed, as opposed to purplish-black. Averages longer-winged and shorter-tailed. Wing: (10) 119-130 (123.3). Tail: (10) 25.5-29 (27.7) (Zimmmer 1953).

VOICE Cricket-like calls typical of genus. Rapidly repeated *sti-sti-stewstewstew* or a shorter *sti-sti-stistew* with the

emphasis on the third *sti*. Other variations noted have included *whoyzi-whoyzi-whoyzi-zi-zi-zi-zi* (Remsen, in Hilty and Brown (1986). All calls have a wheezing quality.

HABITS Gregarious species usually in small flocks. Readily mixes with other *Chaetura* species. In breeding season exhibits an unusual slow wing-beat display, accompanied by much vocalisation. In Ecuador primarily occupies low strata in mixed-species flocks (Marin 1993).

BREEDING Short-tailed Swift makes a typical half-cup nest with average dimensions (8) of 61.8 width, 53.3 front to back and 25.2 depth. Those recorded breeding on Trinidad were all subterranean, mainly in manholes, with one recorded in a concrete walled room (Collins 1967).

REFERENCES Bond (1985), Collins (1967), Haverschmidt (1968), Hilty and Brown (1986), Marin (1993), Parker and Remsen (1987), Meyer de Schauensee and Phelps (1978), Zimmer (1953).

65   ASHY-TAILED SWIFT
Chaetura andrei   Plate 15

Other name: Andre's Swift

IDENTIFICATION Length 13.5cm. The largest *Chaetura*, although rather short-tailed. Belongs to the group of *Chaetura* swifts with contrasting pale rumps and uniform uppertail-coverts, from which Grey-rumped, Band-rumped and Pale-rumped can be distinguished by virtue of their darker uppertail-coverts. Grey-rumped has largely uniform proximal uppertail-coverts and rump. Some contrast is apparent due to darker lateral and distal coverts and tail (which is not cloaked by the tail-coverts). The upperparts of Ashy-tailed are most similar to Vaux's, from which it differs by much darker underparts. Shows greater throat contrast, with underparts being sooty-brown from upper breast to vent, and diagnostically in that undertail-coverts are paler (grey-brown) than belly as opposed to darker than belly in Vaux's. This last difference, although hard to see, separates all other *Chaetura* except Short-tailed. The darker Chapman's differs also in that the throat contrasts less, the rump has a greater contrast and the saddle and wings are much darker and more glossy. Chimney differs in that the rump and uppertail-coverts contrast less, and the underparts are paler, with the throat patch being more extensive. Short-tailed has more contrastingly pale undertail-coverts and otherwise much blacker underparts with no throat patch contrast. In addition the rump of Short-tailed contrasts much more, the saddle and wings being distinctly blacker. Ashy-tailed Swift differs from all other *Chaetura* Swifts, apart from Short-tailed, by virtue of its shorter tail. Tail is intermediate in length between the other species and Short-tailed. Differs further from that species in its less exaggerated wing shape. The flight is intermediate between the fluttering bat-like Short-tailed Swift and the more typical shimmering wing beats of the other *Chaetura*.

DISTRIBUTION Endemic to South America. Two separated populations. A rather scattered, sedentary population in the north of the continent in Venezuela: in Carabobo, Guarico, Sucre, along the middle Orinoco in northern Bolivar and in the Sierra Imataca. The migratory C South American form occurs from E and SE Bolivia:

in Santa Cruz, Chuquisca and Tarija; E and S Brazil in Piauí, Bahia, Rio de Janeiro south to Mato Grosso, Sao Paulo and Santa Catarina, and in N Paraguay and NW Argentina in Salta and Tucuman. Records from Panama, N Colombia, Venezuela and Surinam in austral winter, presumed to be southern migrants (and indeed those taken in N Colombia and Venezuela have proved to be *meridionalis*).

Meridionalis is a fairly common breeder in the south of its range, but its appearance in the north of its wintering grounds appears to involve only small numbers. Sick (1993) notes that this is the commonest Brazilian swift away from the Amazon. *Andrei* is uncommon and local.

MOVEMENTS The southern population is at least partially migratory with winter records from Panama (August), N Colombia (August), Venezuela (September) and Surinam although there is a scarcity of records from these northern areas and it can be presumed that the main wintering grounds are further south, perhaps even within the breeding range of *meridionalis*. Vagrant to Falkland Islands in March 1959 (Woods 1982).

HABITAT Generally a lowland species found to 900m over a wide variety of habitats, including forested and cleared areas. In Venezuela *meridionalis* has occurred at the top of this altitude range over primary cloud forest at Rancho Grande, Miranda.

DESCRIPTION *C. a. meridionalis*
Adult Sexes similar. **Head** Lores, forehead, crown and ear-coverts grey-brown with an olive gloss. Pale grey fringing particularly on forehead and narrow line over eye (reduced when worn). Black eye-patch. Grey-brown ear-coverts paler towards throat. Pale grey-brown throat with some darker grey mottling when worn as result of darker bases. Dark-capped appearance, less marked than Chimney Swift. **Body** Mantle and nape uniform with upper head, though upper back clearly paler grey-brown, becoming progressively paler onto uppertail-coverts. Throat uniform, becoming darker on border of lower throat and progressing to dark grey-brown on lower belly and vent, becoming pale grey-brown again on undertail-coverts. **Upperwing** Typically uniform. Remiges black-brown, darkest on the outer primaries and slightly browner on the secondaries, with marked olive gloss. Innerwebs pale grey-brown. Coverts progressively darker

towards the lesser coverts and alula, with the leading edge coverts grey-brown with slight paler fringing. Tertials and in particular the innerwebs are the palest feathers (grey-brown) in wing. **Underwing** Remiges appear paler and greyer than above, uniform with respective greater coverts. Degree of contrast more marked than on upperwing with darker grey-brown axillaries, and median and lesser coverts clearly darker than remiges and greater coverts. Indistinct pale fringing to coverts is hard to discern in field. **Tail** Grey-brown, paler than the remiges, with paler fringing. Paler on underside. The tail coverts cloak tail, which is visible only when fully spread. Rectrix spines weaker than most *Chaetura*, with little difference between central and outer rectrices. Spines extend beyond web up to 2.1mm on the central rectrices and 1.8 on the outer. **Body/Wing Contrast** The upperwing appears darker than body especially between secondaries and rump, but also to a lesser extent across the saddle. Darkest underwing-coverts are darker than adjoining areas of body.

Juvenile Plumage closely similar to fresh adult, but distinct, narrow white tips to inner primaries, secondaries and tertials.

Measurements Wing: (6) 122-138 (129.8). **Tail:** (6) 36-39 (37.8) (Darlington 1931). Seven males and 11 females from throughout range. Wing: male 125.7-129.7 (127.5); female 126.1-139.6 (131.7). Tail: male 33.8-38.7 (36.6); female 34.7-39.8 (37.4) (Ridgely 1976). Weight: (female Surinam) 21.7 (Haverschmidt 1968). Weight: 19.5 (Belton 1984).

GEOGRAPHICAL VARIATION Two races.
    C. a. meridionalis (migrant form breeding south of *andrei*, in E and SE Bolivia, E Brazil, N Paraguay and NW Argentina, wintering north to Colombia) Described above.
    C. a. andrei (C and N Venezuela) Much shorter-winged than *meridionalis* and a little darker below. Wing: 114.5-117 (Darlington 1931).

VOICE Call (recorded in Paraguay) is a low-pitched *chu-chu-chu-chu* preceding a 'rattling chipper' (Wetmore 1926). Sick (1993) transcribes the call as *tip tip tip* adding as a song a *tli-ti-tit*. Belton (1984) and Brooks *et al.* (1992) described an excited, rapid, chattering *see tsdee ts tsee tsee tsee*.

HABITS Usually encountered in small groups.

BREEDING Ashy-tailed Swift was first recorded nesting in a hollow buriti palm in the Mato Grosso, Brazil. It also breeds commonly in chimneys in more populated areas. The nest is the typical construction of the genus with, in palm-nesting birds, leaf stalks and even fibres from palm leaves shaped into a half-cup adhered to a vertical surface by saliva. Nests have average measurements of: width 85, depth 37 and from front to back 43. Three recorded clutches ranged between 4 and 5 eggs (Sick 1948b).

REFERENCES Belton (1984), Brooks *et al.* (1992), Collins (1968b), Darlington (1931), Hilty and Brown (1986), Ridgely (1976), Meyer de Schauensee and Phelps (1978), Sick (1948b and 1993), Wetmore (1926).

AERONAUTES

This New World genus has three small species, one of which has a population that breeds in North America and migrates to Central America in the winter. All species are strongly piebald in plumage and are not dissimilar to the Old World *Apus* in structure. The inner secondaries are relatively long and decrease in length towards the primaries to produce a rather typical wing shape. The tail shape of White-tipped Swift is shallowly cleft whereas the other species have distinctly forked tails.

Separation from sympatric genera

The relatively small size, distinctive wing shape, striking plumage and rapid flight make this a distinctive genus. Tail shape is important in excluding other sympatric genera. The *Chaetura* and some *Streptoprocne* and *Cypseloides* have square tails. *Panyptila* and *Tachornis* swifts have very long, deeply-forked thin tails that are habitually held tightly closed so that they appear very slim. Those species that also have cleft tails or slightly forked tails are quite different in character (figure 42).

Figure 42. White-throated Swift (left), Chestnut-collared Swift (centre) and Chimney Swift (right). Note differences in structure between the genera. (Not to scale.)

66   WHITE-THROATED SWIFT
Aeronautes saxatalis   Plate 17

IDENTIFICATION Length 15-18cm. Distinctive plumage patterns make this species easily recognisable throughout most of range. The broad white throat and pale upper head contrasting with the otherwise black upperparts give it a very white-headed appearance. In the south of the range Lesser and Great Swallow-tailed Swifts also have very black and white plumage, but differ in both plumage pattern and jizz. This species, like the other members of the genus has a very rapid flight, on shimmering wing beats, similar to the Old World *Apus*.

DISTRIBUTION Extensive North and Central American range. In North America from S British Columbia south through E Washington, C and S Montana, Idaho, Wyoming, far W South Dakota and Nebraska, Utah, Colorado, Nevada, the southern two-thirds of California, Arizona, New Mexico and W Texas. Throughout Mexico, except the Gulf coastal lowlands, south to Guatemala, El Salvador and Honduras where it occurs in the interior mountains. In the north of the range it is apparently steadily increasing its range. First bred in British Columbia in 1907, at Lake Vaseux. Since 1947 small colonies have been found in several areas of British Columbia. Has occurred

as a vagrant in the E United States, Arkansas and Michigan.
  Locally common throughout much of the range. In Mexico believed to be rare. Uncommon in Honduras, whilst in the large north American range it is often a common resident.

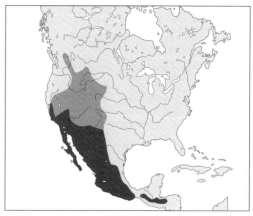

MOVEMENTS The northern populations, *saxatalis*, are migratory. In British Columbia first birds arrive in early

174

April, with main arrivals in late April and early May. Return migration from late August, with few birds remaining after late September. Winter records rare north of C California. In the winter in S California it is harder to find and occurs in smaller numbers, tending to withdraw entirely from colder areas. Larger numbers are only found in some coastal and desert areas. In Nevada, where the species is resident away from the colder montane areas, can be seen in large numbers in winter. The main wintering grounds are further south in Mexico, and as far south as Honduras. In flocks of up to 50 when migrating.

HABITAT Found to 3000m in Mexico and 2450m in Guatemala. In Honduras occasionally descends to 600m. Essentially a species of open country with a preference for canyons and cliff faces, both in the interior and locally on the coast. Often in very arid areas, especially in the south of the range.

DESCRIPTION *A. s. saxatalis*
Adult Male. Head Grey-brown upper head, with paler grey lores, forehead and feathers over eye and along top of ear-coverts. Black eye-patch. All feathers with pale grey fringes, most apparent when plumage fresh. Ear-coverts and side of neck grey-brown, paler than upper head, but uniform with nape. Wholly white throat with darker feather bases that can produce slightly mottled appearance in worn individuals. **Body** Nape paler than upper head though uniform with ear-coverts. Mantle to uppertail-coverts uniformly black-brown. The only exception is the white lateral rump. White on the throat extends broadly to the level of the wings and then narrowly to the legs through the mid belly. The rear flanks are white in a patch contiguous with the lateral rump. The remainder of the underparts is sooty black. **Upperwing** Outerwebs of the primaries black-brown with pale grey fringes and tips. Innerwebs grey-brown, darkest near the shafts becoming paler on the outer vanes. In the outer primary the innerwebs has a narrow pale fringe but is only slightly paler on the outer vane. Descending down the primaries the outer vane becomes progressively paler with the inner primaries having a practically white innerwebs. The secondaries are very broadly and conspicuously white-tipped with black-brown bases. Coverts fairly uniformly black-brown, perhaps darkest on lesser coverts and alula. The coverts are not fringed. The outerweb of the outer primary is white, causing the appearance of a narrow white leading edge to the wing as the leading edge coverts are similarly broadly white-fringed (black-brown bases). **Underwing** Remiges appear paler and greyer than above, with the secondaries pattern still clearly visible. The greater coverts are a little darker than the remiges, the median coverts are clearly dark grey, and the lesser coverts are black-brown and uniform with the axillaries and flanks. The degree of contrast is greater between tracts than on the upperwing, especially between inner and outer primaries. **Tail** Black-brown above (darkest on outerwebs) with paler fringe (white on outer web of outer rectrix), and paler grey below.

Adult Female Averages narrower white fringes on the secondaries and tertials (figure 43). In the male the white is extensive on the outerwebs. In the female the fringes tend to be more uniformly distributed between the webs.

Juvenile Similar to adult but a greater degree of grey-white feather fringing to plumage when fresh. White duller on underparts and dark areas less black (more sooty), causing pattern to be less contrasting.

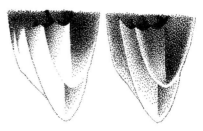

Figure 43. Typical tertial patterns of male (left) and female (right) White-throated Swift.

Measurements Wing: (10 BMNH) 142-153 (147.6). Tail: adult male 55-58; adult female 54.5-57.6. Tail figures are based upon the average figures for birds from the Rockies and California; in both sexes the average figures for California are lower.

GEOGRAPHICAL VARIATION Two races.
    A. s. saxatalis (north of the range, partially migratory) Described above.
    A. s. nigrior (S Mexico, Guatemala, El Salvador and Honduras) Similar to *saxatalis*, but dark areas of plumage darker. Most apparent on the forehead, lores and narrow superciliary line - whitish in *saxatalis*, dark in this race. Wing: (4 BMNH Guatemala) 148-153 (149.5).

VOICE Shrill, descending *skeeeeee* (Gosler 1991) or *kee-kee-kee-kee*, carries for great distances. This call is heard throughout the year.

HABITS Gregarious, can be noted in large numbers depending on local abundance. 1200 were recorded with 200 Vaux's Swifts in Nevada, February 1978 (Alcorn 1988).

BREEDING White-throated Swift is a colonial breeder. Nests are usually situated inaccessibly in rock fissures, under overhangs of steep cliff faces or in holes in houses; will also breed on sea cliffs (Marin pers. comm.). Nesting in old hirundine nests has been recorded. The nest is a cup attached either to a vertical surface or on a ledge, and consists of grasses, moss and seed down with some feathers, bound together with saliva. Lack quoted clutches of 4-6, but one recorded in British Columbia was of only 3 (Campbell 1990, Lack 1956b).

REFERENCES Alcorn (1988), Campbell *et al.* (1990), Edwards (1972, 1989), Garnt and Dunn (1981), Gosler (1991), Lack (1956b), Land (1970), Monroe (1968), Ridgway (1911).

---

67   WHITE-TIPPED SWIFT
Aeronautes montivagus   **Plate 17**

Other name: Mountain Swift

IDENTIFICATION Length 13cm. Distinctive species with piebald plumage pattern and typical *Aeronautes* jizz. Andean Swift is potentially the most confusing species where they occur together. However, White-tipped Swift lacks the nape collar and largely white underparts of that species, and has white-tipped tail feathers (in male). Additionally Andean has a broad white rump patch, but note that White-tipped can appear to have a white rump due to the large white flank tufts. Broad white throat patch, leg tufts, and generally sooty black-brown plumage are a unique combination of features within range. The prominent white tail tips, of male, are shown by no other swift species. Slightly forked tail is longer than any *Chaetura*, and the scythe-shaped wing is quite unlike that genus. Indeed, as is typical of this genus, the inner secondaries appear very broad, thinning on the outer secondaries and the wings becoming very narrow on the outer primaries. *Panyptila* and *Tachornis* swifts are easily separated by their much longer and slimmer tail shapes which, when fully opened (only rarely), are seen to be very deeply forked. The much larger *Cypseloides* and *Streptoprocne* swifts have a different jizz. Flight action is very rapid, much faster than *Chaetura* swifts, appearing very similar to that of Old World *Apus*.

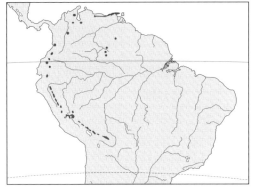

DISTRIBUTION Patchy South American range in Andes and Tepuis. Locally in N Cordilleras of Venezuela from Sucre in east, through to Miranda, and the Andes in Zulia and Merida. In Colombia in the Santa Marta mountains, the west slope of east Andes at Santander, and on the east slope in Meta and Pacific slope in Valle and Narino. Southwards into Ecuador, occurs on both Andean slopes and more locally in central valleys. In Ecuador appears to be commoner in the north than south. Occurs patchily southwards through Peru into N Bolivia south to La Paz.
In S Venezuela occurs in the Tepuis of S Bolivar and S Amazonas and the Sierra de Imeri of N Amazonas, in extreme N Brazil.
Locally common. Found to be abundant but localised in a study of Ecuadorian swifts (Marin 1993).

MOVEMENTS Believed to be resident throughout range.

HABITAT Found from 500-2700m in Andes and from 700-1900m in Tepuis. In Ecuador on west-facing slopes of the inter-Andean zone from 1500-2700m (Marin 1993). Usually over forested areas, both primary and secondary, but also in more open landscapes and along rocky ridges. Although it mainly occurs in mountainous subtropical zones it is frequently encountered in arid mountain valleys. Often seen descending to valley bottoms in times of poor weather, but not recorded in lowlands. Sometimes over human habitations and will utilise derelict buildings.

DESCRIPTION *A. m. montivagus*
Adult Male. **Head** Black-brown upper head, paler grey-brown lores, forehead and line over eye. Black eye-patch. Ear-coverts grey-brown, clearly paler than upper head. Throat entirely white, but grey-brown feather bases can produce some mottling, especially on chin, when worn. White throat patch extends a little onto side of neck behind ear-coverts and onto lower throat/upper breast. **Body** Black-brown from upper head to uppertail-coverts. Un-

derparts black-brown (slightly paler than upperbody) from below throat patch (cut-off can be slightly mottled) to undertail-coverts. White patches around legs often linked. Black-brown body feathers have paler bases and can appear slightly mottled when worn. **Upperwing** Black-brown remiges darkest on outer primaries. Outerweb darkest, innerweb brown. Both webs have indistinct pale fringes. Fringes on secondaries are white and can appear as a pronounced white trailing edge, from above and below. Greater coverts fairly uniform with remiges, median coverts, lesser coverts and alula appearing blacker. Tertials have conspicuous white tips. **Underwing** Remiges paler and greyer appearing darkest on outer primaries and fairly uniform with greater coverts. Median and lesser coverts blacker-brown. **Tail** Black-brown tail with broad white tip, most visible from above, and individually variable in extent.

Adult Female Plumage generally browner than male. Lower back and rump can be contrastingly paler than mantle, white on underparts is duller as is white on tertials. White tail tips often missing.

Juvenile Similar to adults with greater degree of fringing on the body feathers and the tips of the rectrices when fresh.

Measurements Wing: (3 BMNH) 116-119 (117.7). Weight: (26) 17.2-22.9 (19.62) (Marin *et al.* 1993).

GEOGRAPHICAL VARIATION Two races.
A. m. montivagus (Predominantly Andean range, though also occurs in N Cordilleras of Venezuela) Described above.
A. m. tatei (Tepuis of N Brazil and S Venezuela) Upperparts blue-black and very glossy, unlike rather matt black-brown in *montivagus*. Measurements (Type). Wing: 112. Tail: 44. Tail-fork: 8.5 (Chapman 1929). Weight: (female) 21 (Dickerman and Phelps 1982).

VOICE Very vocal, with three recorded calls: a long buzzing, clicking trill not unlike Common Swift that accelerates and decelerates and is of differing durations, usually 4-5 seconds; a very light sibilant squeaking and a short harsh note not unlike alarm call of European Starling *Sturnus vulgaris*.

HABITS Gregarious. Usually encountered in small groups of up to 20. Forages mainly in the lower strata, either when encountered in mixed-species flocks or single-species flocks (Marin 1993).

BREEDING Nests in fissures in steep ravines and also in holes in buildings, although this may be relatively rare. Nesting behind waterfalls has been recorded. No details about nest type are recorded (Fjeldså and Krabbe 1990). Breeding behind waterfalls for *Aeronautes* is considered unlikely by Marin (pers. comm.).

REFERENCES Chapman (1929), Dickerman and Phelps (1982), Fjeldså and Krabbe (1990), Gilliard (1941), Hilty and Brown (1986), Marin (1993), Marin *et al.* (1992), Meyer de Schauensee and Phelps (1978), Zimmer (1953).

68 ANDEAN SWIFT
Aeronautes andecolus **Plate 17**

IDENTIFICATION 14cm. High altitude swift of the sub-equatorial Andes. Distinctive *Aeronautes* shape and plumage patterns make for easy identification, with the only real confusion species being the closely related White-tipped Swift. The shape of the two is rather similar but the slightly larger Andean appears longer and broader-winged with a longer tail. Like White-tipped, tail shape varies considerably with posture, but usually appears narrower than body and slightly forked. Differs in plumage, with Andean showing much more white on underparts, a white nape-collar (often interrupted in the races *andecolus* and *peruvianus*), a broad white rump and lacking the white tail tips shown in some White-tipped Swifts. However, White-tipped can show a narrow indistinct paler stripe from the throat to the belly and often appears to show a narrow and indistinct whitish rump as a result of the silky white flank tufts. Tends to occur at higher altitudes than White-tipped Swift.

DISTRIBUTION Endemic to Andes where three, roughly isolated, populations occur. From Cajamarca in N Peru in the W Andes south to Tarapaca, N Chile. East of this occurs in the Peruvian valleys of Huancavelica south to Cuzco. In the E Andes occurs from Bolivia south to the Rio Negro, W Argentina.

No detailed population studies have been made but apparently common, except in extreme south of range in Chile, where it is scarce.

MOVEMENTS Believed to be resident throughout range.

HABITAT Mainly in the altitudinal range of 2500-3550m in Peru and Bolivia and in the far south of range from 2000-2500m, but it has been recorded from 340-3900m. Mainly over rather bushy semi-arid mountainous landscapes. Occasionally over wooded sites. In some areas in rocky desert areas with scattered cacti.

DESCRIPTION *A. a. andecolus*
Adult male. Head Dark grey-brown cap, on crown and fore-crown. Lores and forehead off-white as are feathers immediately over eye. Off-white ear-coverts, entire throat and whole of nape and side of neck. These white feathers are dark based, appearing slightly mottled when worn.

Black eye-patch. **Body** Mantle to lower back dark sooty-brown, clearly a little darker than upper head. Rump off-white and contrasting strongly with sooty-brown mantle and uppertail-coverts. Underparts white except for a U-shape of dark-brown feathers on flanks and across hind belly, separating off-white central belly from the off-white undertail-coverts (the distal undertail-coverts are pale brown). Below the lower throat/upper belly white becomes a little dirtier, with a greater tendency to show buff mottling. **Upperwing** Black-brown remiges with slightly paler grey-brown innerwebs. Outer primaries appear darkest with secondaries looking browner and with a fairly well-defined white trailing edge caused by white tips. Coverts slightly darker towards leading edge. **Underwing** Remiges appear greyer and paler than on upperwing, fairly uniform with greater coverts, but clearly paler than black-brown median coverts and lesser coverts. Leading edge coverts grey-brown with broad grey-white fringes. **Tail** Black-brown, uniform with, or slightly lighter than, up-pertail-coverts, these do not entirely cloak the tail. Undertail, like the remiges, is paler and greyer than above and is not cloaked by tail-coverts.

Adult Female Paler-backed than male.

Juvenile Darker on forehead than adults and more buffy below.

Measurements Wing: (4 BMNH) 136-148 (143.6). Wing: (8) 138-144.5 (141.3). Tail: (8) 61-67 (65). Tail-fork: (7) 19-23 (21.1) (Chapman 1919).

GEOGRAPHICAL VARIATION Three races.
  A. a. andecolus (Most southerly race occurring from C Bolivia to Rio Negro in W Argentina) Described above.
  A. a. parvulus (W Andes of Peru into N Chile) A little smaller than *andecolus*, and the most contrastingly plumaged race, in adult plumage, which is almost black on face, upperparts and flanks, with central underparts white, as is the broad, complete nape collar. Wing: (7 BMNH) 127-143 (135.9).
  A. a. peruvianus (Peru east of *parvulus*) Smaller than *andecolus* with shorter more shallowly forked tail, differing in plumage as lacks buffy tints on white areas of plumage, has darker forehead and far less white on undertail-coverts. From somewhat darker *parvulus* can be further recognised by having a more interrupted collar. Wing: (8 BMNH) 131-146 (140.3). Wing: (5) 131-139 (134.8). Tail: (5) 54-56 (54.4). Tail-fork: (5) 12.5-14 (13.3) (Chapman 1919).

VOICE Two shrill screaming calls are commonly heard: a *zeezeezeezeeer* and a slightly weaker *trritrrritrri.....* A nocturnal flight call, a low *trp-rrie* has been recorded (Fjeldså and Krabbe 1990).

HABITS Gregarious. Feeds in large flocks in one area and then quickly moves on.

BREEDING Nests in holes in steep rock faces and road cuttings, but no details about nest form are recorded (Johnson 1967).

REFERENCES Chapman (1919), Fjeldså and Krabbe (1990), Johnson (1967), Narosky and Yzurieta (1987), Wetmore (1926), Zimmer (1953).

TACHORNIS

This group of small to tiny species comprises three exclusively Neotropical species, all of which are very distinctive. Their very rapid flight, like their Old World equivalents, is rather 'helter-skelter' with numerous direction changes. All three habitually occur close to palm trees and are therefore often encountered in town parks and gardens.

Separation from sympatric genera

Structurally the two South American species with their long deeply forked tails with very thin outer rectrices are unique, with the exception of the two very differently plumaged *Panyptila* species. Additionally, even the smaller of the two *Panyptila* species is larger than the largest *Tachornis*. The tails of this genus are usually held closed and appear rather spiky, although the tail is seldom held so tightly closed to prevent some of the fork being evident. Antillean Palm Swift is relatively shorter-tailed and has a much less deeply forked tail, although it is considerably more forked than the only other Caribbean species with forked tails, Black and White-collared Swifts. The wings are similarly attenuated in all *Tachornis*, being long and very thin. The bodies are also very thin and streamlined.

69   ANTILLEAN PALM SWIFT
Tachornis phoenicobia   Plate 18

IDENTIFICATION Length 9-10cm. Within its range this very small and highly distinctive swift presents no identification problems. Unusual plumage pattern with wide white throat patch, dark brown breast band, white belly patch, broad white rear flanks and rump separated from the belly by dark line of feathers between dark vent and uppertail-coverts and broad blackish flanks. Unlike other *Tachornis* species the tail is rather short and the fork is shallow. Only sympatric with the very different White-collared and Black Swifts. The very dark and very different Chimney Swift is the only other regularly occurring species. The flight has been described as tortuous and bat-like (Bond 1985).

DISTRIBUTION Endemic to the Greater Antilles. Found on the Greater Antillean islands of Cuba (including Isle of Pines), Hispaniola (including Saona, Beata and Île à Vache) and Jamaica. Recorded as a vagrant in Key West, Florida and Puerto Rico.

A rather common species. Can be observed in large numbers in suitable habitat, either in rural or in urban settings.

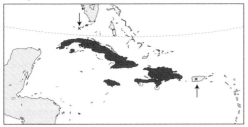

MOVEMENTS Believed to be resident but perhaps dispersive to some extent, as shown by vagrant records.

HABITAT Like other *Tachornis* species found primarily around palms in lowland settlements, including large cities. Sometimes found in mountainous regions.

DESCRIPTION *T. p. phoenicobia*
Adult Sexes similar. **Head** Brown upper head with some pale fringing when fresh. Ear-coverts slightly paler and edge of ear-coverts and side of neck whitish. Pure white

throat patch contrasts strongly with ear-coverts and extends to gape along its length and slightly onto side of neck, where it is less clearly white, and onto upper breast. **Body** Upperparts uniformly brown from head to lower back, with some paler fringing when fresh. Rump is contrastingly pure white, with uppertail-coverts being brown uniform with saddle. Underparts mid brown (slightly paler than the upperparts) breast band across mid breast contrasts with white throat and white belly patch. Breast band narrowest at its centre as flanks are extensively mid brown, narrowest on rear flanks where belly patch extends onto upper rear flanks and white rump patch extends slightly (though visible from below) onto lower rear flanks. However, the mid brown flank is not broken and extends onto undertail-coverts. The belly patch is therefore rather rhombus-shaped. **Upperwing** Upperwing blackish-brown, slightly darker than saddle, with remiges darkest on the outer primaries and outerwebs. Innerwebs are browner and paler. Greater coverts uniform with remiges, the smaller coverts appearing slightly darker. Wing pattern is, however, one of great uniformity. **Underwing** Remiges paler than above and contrast throughout underwing is far greater. The remiges are uniform with the greater coverts, the median coverts and lesser coverts are blacker. Outer primaries also appear comparatively darker than in upperwing. The darkest underwing-coverts appear much darker than underbody. **Tail** Tail similar in colour to remiges, above and below, not contrasting noticeably with tail-coverts.

Juvenile Basic pattern as adults but white on underparts duller. Flanks and undertail-coverts paler.

Measurements Wing: adult male 100.5-102.5 (101.7); adult female 97-102 (100.7). Tail: adult male 39-40 (39.3); adult female 38.5-44.5 (41.5) (Ridgway 1911). Wing (Jamaica 13 BMNH): 88-107 (102.2).

GEOGRAPHICAL VARIATION   Two races.
T. p. phoenicobia (Hispaniola and Jamaica) Described above.
T. p. iradii (Cuba and the Isle of Pines) Larger than *phoenicobia*, with a deeper tail-fork. Upperparts sootier and rather less black. Flanks much paler and sides of head more extensively grey-brown. Wing: adult male 94.5-106 (100.6); adult female 98.5-110 (106). Tail: adult male 38-47 (42.5); adult female 41-48 (45.9) (Ridgway 1911).

VOICE The call has been described as a weak twittering (Bond 1985).

HABITS Usually seen in small groups, sometimes with hirundines. Associates with other swift species less regularly than most swifts.

BREEDING Antillean Palm Swift is a colonial palm breeder. The nest is sited in the same way as Fork-tailed Palm Swift and is a globular construction of soft materials bound together with saliva with the cup close to the bottom. Clutch size is 3-5 (Bond 1985).

REFERENCES Bond (1985), Kepler (1971), Ridgway (1911).

70 PYGMY SWIFT
Tachornis furcata Plate 18

IDENTIFICATION Length 10.2cm. Tiny version of the small Fork-tailed Palm Swift. Differs from that species by distinct upper breast band, separating white throat from white central belly, and white outerwebs to the base of the tail feathers that can be seen from above when the species banks. The breast of Fork-tailed Palm Swift can appear rather mottled, but never as a distinct band. The tail appears a little shorter and thinner than Fork-tailed Palm Swift and wing beats are more rapid. The two species have not been recorded from the same areas: Fork-tailed Palm Swift is found south and east of the Andes. All other sympatric species easily separated on jizz alone. Like the other New World palm swifts mainly found around palms.

DISTRIBUTION Very restricted NW South American range. Known only from NE Colombia, in Petreola in Norte de Santander, and W Venezuela, in Alturita and Guachi, Zulia, Las Mesas in Tachira, El Vigia in Merida, and Betijoque in Trujillo. Unauthenticated records away from these areas in the cordilleras of NC Venezuela probably concern Fork-tailed Palm Swift.

With the exception of White-chested Swift this is apparently the scarcest South American swift. Within small range distinctly uncommon with very few sites.

MOVEMENTS No evidence of any movements.

HABITAT Found over a variety of habitats, including primary forest, secondary growth, plantations, clearings and open land with scattered trees. However, palms usually make up a considerable part of the habitat and it is dependent on palms for nesting sites.

DESCRIPTION *T. f. furcata*
Adult Sexes similar. **Head** Sooty-brown uniformly across upper head. Ear-coverts slightly paler though still clearly darker than grey/off-white upper throat. Lower throat pale grey-brown clearly darker than upper throat and chin.

Area of intergradation between lower and upper throat is indistinct. Black eye-patch appears reduced compared to other swifts. **Body** Nape to uppertail-coverts uniformly sooty-brown as head. On underparts the grey-brown on the lower throat extends to the mid breast forming an indistinct breast band, contrasting with off-white upperthroat and lower breast to lower belly. Flanks and vent uniform with breast band, with the undertail-coverts being sooty-brown and only slightly paler than the upperparts. **Upperwing** Remiges black-brown on outerwebs appearing browner and paler on the innerwebs. Coverts progressively darker towards lesser coverts and alula but the wing is nevertheless rather uniform in character. Remiges glossy and delicately bordered white, whereas the coverts are less glossy and have grey-white fringes. Fringing not obvious. **Underwing** Remiges greyer and slightly paler than on the upperwing and uniform with the greater coverts, with the lesser and median coverts being significantly darker sooty-blackish, contrasting clearly with the underbody. **Tail** Tail appears sooty-brown, fairly uniform with tail coverts above and below (tail, like remiges, paler below). The bases of the tail feathers are white. This is hard to see except when the bird banks, fanning its tail, as the patches are obscured by the tail-coverts.

Measurements (2, type male and female): wing 90 + 91. Tail: 53 + 55. Tail-fork: 26.5-27 (Sutton 1928). Two measured by Avelado and Pons (1952): wing female 89 + male 94; tail female 51 + male 51.

GEOGRAPHICAL VARIATION Two races.
    T. f. furcata (NE Colombia and adjacent areas in NW Venezuela, in the area to the south of Lake Maracaibo) Described above.
    T. f. nigrodorsalis (W Venezuela) Differs from *furcata* in pure black upperparts and whiter throat. Similar size to *furcata*. Wing: adult male (5) 89-92 (90.6); adult female (3) 92-96 (93.3). Tail: adult male 46-51 (49.4); 50-54 (52) (Avelado and Pons 1952).

VOICE Not known.

HABITS Typically seen in small, highly active groups.

BREEDING Apparently has a similar nest and nest site to Antillean Palm Swift (Bond 1956).

REFERENCES Avelado and Pons (1952), Bond (1956), Hilty and Brown (1986), Meyer de Schauensee and Phelps (1978), Sutton (1928).

71 FORK-TAILED PALM SWIFT
Tachornis squamata Plate 18

Other name: Neotropical Palm Swift

IDENTIFICATION Length 13.2cm. Distinctive species closest in character to the similarly plumaged Pygmy Swift, which occurs in the Andes of E Colombia and W Venezuela, but does not overlap in range with this species. Rapid flight, small size and slight build with very long thin (but not needle-like) tail and narrow wings add up to a distinctive jizz. Often close to palm trees. On the upperparts rather dark brown and indistinct, but below shows a pale throat and central underbody, with the flanks and undertail-coverts being brown with some fringing. The throat is usually quite clear white but the rest of the white on the underparts can be variably mottled appearing rather dirty. Tail shape alone separates the species from all other New World swifts, with the exception of the *Panyptila*. The sympatric Lesser Swallow-tailed Swift is easily separated on plumage - very black with white patches on lores, throat (including nape collar) and flank patches. Additionally, the tail is not quite as long and appears thinner, the head appears bigger, and the flight and general jizz is far more graceful.

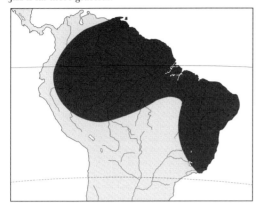

DISTRIBUTION Extensive tropical South American range east of Andes. Throughout Colombia east of Andes south from Meta and Vichada, into lowland E Ecuador and NE Peru, at Loreto. Occurs throughout Venezuela south of Andes and N Cordilleras, but including Sucre, and also in the Guianas. Found throughout Amazonian Brazil, and in the east and interior through Ceara, Paraiba and Pernambuco and into Espirito Santo, W Minas Gerais and Goias. Also throughout Trinidad.

Locally common, seldom seen in large numbers.

MOVEMENTS Presumed to be resident throughout range.

HABITAT Closely associated with Moriche *Mauritia* palms and seldom seen far from them. Accordingly the presence of these plants has a great bearing upon the habitat this species will occur in. Often in an urban setting, particularly the parks of towns or cities. Away from towns tends to be found in rather open country (including llanos) or in clearings within forests. Found to 1000m, although usually at much lower altitudes. On the eastern slope of the Ecuadorian Andes occurs between 200-1000m (Marin 1993).

DESCRIPTION *T. s. squamata*
Adult Sexes similar. **Head** Black-brown upper head slightly green-glossed and with pale grey fringes, less apparent when worn. Ear-coverts paler grey-brown with some pale fringing, darker than throat. Throat patch very pale brown-white appearing rather mottled, uniform in shade across throat and down towards upper breast. **Body** Upperparts uniformly black-brown and slightly green-glossed, as upper head, from nape to uppertail-coverts, perhaps appearing slightly paler across rump. The underparts appear rather mottled light brown, palest on throat and centre of body as far as the undertail-coverts. Flanks progressively darker and browner towards the wings. Undertail-coverts are the darkest area of the underparts being black-brown with prominent pale fringes. All underpart feathers are lightly pale grey/off-white fringed. **Upperwing** Rather glossy black-brown. The remiges are

darkest on the outerwebs with innerwebs appearing paler brown. Outer primaries appear darker than inner remiges and all remiges appear uniform with respective greater coverts. Median and lesser coverts become blacker towards leading edge. All coverts are lightly pale-fringed. **Underwing** Underside of remiges appears paler than the upperside with greater coverts appearing slightly darker and median and lesser coverts considerably darker. **Tail** Like the remiges, black-brown with obvious gloss, appearing paler from below.

Juvenile Has more extensive buffy fringes to the upperparts when fresh and head is rather buff tinged.

Measurements Wing (10 BMNH): 97-107 (100). Weight: (10) 10.5-13.6 (11.7) (Marin *et al.* 1993).

GEOGRAPHICAL VARIATION Two races.
   T. s. squamata (E Peru to Venezuela) Described above. As Zimmer (1953) noted although typical examples of each race are relatively easily separated there is considerable variation with some museum specimens well within the range of *squamata* being closer to *semota*. In addition, Zimmer noted that most birds from the central reaches of the Orinoco in Venezuela are considerably lighter than typical *squamata*.
   T. s. semota (Trinidad, the Guianas, south into C and E Brazil) Typically much darker than *squamata*, having steely-black upperparts with much less feather fringing than *squamata* and an almost uniformly black upperwing. On the underparts the undertail-coverts are black with a variable black fringe and like *squamata* the breast can appear variably mottled due to the visibility of blackish feather bases. Measurements (type). Wing: 106. Tail: 71. Tail-fork: 39.5 (Riley 1933).

VOICE Highly distinctive insect-like buzzing *d-z-z-z-z-z* and a trilling *trrrrreeeeee* (Snyder 1966). Sick (1948a) describes call as a thin screech on an E overtone, like *gs-gs*. However, usually rather quiet, at least away from nest site.

HABITS Usually in small groups (up to 10 birds) or alone. Tends to associate less with other swift species than most. Sometimes loosely with Blue-and-white Swallow *Notiochelidon cyanoleuca*. Usually feeds just above canopy or low over ground. In mixed species flocks forages in lower strata than other swift species (Marin 1993).

BREEDING Fork-tailed Palm Swift is a solitary breeder that nests on the outward facing leaf of a dried buriti palm *Mauritia* spp.

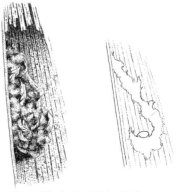

Figure 44. Nest of Fork-tailed Palm Swift on palm frond (and cross-section on right). (Based Sick 1991.)

When withered these large palm leaves crack at the base and hang vertically close to the trunk. The nest is situated in the middle of the leaf where it folds deeply, about 16cm from the cracked base. The nest is a long pouch attached with saliva at the top with the long bulky wall of the nest curving gently away from the palm and forming at its base, on the inside of the wall, an egg trough and an entrance between this trough and the leaf. The whole construction is about 13cm in length, outward diameter at its base (i.e. the outerwall of the egg trough) is 6-7cm in diameter and the egg trough has a diameter of 3.5cm and a depth of 1.5cm. The entrance is about 3cm in diameter and the walls of the nest are mainly .5cm wide, except the base below the egg trough where they are up to 2cm thick. The nest is made of downy body feathers and downy seed cases bound together with saliva. The interior walls are mainly downy and the outer walls are very feathery and appear rather disorderly. Sick recorded three eggs in the nest he studied and one of these measured 10.0 x 15.5 (Sick 1948a) (figure 44).

REFERENCES Hilty and Brown (1986), Marin *et al.* (1992), Marin (1993), Riley (1933), Meyer de Schauensee and Phelps (1978), Sick (1948a), Snyder (1966), Zimmer (1953).

PANYPTILA

The genus *Panyptila* contains two very distinctive Neotropical species, one endemic to Central America, the other having a large Neotropical range. The two species are remarkably similar in their very piebald plumage and they differ primarily in size. Both species are very graceful with long pointed wings and tails with the latter being deeply emarginated on the outer rectrices and having a deep fork. The tail is habitually held tightly closed. The body is slim and streamlined although the head is quite large and appears rather protruding (figure 45).

Figure 45. Fork-tailed Palm Swift (left) and Lesser Swallow-tailed Swift (right).

Separation from sympatric genera

Easily separated from all other Latin American genera. Shares long deeply forked tails with the two South American *Tachornis* species, but Lesser Swallow-tailed Swift can be separated easily from these two species by the striking piebald plumage (as opposed to brownish upperparts and indistinct pale underparts), far more graceful flight, larger head and generally being a somewhat sturdier species.

72   GREAT SWALLOW-TAILED SWIFT
Panyptila sanctihieronymi   Plate 18

Other name: Geronimo Swift

IDENTIFICATION Length 18-20cm. Large swift with striking piebald plumage. It can be separated from the larger sympatric species, White-collared and White-naped Swifts, by its white flank patches, white forehead patch, white trailing edge to the wing and from White-naped Swift by its complete white collar. Best separated, however, by the very different tail shape, which is very long and deeply forked (although usually held tightly closed when it appears thin and spike-like), as opposed to shorter and slightly forked in White-collared and shorter and square in White-naped. The wings are considerably narrower and the whole appearance of the bird is more rakish than in the *Streptoprocne*. Remarkably similar in plumage to Lesser Swallow-tailed Swift but much larger and more powerful in appearance. Two marginal plumage features might be apparent in the field. The white trailing edge to wing may appear broader and the white collar on the nape may be better defined in most birds (some individuals of Lesser Swallow-tailed Swift have very restricted white or even an absence of white on the nape - figure 46).

Figure 46. Heads of *Panyptila* swifts. Great Swallow-tailed Swift (two left) typically has a broader neck collar with less brown-grey than Lesser Swallow-tailed Swift (two right).

181

Direct, powerful flight, described by many authors as being exceedingly acrobatic, is rather different from the light and comparatively fluttering action of Lesser Swallow-tailed Swift.

DISTRIBUTION Rather restricted Central American range: the mountains of S Mexico: Michoacan, Guerrero, Oaxaca and Chiapas, and the central highlands of Guatemala at San Geronimo and Puebla Vieja in Vera Paz and at Volcan de Fuego, near Antigua and throughout the mountains of S Honduras. Recorded in Nicaragua and Costa Rica (5 occasions). These records represent the most southerly sightings and the Costa Rican dates of March, April, September and November suggest that they may refer to migrants.

One of the rarer species. Edwards (1989) states that it is very rare in Mexico, and in Guatemala it is known from only three localities and is thought to be rare. However, Land (1970) states that the species is apparently fairly regular in Antigua. Monroe (1968) considered that this was a common resident in the interior mountains of Honduras.

MOVEMENTS Rather intriguingly, although this species is believed to be resident, the Costa Rican records suggest that some occasional wanderings or genuine migrations may occur.

HABITAT At considerable altitude in Mexico between 1200m and 1500m, and in the Guatemalan highlands between 900m and 1850m; in Honduras mainly over 1000m, but has been recorded to 500m in mountain valleys. The Costa Rican records include some from La Selva, a lowland site, suggesting rather different habitat utilisation away from the breeding range. Mainly seen over rugged terrain with canyons and large cliff faces. On occasions, when it descends into valleys, may even be seen over large cities.

DESCRIPTION *P. sanctihieronymi*
Sexes similar. **Head** Upper head uniformly black (becomes greyer towards collar on nape) apart from white supraloral spot, in front of black eye-patch, reaching the bill and spreading to the gape. Ear-coverts similarly black. Throat to upper breast white, extending in collar around the back of the nape. The collar is wide at the back of the ear-coverts and narrowest on the nape. Some individuals show buff feathers on white area of nape. **Body** Below the white on the nape the upperparts are uniformly black from the mantle to the uppertail-coverts. Below the white breast the underbody is deeply black (as dark as the upperparts), except for a narrow white patch (rather squarish) on the rear flanks. **Upperwing** Very uniformly black with the remiges appearing slightly paler than the coverts by virtue of their paler innerwebs (also the outer primaries appear a little darker). The secondaries and inner primaries are white-tipped, broadest on the secondaries, and this forms a distinctive feature when seen from above. **Underwing** The remiges appear paler below but the contrast is less marked than in other species with the effect still being rather sooty-brown. The underwing-coverts are far blacker than the remiges, appearing uniform with the underbody. **Tail** Similar to remiges above and below.

Measurements Wing: 185-190.5 (188). Tail: 86.5-88 (87.2) (Ridgway 1911). Wing: (2 BMNH) 182 + 188.

GEOGRAPHICAL VARIATION None. Monotypic.

VOICE Described as a plaintive *tyee-ew* (Edwards 1989).

HABITS Apparently a rather solitary species seldom in large groups or with other species. Usually seen flying at great height and speed.

BREEDING Great Swallow-tailed Swift has a nest that is rather similar to its smaller relative. However, the nest described by Salvin (1863) had a false entrance on the side of the tube and was considerably larger, being 660 x 150 in dimensions (Sick 1958).

REFERENCES Edwards (1972 and 1989), Howell and Webb (1995), Land (1970), Monroe (1968), Ridgway (1911), Salvin (1863), Sick (1958).

73   LESSER SWALLOW-TAILED SWIFT
Panyptila cayennensis   Plate 18

Other name: Cayenne Swift

IDENTIFICATION Length 13cm. Small size and slender graceful aspect, coupled with black and white plumage, render this species easily identifiable throughout range. Structurally has long thin wings, a long tail that, although deeply forked, is usually held tightly closed appearing thin and spike-like, and a rather protruding, bull-headed aspect. Remarkably similar in plumage to Great Swallow-tailed Swift, but considerably smaller with a less powerful flight action. In South America most likely to be mistaken for Fork-tailed Palm Swift and the much rarer (and far more restricted) Pygmy Swift which have similar structures. However, the plumage of these two species is very different with the dingy brown and pale-bellied *Tachornis* swifts differing from the deeply black Lesser Swallow-tailed Swift with its prominent white flank, loral and throat patches, white collar and white trailing edge to the secondaries and inner primaries. Furthermore the flight action of Fork-tailed Palm Swift is more frantic with more flickering wing beats, the tail appears proportionally longer and the species is also found usually at lower altitudes invariably around palms, often in an urban setting. White-tipped Swift has a rather similar plumage although it lacks the white collar and the male has white tail tips. Most importantly White-tipped Swift has a rather square or slightly notched tail that is considerably shorter and more bulky than Lesser Swallow-tailed. Tail shape alone separates this species from all other sympatric species. The flight action is very graceful and rather hirundine-like, with much gliding interspersed with rather fluttering wing beats.

DISTRIBUTION Extensive Neotropical range. Central

182

America, on the Caribbean slope, from S Mexico, in Veracruz, Oaxaca and Chiapas, into Belize, Honduras, Nicaragua and Costa Rica, including the Pacific lowlands. South into Panama it occurs on both the Caribbean and Pacific slopes. The distribution in N South America is rather broken and it is likely that the species' scarcity disguises the true range, and it is probably more widespread than currently thought. In Colombia there are sightings from Valle on the Pacific slope, the central Andes at Caldas, Vaupes, Leticia and in NE Guiania. In Ecuador it occurs in both the western (S to W Loja) and eastern foothills and lowlands. Also in E Peru and N Bolivia, in Pando, Beni and N La Paz. The range in Venezuela is even more scattered with records from around Zea in Zulia, Henri Pittier National Park in Miranda and Cerro Golfo. East of Venezuela it occurs through the Guianas and Surinam. In Brazil occurs in Amazonia, Para, Mata Grosso, Maranhao, Bahia, Espirito Santo and Sao Paulo.

Generally rather local and uncommon, although quite common at Limoncocha in E Ecuador. Monroe (1968) describes the species as uncommon to rare anywhere north of Costa Rica. Noted to be most abundant between 300-500m altitude on both slopes of the Ecuadorian Andes (Marin 1993).

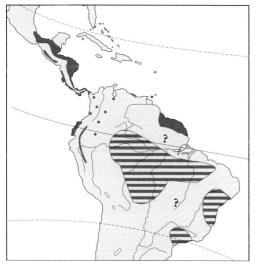

MOVEMENTS Believed to be resident throughout the range.

HABITAT Occurs over a variety of habitats both primary and secondary forest, mainly in large clearings or at forest edge. Also over open cultivated land, and will occur over towns and large rivers. Seems to prefer rather humid areas and is generally scarce or absent from more arid parts. Tends to be a lowland and foothill species with records to 800m on the Caribbean slope and 1000m on the Pacific slope in Costa Rica, to 1500m in Guatemala, 600m in Mexico, to 1400m in Colombia, 1500m in Ecuador (though mainly below 900m) and to 1000m in Venezuela. In Honduras it is only found below 300m. In Ecuador occurs over a broader altitudinal range on the west slope of the Andes (300-1500m) than the east slope (200-600m) (Marin 1993).

DESCRIPTION *P. c. cayennensis*
Adult Sexes similar. **Head** Upper head uniformly black apart from white supraloral spot, in front of black eye-

patch, reaching the bill and spreading to the gape. Black cap is paler and greyer towards nape. Ear-coverts similarly black. Throat to upper breast white, extending in collar around the back of the nape. Collar widest at the back of the ear-coverts and narrowest on the nape. Most show grey-brown feathers in collar on nape. Rarely individuals lack white on nape. **Body** Below nape the upperparts are uniformly black from mantle to uppertail-coverts. Below the white mid breast the underbody is deeply black (as dark as upperparts), except for a narrow white patch (rather squarish) on the rear flanks. **Upperwing** Very uniformly black with the remiges appearing slightly paler than the coverts by virtue of their paler innerwebs (also the outer primaries appear a little darker). The secondaries and inner primaries are white-tipped, broadest on the secondaries, forming a distinctive feature when seen from above. **Underwing** The remiges appear paler below but the contrast is less marked than in other species with the effect still being rather sooty-brown. The underwing-coverts (including the greater coverts) are far blacker than the remiges and appear uniform with the underbody. **Tail** Similar in plumage to remiges, above and below.

Measurements Eleven males and 17 females. Wing: adult male 116.5-122.4 (119); adult female 116.3-124.6 (119.1). Tail: adult male 52.4-58.8 (55.3); adult female 52-59.3 (56.4) (Ridgely 1976). Wing: (15 BMNH) 116-129 (120.7). Weight: 18.1. (Marin 1993).

GEOGRAPHICAL VARIATION Two races.
    P. c. cayennensis (S Honduras southwards to SE Brazil) Described above. Monroe (1968) considered that if a large series of skins was available for Honduras the variation in size between the two races might show intermediate individuals. If this were the case the validity of these races would be called into doubt.
    P. c. veraecrucis (Mexico to N Honduras) Similar in plumage to *cayennensis* but rather larger. The type from Veracruz has a wing of 126.8 (Monroe 1968).

VOICE Soft chattering notes have been recorded in flight, whilst at the nest a light *chee-chee-chee* has been recorded (ffrench, in Stiles and Skutch 1989). Sick (1958) described the call as a high-pitched *djip-djip-djip*.

HABITS Less gregarious than most swift species, usually seen singly or in pairs. Groups of up to six have been recorded. Often associates loosely with *Chaetura* and *Cypseloides* species. Forages in higher strata when noted with other swift species, except if White-collared and *Cypseloides* swifts are present when it moves to comparatively lower strata (Marin 1993).

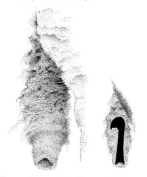

Figure 47. Nest of Lesser Swallow-tailed Swift (with cross-section on right).

183

BREEDING Solitary breeder that builds two main nest types. One type Sick (1993) likens to a thick woollen stocking fixed by its sole to a branch; the entrance is a narrow sleeve hanging down from the sole. The other type is a long sleeve attached to a vertical wall or trunk for the whole of its length. In this latter type the entrance is sometimes slightly outward-turned. Eggs are laid in a pocket on a ledge in the enlarged upper area of the nest. A great variation in the dimensions of these nests has been noted with a range between 240 x 90 and 355 x 75 to 480 x 165.

The wall of one was 10cm wide at the top but the sleeve entrance is usually 1-5cm wide. The nest is made of plant down bound together with saliva into a felt-like material. Feathers are often placed on the outside of the nest. Clutch 2-3 (Sick 1993).

REFERENCES Edwards (1972), Hilty and Brown (1986), Land (1970), Marin (1993), Marin *et al.* (1992), Monroe (1968), Ridgway (1911), Ridgely (1976), Meyer de Schauensee and Phelps (1978), Sick (1993), Stiles and Skutch (1989).

CYPSIURUS

A small genus consisting of two closely related species that occupy very similar niches. One has an extensive sub-Saharan African range (some authors consider the population on Madagascar to constitute a third species), the other occurs throughout the Oriental region from the Indian subcontinent, eastwards to the Greater Sundas and the Philippines. The highly attenuated appearance of both species is a result of the long thin wings and tails which are deeply forked. The outer rectrices of African Palm Swift are heavily emarginated and can appear streamer-like when the tail is spread.

Separation from sympatric genera

Superficially the structure is reminiscent of the *Apus*. However the much paler, more nondescript plumage and the association with palms, around which both species are regularly seen flying in an erratic manner at great speed, serves to ease separation. In Africa *Schoutedenapus* can be a problem but flight action is very different (invariably more fluttery, less rapid than the Palm Swifts), as is general jizz and wing shape: very straight trailing edge with a sharply tapering leading edge to the primaries in *Schoutedenapus* compared with a thin but more *Apus*-like shape in *Cypsiurus*.

74 AFRICAN PALM SWIFT
Cypsiurus parvus   Plate 19

Other names: Madagascar Palm Swift (race *gracilis*); Old World Palm Swift

IDENTIFICATION Length 16cm, including tail of up to 9cm. Tiny, rather featureless swift best recognised by its highly distinctive jizz. Superficially *Apus*-like in shape but obviously smaller and, in particular, slimmer with very narrow long wings and tail. The tail is held tightly closed appearing as a needle-thin spike. Only when the tail is infrequently fanned will the deep fork and needle-thin outer rectrices be appreciated. Juveniles show more rounded outer rectrices and slightly shallower forks. Body and head also appear very thin and streamlined. The plumage is very uniform grey-brown or sandy-brown, lacking an obvious throat patch, with the upper head usually appearing as the darkest area of the body. The outer wing and the underwing appear darker than the body. The darker upper head can give a slightly capped appearance and the reduced eye patch (cf *Apus*) gives the face a rather open or bare look. The flight is a frantic dash, often in tight circles, around its favoured palms, invariably calling loudly. Lacks the power either in appearance or in flight action of *Apus*. Separation from Asian Palm Swift is dealt with under that species.

DISTRIBUTION Widespread sub-Saharan range including Madagascar; also SW Arabia. South of the Sahara from Senegal in the west to NW Ethiopia (though not in C and N Ethiopia or the Horn of Africa), throughout the continent as far south as extreme SW Angola in the west and in a line eastwards through NE Namibia, N Botswana and S

Zimbabwe and Mozambique. Range extends south along coastal strip of Mozambique into South Africa as far as the Transkei, and inland into the Drakensberg range, widely in Transvaal and N Cape Province (north of the Orange River). Found in lowland Madagascar and on the Comoros. In Arabia occurs along the coastal strip and Red Sea mountains in the region of Asir in SW Saudi Arabia and into Yemen. Formerly in Egypt, the last record being at Abu Simbel in 1928. Bred as far north as Wadi Halfa in the last century. The creation of Lake Nasser has destroyed potential breeding habitat.

A familiar species throughout range. Often abundant.

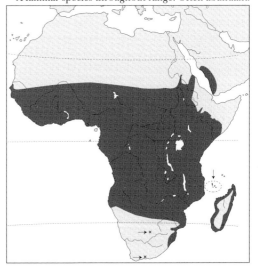

MOVEMENTS Resident throughout range, and extralimital records from East Cape (apparently bred) may be the result of range expansion rather than dispersal or localised migrations.

HABITAT Close association with palms, particularly fan palms which are used for nest sites, and the species is accordingly found in any areas in which they occur. This includes some very arid regions, through to rainforest zones. Occurs in towns and large cities particularly favouring parks. Washington palms (exotics from SW USA) are also much used and are a major factor in range expansion. The species will also freely use bridges for nesting if suitably constructed. Primarily a lowland species occurring to 1100m in Madagascar.

DESCRIPTION *C. p. parvus*
Adult. Head Forehead, crown and nape brown-grey. Lores and feathers around nape grey. Ear-coverts browner towards crown. Feathers of upper head fringed pale grey when fresh. Throat grey-white (palest on chin and central throat) usually clearly the palest area of the very uniform underparts, though often hard to discern. Throat never appears as a defined patch as in *Apus* and often shows dark grey shaft streaking. Bristle feathers in front of eye much reduced compared to *Apus*, never appearing as a patch. Face appears more open as a result. Head usually appears dark-capped (indeed crown can appear as the darkest area on upperbody in some individuals). **Body** Upperbody brown-grey across saddle with rump usually uniform or slightly paler, and head uniform with mantle or slightly darker. Some paler grey fringing evident across upperparts when fresh. Individual feather patterning far more uniform than in *Apus*. Underbody mouse-grey, palest on throat. Some whitish fringing when fresh. When moulting can look very untidy and mottled. **Upperwing** Dark olive-brown darkest on outer primaries (innerwebs paler), becoming paler towards inner primaries. Inner primary webs become progressively paler grey and have narrow white fringes. Secondaries similar to inner primaries but fringed greyish-white. Greater coverts and greater primary coverts fairly uniform with respective remiges. Median, median primary and lesser coverts and alula darker uniform olive-brown, fairly uniform with outer primaries and uniform with, or little darker than, saddle depending on wear. Leading edge coverts uniform with lesser coverts. All coverts narrowly edged paler though least distinctly on smaller darker coverts. **Underwing** Similar to upperwing though remiges slightly paler. Median and median primary coverts fairly uniform with respective remiges, other coverts slightly darker grey, typically appearing slightly darker than body. **Tail** Olive-brown as remiges.

Sexual Differences Male averages paler with more whitish on the throat.

Juvenile Similar to fresh adult though fringes more rufous. Structural differences as mentioned previously. A subadult plumage is recognised as a result of the intermediate nature of the fifth rectrix of birds that have undergone their post-juvenile moult. This is less rounded than in juvenile, but not as highly emarginated as adult. Delta-length also intermediate. This situation applies to all races with the exception of *gracilis* (figure 48).

Measurements Wing: male 128-138 (133); female 129-136 (132); juvenile male 123-133 (128); juvenile female 119-128 (124). Tail: male 89-96 (92.4); female 87-94 (90.8).

Tail-fork: male 55-63 (59.2); female 54-62 (58.7) (Fry *et al.* 1988). Average weight: 13.0 (Cramp 1985).

Figure 48. Tails of juvenile (left) and adult (right) African Palm Swifts.

GEOGRAPHICAL VARIATION Eight races. The form on Madagascar, *gracilis*, may be a separate species.
C. p. parvus (From Senegambia to Ethiopia and south in Sudan as far as Juba. Also SW Arabia) Described above.
C. p. laemostigma (S Somalia to Mozambique, in coastal lowlands) Darker and throat more mottled than *parvus*. Smaller than *parvus*. Wing: male (28) 122-135 (127); female (24) 119-130 (125). Tail: 76-105 (96.7). Average tail-fork 65. (Fry *et al.* 1988). Weight: (19) 10-13.5 (11.76) (Brooke 1972a).
C. p. myochrous (S Sudan south through E Africa to S Africa. Prefers highlands) Similar to *laemostigma* but with more marked greenish gloss. Slightly larger than *laemostigma*. Wing: (123) 123-148 (133). Weight: (14) 11.0-16.3 (14.16) (Brooke 1972a).
C. p. brachypterus (From Sierra Leone east to NE Zaïre and south to Angola. It is believed that this is the race occurring on islands in Gulf of Guinea) Darker than *parvus* or *laemostigma* with throat mottling like *parvus*. Wing (219): 112-138 (126). Weight: (7, north of Rovuma River) 12-14 (12.57); (5, south of the Rovuma River) 14.2-15.6 (14.96) (Brooke 1972a).
C. p. hyphaenes (N Namibia and N Botswana) Paler than *parvus*, with least mottled throat. Apparently closest in size to *myochrous*. Wing: 123-142 (133.4). Weight: (22) 10.0-18.1 (13.74) (Brooke 1972a).
C. p. celer (Mozambique to Natal) Similar to *myochrous* in shade though warmer in tone, less marked pale fringing and very fine throat streaking restricted to upper throat. Large size: wing (8); 130-141 (136) (Fry *et al.* 1988).
C. p. griveaudi (Grand Comoro) Very dark race with heavily streaked throat and upper breast. Wing: male (5) 125-132 (128.8); female (7) 123-142 (124.6) (Brooke 1972a).
C. p. gracilis (Madagascar) Small race, darker than continental African races. Like *griveaudi* has heavily streaked throat, but does not extend onto breast in this race. Juvenile has deeper tail-fork than juveniles of other races. Wing: (25) 116-132 (123) (Brooke 1972a).

VOICE Very vocal. Frequently heard uttering a call not dissimilar to Little Swift, a rapid chattering *si-si-si-soo-soo*, varying in duration, but usually repeated again and again. This chattering is heard whilst feeding, but particularly vigorously around nest sites. Variations of this call can be

185

heard around the nest site during breeding season.

BREEDING African Palm Swift is a solitary or colonial nester. Palms are the most frequent nest sites, but artificial structures, especially tall bridge girders, are not infrequently used. The nest is constructed largely of seed-down, often up to half of the material, or even more when cotton floss is employed, bound together with saliva. Normally a variety of soft feathers will also be stuck to the nest. The nest consists of a shallow cup standing away from a wall by 4-15mm and being 45-55mm wide. The back wall is between 45-120mm high and is adjoined to either the upper surface of the frond (either withered or fresh fronds are utilised) or on *Washingtonia* palms next to the trunk between the base of two fronds. Clutch usually consists of 2 eggs, but 1 and 3 have been recorded. Eggs average (7) 19.2 x 12.6 (Fry *et al.* 1988).

Figure 49. Nest of African Palm Swift.

REFERENCES Brooke (1972a), Chantler (1993), Cramp (1985), Fry *et al.* (1988), Goodman and Meininger (1989).

---

75   ASIAN PALM SWIFT
Cypsiurus balasiensis   Plate 19

IDENTIFICATION Length 13cm. Like its African counterpart a tiny, highly attenuated, rather featureless species, more easily recognised by distinctive jizz than any plumage features. Usually seen in a high speed flight with frantic wing beats. Underparts appear very uniformly pale, with a slightly paler throat and usually darker underwings, especially the median and lesser coverts. Darker upper head gives a slightly capped appearance and when seen from above head, mantle and upper back (often head is darkest) are darker than the rump and uppertail-coverts, though uniform with or slightly paler than the lesser and median coverts and outer primaries. The generally grey-brown plumage is paler than any sympatric *Apus*. All other sympatric species can be excluded on structure: the long thin and highly forked tail (usually held closed and appearing as a thin spike), long thin wings and very slender body, coupled with the flight action and regular calling, create a very distinctive jizz. Regularly seen in the proximity of fan palms. If an extralimital Palm Swift is encountered separation from African is best achieved by that species' slightly paler plumage, less marked saddle, larger size and longer sharply pointed outer rectrices. This last feature is particularly useful if sustained views can be achieved when the incredibly thin streamer-like outer rectrix tips can be observed. Only an adult African Palm Swift will show this to any great extent.

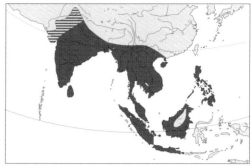

DISTRIBUTION Extensive S and SE Asian range, extending throughout the Greater Sundas and the Philippines. Found in the Indian subcontinent south of the Himalayas and eastwards from Rajasthan and Gujarat where it has a patchy distribution; also on Sri Lanka. The range is continuous eastwards through Bangladesh, Burma, Thailand and Indochina, south of the highlands of Yunnan. Also on Hainan. Southwards it occurs throughout the Malaya Peninsular, Sumatra (including the Lingag archipelago and Belitung island), Java, Bali, Borneo and the Philippines.

Common resident throughout most of range, becoming more locally common in areas where fan palms are less abundant, especially dense primary forest areas.

MOVEMENTS Entirely resident throughout range. Low dispersal rate is indicated by low incidence of extralimital records.

HABITAT Very dependent on fan palms (*Livistona, Borassus, Corypha* and *Areca* species) which are the main (in some areas exclusive) nest sites. These trees seem to be essential for the utilisation of a habitat, though in some areas this dependency is reduced by adaptation to nesting and roosting in thatched roofs. This species is therefore often found in close proximity to man, even in large Asian cities. The preference for fan palms means that it is found primarily in the lowlands, but it can be seen in hill areas up to 1000m in India and 1500m in Java and Burma. Seldom far from palms but will utilise adjacent areas for feeding, especially mangroves and paddy fields. Less common in forest areas.

DESCRIPTION *C. b. balasiensis*
Adult Sexes similar. **Head** Forehead, crown and nape sooty brown-grey, with lores and feathers around nape slightly paler grey. Ear-coverts become darker towards crown. Pale grey fringing on upper head when fresh. Grey throat slightly paler than very uniform underparts (especially chin and central throat) though never showing as a clear patch. Black eye-patch smaller than in *Apus*, restricted to a narrow semicircle in front of the eye, giving the head a very bare appearance. Dark upper head creates slightly capped appearance. **Body** Brown-grey upperbody darkest on the head, mantle and upper back (in many individuals the upper head is darker than the saddle). Rump and uppertail-coverts are usually paler grey. Some paler fringing apparent in fresh plumage on the generally very uniformly marked feathers. Mid grey-brown underbody palest on the throat and lightly white-fringed when fresh. Appearance can be very untidy when moulting and almost spotted in extreme cases as new darker feathers replace old pale feathers. **Upperwing** Dark brown outer primaries, with pal-

186

er innerwebs that are paler than grey-brown inner primaries and secondaries. The primary innerwebs are
progressively paler grey with narrow white fringes. The
secondaries have grey-white fringes. The greater primary
coverts and greater coverts appear uniform with respective remiges, but the smaller coverts and alula are clearly
darker brown, being uniform with the outer primaries and
usually a little darker than the saddle. All coverts narrowly pale-fringed. **Underwing** Remiges paler grey on
underwing than upperwing. Pattern similar to upperwing
with the greater coverts fairly uniform with the remiges,
and the smaller coverts darker towards leading edge. All
coverts have whitish fringes most prominent on the greater coverts. Typically the darkest underwing-coverts look
darker than the body. **Tail** Grey-brown slightly darker than
uppertail-coverts, but more uniform with undertail-coverts.

Juvenile Similar to the fresh adult though perhaps fringing more apparent. Better aged by the length of the outer
tail and the relative depth of the tail-fork. Adults have clearly longer, more sharply pointed outer rectrices and a
deeper fork than juveniles.

Measurements Wing: (8 Sri Lanka) 113-123 (118.4). Tail-
fork: (8) 30-39 (33.63) (Benson, in Brooke 1972a). Wing:
(5) 107-122 (Baker 1927). Tail: central rectrices 30-33 and
outer rectrices 60-68 (Ali and Ripley 1971).

GEOGRAPHICAL VARIATION Four races.
  C. b. balasiensis (Throughout the Indian subcontinent
  east of Pakistan and west of the Assam hills) Described
  above. Within its range clinal variation can be noted
  with the palest birds in the north-west and the darkest in Sri Lanka.
  C. b. infumatus (East of the range of *balasiensis* from
  the Assam Hills eastwards to Indochina, and then
  south through the Malay Peninsula, Sumatra and
  Borneo) The intergradation zone with *balasiensis* is
  apparently very narrow (Brooke 1972a). Differs from
  balasiensis in its darker plumage especially the rump,
  which is more uniform with the mantle. The outer
  remiges and the whole of the tail are black-brown and
  slightly glossed. Tail averages shorter with a less pronounced fork that is less than 26mm as opposed to
  more than 28mm for *balasiensis*. Wing: 113-126 (Baker 1927).
  C. b. bartelsorum (Java and Bali) Differs from other
  races in having a lightly streaked throat in adult plumage and having no pale fringes on the juvenile
  plumage. Juvenile also paler than adult. Rump paler
  than mantle. Tail longer and more forked than *pallidor*. Wing: 110-122 (116.6). Tail (type): 54. Tail-fork:
  18-25 (22.7) (Brooke 1972a).
  C. b. pallidor (Philippines) Like *balasiensis*, much paler than *infumatus* but rump more uniform with mantle
  like *infumatus*. Wing: 112-119 (115.3). Tail-fork: 15.5-
  21 (18.1) (Brooke 1972a). Type measurements: wing
  119; tail 62; tail-fork 19 (Hachisuka 1935).

VOICE Frequently heard call is a high-pitched reedy trill
sisisi-soo-soo. Much variation can be heard, mainly in duration. Call is often the first evidence of the species.

HABITS Gregarious. Often seen feeding in groups of up
to 40 birds, typically around palms. In Burma feeding
flocks over water of up to 1000 individuals have been recorded. Readily joins mixed flocks with other swifts and
hirundines.

BREEDING Asian Palm Swift is a solitary breeder that habitually nests on palm fronds. However, it will, mainly in
the absence of these trees, nest amongst the eaves of
thatched roofs of buildings. The nest is a rather flimsy
half-cup, usually about 10mm deep, with a diameter in
the region of 40mm. Constructed primarily of seed down
bound together with saliva and attached in the manner of
a bracket, on a furrow of an overhanging frond, with a
supporting apron stretching up to 8cm above the nest.
Sometimes located near the frond mid-rib or under the
palm leaves. Two eggs are the usual clutch although three
is often recorded. Average size of nominate races eggs:
(50) 18.2 x 11.5 (Ali and Ripley 1970).

REFERENCES Ali and Ripley (1970), Baker (1927),
Brooke (1972a), Chantler (1993), Hachisuka (1935),
MacKinnon (1988), Smythies (1953).

TACHYMARPTIS

This Old World genus has only two species, both of which are large. One is endemic to Africa, the other occurs in the Western Palearctic, Africa, India and Madagascar. Both have large amounts of white in the feathers of the underparts, forming a white belly patch in one and a variable amount of white mottling in the other. Structurally, both species have medium length tails that are distinctly forked. The wings are rather broader and less sharply pointed than the *Apus* swifts. The feet are heterodactyl.

Separation from sympatric genera

There should be little difficulty in separating this genus. The impressive bulk of these species, with their broad wings, large heads and medium forked, rather short tails (in comparison with the *Apus*), make them easily identifiable.

76   ALPINE SWIFT
Tachymarptis melba   Plate 20

Other name: *Apus melba*

IDENTIFICATION Length 20-22cm. Large size, powerful broad body and wings, relatively short forked tail (cf *Apus*) and deep powerful wing beats typical, in the Old World, only of *Tachymarptis* swifts. On brief initial views profile easily mistakable for small falcon. Distinctive underpart markings enable easy identification, once size and structure have been gauged and the possibility of an albino *Apus* has been excluded (check shape and extent of belly patch, as this should not extend onto the undertail-coverts as has been shown in some literature, but in fact can appear pointed distally contrary to some other references - figure 50).

Figure 50. Variation in shape of belly patch on Alpine Swift. Left hand bird is more typical.

White belly patch always highly visible, but throat patch far less so and the fact that this latter feature cannot be seen on a suspected Alpine should not rule out this species. Within Western Palearctic it has the palest upperparts of any swift, though the southern populations are much darker. For elimination of Mottled Swift see that species. Flight very powerful and rapid with deep slow wing beats rather deceptive as the speed is usually far greater than any *Apus*. Appears less manoeuvrable than *Apus* swifts.

DISTRIBUTION Extensive Old World range. Northern (and southern African) populations winter in tropics, southern populations resident. In the Western Palearctic throughout Mediterranean basin, and mountains of S Europe north to the N Alps and S Carpathians. Scattered distribution in NW Africa, throughout Atlas mountains with an isolated population in N Libya. Throughout Turkey and northwards through the Caucasus and along the east coast of the Black Sea to the Crimean peninsula. Extends east from Turkey, through Iran and Afghanistan to Baluchistan in W Pakistan, and northeastwards to Turkestan. The status in the W Himalayas is uncertain, with isolated populations likely to exist. Scattered populations exist in the mountains of the eastern Mediterranean extending to the Dead Sea, into the northern Rift Valley and the northern Red Sea. Northern populations winter primarily in sub-Saharan Africa. There are winter records from NW India, the Dead Sea, Sinai, Egypt, Cyprus, Iran and Yemen. There are also presumed Western Palearctic wintering records from W Arabia, although a resident population also occurs there. Wintering possibly occurs throughout N Africa, although the picture is clouded as separation of the resident races is only safely achieved in museum collections. No evidence of wintering south of the equator. Wintering has been proved in W Africa from Guinea and Liberia in the west, to NE Cameroon in the east, and as far north as the Bandiagara plateau, and in E Africa throughout Ethiopia and S Sudan, N Zaïre and Uganda. There are scattered records presumed to be Palearctic passage migrants throughout Africa north of the wintering range. In S Africa the populations in S Zimbabwe, South and East Cape, Transvaal, Orange Free State, Lesotho and Natal are mostly summer visitors. The wintering grounds are unknown. Passage is recorded from Zimbabwe (May - June and August - October), SE Zambia and S Malawi, so wintering perhaps occurs in E Africa. Resident African populations occur in C Mali on the Bandiagara escarpment, W and SE Ethiopian highlands, N Somalia, NE Zaïre and Uganda in the Ruwenzori Mountains, W and C Kenya and N Tanzania. In SW Africa occurs in SW Angola, Namibia and NW Cape. All of these populations are to some extent dispersive. A resident population exists in mountains of NE and C Madagascar. In the Indian subcontinent found south from the Himalayan foothills, throughout the peninsula, except the east and west coastal strips. A population exists in the interior of Sri Lanka. Annual visitor to Britain from March to October, mainly in the spring, including regular small flocks. Rarer elsewhere north of breeding range in Europe. Accidental in Scandinavia. Three New World records from the Caribbean: Barbados, Des Echeo Island off Puerto Rico and Moulea Chique, St. Lucia.

Few accurate population counts, even in Europe. Locally fairly common to common throughout Western Palearctic range. In Switzerland, in the period 1970-4 there was a minimum of 1250 pairs (at least 760 of these were nesting on buildings). Severe weather in autumn 1974 caused mass fatalities, with a resultant decrease in pairs breeding on buildings to no more than 320 in the period 1975-8. This population has increased since. Locally abundant in Turkey, especially Taurus mountains, southern republics of former USSR and throughout Indian subcontinent though no precise numbers recorded. In Africa common to abundant throughout range of resident population, and locally common in range of southern migrant population.

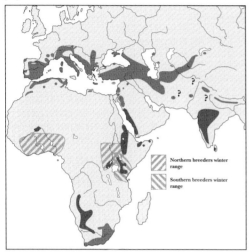

Northern breeders winter range

Southern breeders winter range

MOVEMENTS Migration from Palearctic on broad front, without noticeable concentrations at natural land bridges, and through higher airspace than other species. Scarcity of records from W Africa suggests that trans-Saharan migration is preferred avoiding Atlantic coast. Migration from breeding range occurs primarily September to mid October, with many juveniles moving earlier in August. Present in African wintering grounds from October to March. First returnees appear in Mediterranean basin from second half of February, abundant from March in north-west Africa. In Switzerland arrives late March to early April, with some migration still apparent until mid May. Migratory populations of southern Africa leave the breeding grounds from May-August. Migration through Zimbabwe is noted during May-June and August-October. Resident populations of Africa are to some extent dispersive in the non-breeding season. The situation within India is particularly confused, with populations resident, though local migrations occur particularly in the monsoon. The race *nubifuga*, from the Himalayas, is thought to winter in central India.

HABITAT Within Western Palearctic in temperate and Mediterranean habitats, breeding mainly in mountainous regions, also occasionally in lowlands. Large daytime wanderings for food often take birds through a great variety of habitats. In sub-Saharan Africa found to breed in a variety of habitats from subdesert steppe to equatorial mountains. In Turkmenistan found to 2500m, in Africa can be found at much higher altitudes being common at 4000-4300m in the Ruwenzoris and at 4600m on Mount

Stanley. Migration occurs over a variety of habitats.

DESCRIPTION *T. m. melba*
Adult Sexes similar. **Head** Broad, white throat patch extending to gape at eye and down to lower throat, where bordered by olive-brown chest band. Patch often ill-defined. Dark shaft streaks most apparent when worn. Pale grey-brown forehead contrasts slightly with black eyepatch, darkening to mid brown crown and sides of head/neck and feathers around gape and lores. When fresh, all feathers narrowly fringed white. **Body** Broad mid brown chest band below throat patch narrows slightly at centre, and is white fringed when fresh. Whole of breast and belly white. Vent, undertail-coverts and narrow line of flank feathers mid brown and contrasting sharply with white patch. These feathers when fresh have white fringes, narrow dark brown subterminal crescents and pale grey bases (these bases are particularly apparent on vent/undertail-coverts). Fringes abraded when worn. Nape and saddle uniform with mid brown crown, rump and uppertail-coverts slightly paler greyer-brown. All feathers white-fringed with narrow dark brown subterminal crescents. Fringes abrade when worn and pattern most distinct on rump/uppertail-coverts. **Upperwing** All rectrices black-brown on outerwebs and tips, inners grey-brown. Outer primaries darkest. Outerweb of tenth primary and innerwebs and tips of secondaries and inner primaries narrowly fringed white when fresh. Primary coverts as primaries, greater and median coverts slightly paler than secondaries and dark grey-brown lesser coverts. Leading edge coverts black-brown and broadly fringed white. **Underwing** Rectrices paler, contrasting with darker greater coverts. Greater primary coverts and greater coverts fairly uniform with rectrices, median and lessers clearly darker, uniform with axillaries and flanks. All underwing-coverts fringed white. **Tail** Olive-brown, narrowly fringed off-white, palest below. **Wing/Body Contrast** Outer primaries appear as the darkest areas of the upperparts with the upperwing-coverts and tertials appearing fairly uniform with adjoining parts of the body. Underwing darker than underbody.

Juvenile similar to fresh adult, plumage being extensively fringed white, more pronounced especially on wing coverts. Upperparts appear slightly darker.

Measurements Switzerland. Wing: male (11) 220-240 (227); female (16) 214-230 (221). Tail: male (10) 79-94 (84.4); female (17) 75-89 (80.7). Tail-fork: male (10) 24-28; female (17) 18-27. Weight (adults Switzerland April-August): 76-120 (Cramp 1985).

GEOGRAPHICAL VARIATION All ten subspecies share distinctive plumage pattern of the species and are therefore easily identified specifically. Racial separation is more troublesome. Main differences are in size, relative darkness of dark plumage areas, and size of breast band and throat patch. The darkest populations occur in areas of high rainfall, and the palest in arid habitats.
T. m. melba (From Spain through N Mediterranean basin and Asia Minor east to the W Himalayas) Variation in plumage darkness occurs within the range, especially in the area of intergradation with *tuneti* where paler individuals are more commonly encountered. Those breeding in western Himalayas often as dark as the Indian subspecies *bakeri* or *nubifuga*.
T. m. tuneti (South of *melba* from NW Africa, through S Mediterranean basin and Middle East, and east to Iran) From *melba* by paler plumage, more grey-brown

than olive-brown. Boundary between the two forms is not easily drawn and an area of intergradation occurs. Furthermore, considerable variation within each form has been noted, with individuals closely resembling either form being noted well within the other's range. Size similar to *melba*. Wing: male (12) 222-231 (227); female (12) 216-227 (222). Tail: male (10) 79-86 (82.5); female (9) 75-83 (79.7) (Cramp 1985). Weight: 95-110 (Brooke 1971c).

T. m. archeri (Somalia and SW Arabia) Similar in plumage to *tuneti*, in some cases paler. Main difference is smaller wings. Wing: (9) 193-208 (197.1) (Brooke 1971c).

T. m. maximus (Ruwenzori Mountains, Uganda and Zaïre) Very dark. Wing, other than brown inner greater wing-coverts, practically black. Head, tail and broad breast band similarly blackish. Largest race. Wing: (9) 221-230 (226.2) (Brooke 1971c).

T. m. africanus (E and S Africa) Plumage distinctly darker than *melba*, especially upperparts and breast band (also broader). Throat patch considerably reduced and shaft streaks blacker as they are on the belly patch. Smaller than *melba*, but variation within range. Wing: Ethiopia 199-218 (207.6); south-east Africa 200-228 (210.9); Angola 190-207 (199.8). Longest wing lengths recorded for South African birds. Weight (south-western Angola, 12) 67-87 (76) (Brooke 1971c).

T. m. marjoriae (Namibia to NW Cape) Closer to *tuneti* than *melba* being paler and greyer. From *africanus* by much paler plumage, with larger throat patch (unstreaked as is belly patch) and narrower breast band. Size similar to Angolan *africanus*: wing (15) 192-209 (202.6), weight 91 (Brooke 1971c).

T. m. willisi (Madagascar) Small race, with darkest plumage and broad breast band. Closest to *bakeri*. Wing: (36) 186-203 (194.1) (Brooke 1971c).

T. m. nubifuga (Himalayas, wintering C India) Darker than *melba*, with smaller throat patch and slightly broader breast band. Resembles *africanus*. Smaller than *melba*. Wing: (7) 212-217 (214).

T. m. dorobtatai (Mountains of western peninsular India south of *nubifuga* breeding range. Separated, in the breeding season, by the Ganges Valley and the Punjab lowlands) Darker than *nubifuga*, with broader breast band and shorter wings. Differs from *bakeri* in lighter plumage and broader breast-band. Wing: 194-207 (201).

T. m. bakeri (Sri Lanka) Smaller and darker than *nubifuga*, with narrower breast band. Wing: 194-207 (201.8).

VOICE Only likely to be heard in breeding season. Most common call is a canary-like (*Serinus*) trill, that accelerates and then decelerates, *trit-it-it-ititit-it-it-it*. Variations in duration and pitch often heard, especially around nest-site, and may be threat or alarm calls. Notably a single *zri* or *ziiu* call commonly recorded for solitary birds.

HABITS Gregarious, often in large flocks at any time of year and also in mixed species flocks. Usually seen at greater height than other species, though will feed at low level, and over water, especially in bad weather.

BREEDING Alpine Swift is a colonial breeder, nesting on ledges or in holes on cliff faces or tall buildings. Most nests are made on flat surfaces and are saucers with diameters

averaging (12) 12.5 x 13.0cm, with shallow depressions averaging 2.8cm. Average overall height varies between 1.5-6.5cm, averaging (8) 3.9cm on level surfaces and (4) 4.9cm on angled surfaces. Clutches vary between 1-4, but are normally 3 (64% of 2661 clutches in Switzerland, Arn-Willi 1959) and average (115) 30 x 19mm in size. Cramp (1985).

REFERENCES Ali and Ripley (1970), Arn-Willi (1959), Brooke (1971c and 1972), Chantler (1993 and 1995), Cramp (1985), Dean (1994), Dement'ev and Gladkov (1966), Fry *et al.* (1988).

77   MOTTLED SWIFT
Tachymarptis aequatorialis   Plate 20

Other name: *Apus aequatorialis*

IDENTIFICATION Length 23cm. Large size and bulky appearance (broad wings and body), with rapid flight on powerful, deep wing beats, create distinctive *Tachymarptis* jizz. In structure, from Alpine by comparatively longer and more deeply forked tail. Like Alpine upperpart colour racially variable. Best separated from Alpine by underparts. No Mottled Swift races show the clear-cut white belly patch of Alpine, but *furensis*, and to a lesser extent *lowei*, show bellies whiter than the rest of the underparts but barred with dark subterminal bands. Effect, however, is not clear-cut. Greatest problem in identifying Mottled Swift is when single birds are encountered and therefore the relative size and structure are harder to assess. Due to the lack of any obvious features, theoretically any uniform *Apus* could cause confusion, though even with single birds, once familiar with the jizz, observers should have no problem. Dark plumage, uniform upper head (not contrasting greatly with saddle) and restricted throat patch of *aequatorialis* and *gelidus* exclude Pallid and Forbes-Watson's Swifts. When seen well the heavily marked underparts of these two races, exclude Bradfield's, Nyanza, Common and Bates's Swifts. The most troublesome species is African Swift, which is close in plumage to *aequatorialis* and *gelidus*. In this species the plumage of the upperparts is very uniformly black-brown with a less clearly defined saddle, but a more marked pale forehead, lores and secondaries. The West African race, *lowei*, and the west Sudanese race, *furensis*, as mentioned above, have whitish bellies that further ease identification.

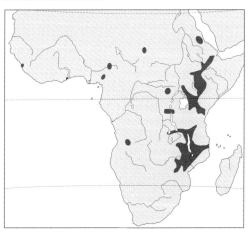

DISTRIBUTION Endemic to sub-Saharan Africa. Breeds in an extensive East African range, and a disjointed West African range. In W Africa occurs in W Sierra Leone, SE Ghana, EC Nigeria and at Bamenda, Cameroon. An isolated population breeds in Darfur, W Sudan. In E Africa has large range from Eritrea, south through the mountain ranges of the Ethiopian highlands, Kenya, N Tanzania and E Uganda. In the Ruwenzoris, NE Zaïre and Uganda, and the mountains of Burundi and E Zaïre. Continuous range from SE Zaïre, S Tanzania, Malawi, N Zambia, N Mozambique (from the Zimbabwe border to the Indian Ocean) and S Zimbabwe. An isolated population occurs at Quela, Angola, and there are records from Luanda. Recorded as a vagrant from southern Cape province.

Very local throughout scattered West African range. However, can be encountered in large numbers, with up to 1000 encountered in Nigeria, at Wase Rock. Presumed breeder only throughout West African range. Common to abundant throughout East Africa. Common within isolated Sudanese breeding population. No details of relative abundance in Angola or Bamenda, Cameroon where the species probably breeds.

MOVEMENTS Poorly understood. Probably dispersive resident, with little evidence of intra-African migration. Broadly scattered records away from known breeding areas during breeding season suggest that non-breeding populations wander extensively. Common non-breeding summer visitor to western Sudan (probably *aequatorialis*), with records further east into Chad.

HABITAT Due to highly specialised nest site requirements, breeding range is restricted to rugged granite highlands of up to 3000m. Travels widely, in search of food, and can be seen over many habitats. Wide African range shows adaptation to wide range of climatic zones from arid Eritrean highlands to equatorial Ruwenzori range.

DESCRIPTION *T. a. aequatorialis*
Adult Sexes similar. **Head** Small rounded, with grey-white or pale grey throat patch. Most marked on chin and upper throat, merging with black-brown feathers (pale grey-tipped and based) of lower throat and ear-coverts. Black eye-patch contrasts slightly with brown lores, forehead and slightly darker crown. Head feathers lightly pale brown-fringed, most marked on ear-coverts. **Body** Subterminal bands become blacker, and the pale tips broader and whiter, from below throat to undertail-coverts. The belly feathers are most broadly fringed white, the upper breast and undertail-coverts have the least fringing. Upperparts darkest on the black-brown saddle, which is restricted to the mantle. Contrasts slightly with the marginally paler head and more markedly with olive grey-brown rump and uppertail-coverts. When fresh all upperpart feathers are lightly fringed paler, most markedly on rump and uppertail-coverts and least on mantle. **Upperwing** Outer primaries black on outerwebs, with innerwebs black-brown next to shafts becoming paler on outer margins, which are fringed pale brown. Primaries progressively paler towards inners. Greater primary coverts, median primary coverts, lesser coverts and alula black-brown, contrasting with browner tertials, secondaries, greater and median coverts. **Underwing** Remiges paler (dark grey-brown) than on upperwing. Outer primaries similarly darker than inners. Greater coverts appear quite uniform with remiges, with smaller coverts markedly darker. Underwing-coverts pale-tipped and with darker

subterminal bands. Small coverts on leading edge of wing black-brown with pale grey fringes. **Tail** Black-brown above, darker than uppertail-coverts. Undertail slightly paler.

Juvenile Similar to adult but white tips to 3 outer primaries and shorter outer rectrices.

Measurements Wing: male (45) 188-213 (203); female (26) 192-210 (203). Weight: male 84-104 (92.3); female 83-105 (95.8) (Fry *et al.* 1988).

GEOGRAPHICAL VARIATION Four races, differing in size, clarity and extent of throat patch, degree of white on belly feathers and relative warmth of upperparts.
  T. a. aequatorialis (Ethiopia to Angola, Zimbabwe east of *gelidus*. An isolated population occurs at Bamenda in Cameroon) Described above.
  T. a. gelidus (W and C Zimbabwe) Similar to *aequatorialis*, differing in whiter throat, greyer upperparts and blacker subterminal bands on breast feathers. *Gelidus* is slightly smaller. Wing: male (9) 194-206 (201); female (7) 190-205 (198). Tail: (5) 85-90. Weight: (5) 83.8-92.6 (87.9) (Fry *et al.* 1988). An area of intergradation occurs between *gelidus* and *aequatorialis*; the boundary between the two is thought to be east or west of the 32°30'E line.
  T. a. furensis (W Sudan, Darfur) Very uniformly grey-brown above with indistinct saddle. White throat patch more extensive than in *aequatorialis*, extending onto lower throat. In some individuals the throat patch extends onto upper breast, though in these cases the white feathers are streaked and barred brown. Belly feathers appear very white, with narrow but distinct brown subterminal bands. Large, wing: (15) 197-215 (207) (Fry *et al.* 1988).
  T. a. lowei (W Africa, from Sierra Leone to Nigeria) Grey-brown upperparts resemble those of *gelidus*, throat patch closer to *furensis* in some, extending towards breast with streaking in centre and barring at sides, although smaller and greyer. In at least one specimen the throat is pure white. Wing: (3) 196-211 (203.3) (Fry *et al.* 1988).

VOICE Scream call not unlike Little Swift, shrill but disjointed. Loudest and harshest when fighting for nest sites. A deep harsh trill can also be given in flight. High pitched squeaking alarm notes can be heard when disturbed from the nest. The begging note of the young from the nest is an insistent *seep* note, likened more to a quiet yelping as the young get older. Most unusual, for a swift, is the mechanical drumming note, recorded from East Africa by birds diving to drink from a lake surface. The sound, *prrpt-prrpt-prrpt*, is made by the slight depression of the tail and its movement (apparently always to the right), causing outer rectrix to vibrate (Fry *et al.* 1988).

HABITS Loose association with Alpine and other swifts. Can occur in huge flocks of up to 1000, but usually 10-50. Very distinctive drinking technique has been likened to that of the feeding action of the skimmers *Rynchops* due to habit of flying in low shallow flight over the water with lower mandible in the water, leaving a considerable wake (Fry *et al.* 1988).

BREEDING Primarily a colonial breeder, although solitary breeding is not unknown. Nests are sturdy half-cups on vertical faces or on 45 degree overhangs consisting of leaves, wind-dispersed seeds and feathers, glued together with saliva. They are 80-87mm wide, protrude 60-65mm and are 25-48mm deep. Sites require plenty of airspace in the approach (at least 6m) and are always dry in fissures on cliff faces of granite. One or two eggs are laid averag-ing (3) 29.6 x 19.2 (Fry *et al.* 1988).

REFERENCES Brooke (1967), (Fry *et al.* 1988).

APUS

A large genus with 15 small to medium-sized species, distributed throughout the Old World. Most species occur in Africa, where there are six endemic species and five others which winter or breed there (and which are also found elsewhere). Five species breed in the Western Palearctic and migrate, mainly, to sub-Saharan Africa. Five species breed in the Eastern Palearctic and migrate either to the Oriental region or to Africa. Four species, including three endemics, breed in the Oriental region.

Plumages fall into two groups: blackish with whiter throats and white rump bands, or black-brown to mid brown with paler throats (figure 51).

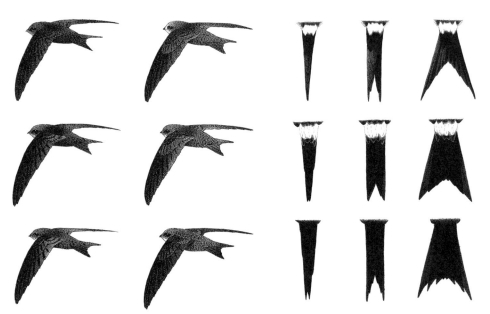

Figure 51. Larger Western Palearctic and African *Apus* swifts: Common Swift (top left), Pallid Swift (top right), African Swift (centre left), Forbes-Watson's Swift (centre right), Nyanza Swift (bottom left) and Bradfield's Swift (bottom right).

Figure 52. Tails of three Apus species with forked tails: White-rumped Swift (top), Pacific Swift (centre) and Dark-rumped Swift (bottom). Left hand figures are tightly closed, central figures are open and right hand figures are fully spread. Note the deeply emarginated outer rectrices of White-rumped and Dark-rumped Swifts in comparison with Pacific Swift.

All species show some degree of gloss in the plumage (from an oily sheen in some to a blue-green gloss in others) and in fresh plumage a variable amount of whitish fringing can be seen on the dark body feathers. Structurally, most species have long deeply forked tails (one species has only a rather cleft tail and another is practically square), sharply pointed wings (with either the ninth or tenth primary longest) and feet that appear to be pamprodactyl but are actually functionally heterodactyl (figure 52).

In-the-hand identification of larger plain *Apus* swifts.

The larger Western Palearctic and African *Apus* swifts can present identification problems in the hand. Common and Forbes-Watson's Swifts can be separated from most African, Nyanza, Bradfield's and Pallid Swifts by the delta-length measurement (see table 8). This is the distance between the outermost (fifth) and penultimate (fourth) tail feather. Furthermore, typically, the outermost (tenth) primary in Common Swift is shorter than the ninth primary. However, many Pallid Swifts share this feature and the habit of arresting moult at the tenth primary means that this feather is often more abraded than the ninth primary with a resultant measurement difference.

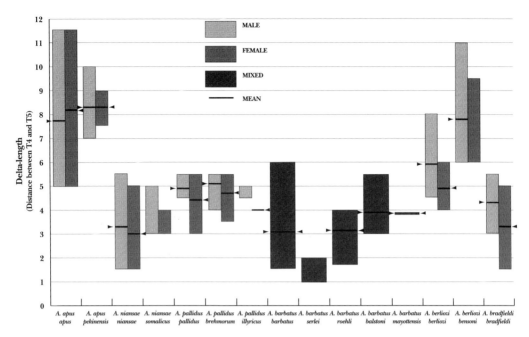

Table 8. Delta-length of large plain *Apus* swifts.

Nyanza averages much the shortest-winged, with a maximum wing length for *niansae* of 162 and for *somalicus* of 160. There is overlap, therefore, only with the shortest-winged Pallid Swifts of the nominate race and *brehmorum*, Common Swifts of both races and the nominate race of Forbes-Watson's Swift. Short-winged examples of the latter two species should be excluded in most cases by reference to the delta-length.

Structurally the other members of this group are very similar with relatively shallow tail-forks, short delta-lengths and long wings. Their identification, as with the other species, can be clinched in the hand by reference to the plumage descriptions.

78 ALEXANDER'S SWIFT
Apus alexandri   Plate 21

Other name: Cape Verde Swift

IDENTIFICATION Length 13cm. Distinctly smaller (10%) than Plain Swift and shorter-winged, with shallowest tail-fork of the *Apus*. Compact shape combined with weak fluttering flight lead to distinct appearance. Closest to Plain Swift in plumage, but is less black. This is apparent particularly on the innerwing, pale grey-brown rump (saddle contrast more prominent) and lightly marked pale grey-brown underparts. The structure and flight action of Plain Swift is also distinct. Common and Pallid Swifts are the only other species perhaps likely to occur on the Cape Verde Islands but both can be easily excluded on structure.

DISTRIBUTION Endemic to the Cape Verde islands where it is known from most islands. No detailed data is available regarding population, but it is apparently common.

MOVEMENTS Thought to be resident. However, it is likely that this species could occur in West Africa, and it is notable that White (1960) suspected absence in early July (dry season), after being unable to find any during his visit in this period.

HABITAT Like Plain Swift it is found from sea-level to the highest peaks and prefers deep gullies, preferably on the coast. Can be seen over any habitat on the islands.

DESCRIPTION *A. alexandri*
Adult Sexes similar. **Head** Indistinct broad throat patch,

193

very pale grey-brown or pale grey. Whole of throat pale becoming progressively darker towards pale grey-brown breast from palest area on chin. Forehead/lores grey-brown and lightly fringed paler. Becoming darker to dark grey-brown on crown. Ear-coverts and sides of neck slightly paler than nape. Sides of throat and feathers around gape becoming darker away from throat. Whole of head lightly fringed when fresh. **Body** Underparts progressively darker away from throat to darkest point on grey-brown belly; undertail-coverts slightly paler. Feathers grey-brown with paler fringes and grey bases; when worn fringes abrade and bases can be seen. Dark grey-brown saddle extends to back and contrasts clearly with grey-brown rump and uppertail-coverts. Feathers of upperparts more narrowly fringed and bases less visible when worn. **Upperwing** Darkest on black-brown outer primaries, alula, outer primary coverts, median primary coverts and lesser coverts. Dark grey-brown inner primaries and secondaries, uniform with or slightly paler than corresponding greater coverts, with median coverts darker, closer to lesser coverts. **Tail** Grey-brown, no clear contrast with tail-coverts. **Wing/Body Contrast** Lesser coverts quite uniform with saddle, other coverts and tertials slightly paler. On underwing darkest coverts fairly uniform with body, rest of wing paler.

Juvenile Closely similar to fresh adult; inner primaries and secondaries narrowly edged white at tips.

Measurements Wing: 139-141. Tail: 56-60. Tail-fork: 16-19 (Cramp 1985).

GEOGRAPHICAL VARIATION None. Monotypic.

VOICE Notably higher pitched than Common Swift with a less fierce and more articulated delivery. The call has a somewhat reeling quality (Hazevoet 1995).

HABITS A gregarious species, often seen in large flocks.

BREEDING Alexander's Swift is not well known in its nesting habits. It nests in both houses and fissures and caves in cliffs from sea-level to 1600m. This species is unique in the Apodidae in having finely red-brown freckled (most densely clustered at broadest end), not pure white eggs. The clutch (two nests) is two (Bannerman and Bannerman 1966 and Cramp 1985).

REFERENCES Bannerman and Bannerman (1966 and 1968), Cramp (1985), Chantler (1993), Hazevoet (1995), White (1960).

---

79 COMMON SWIFT
Apus apus   **Plate 21**

Other names: Eurasian Swift

IDENTIFICATION Length 16-17cm. For identification in sub-Saharan Africa, see accounts of Scarce, Schouteden's, African, Forbes-Watson's, Bradfield's and Nyanza Swifts. For identification from Plain and Alexander's Swifts see under those species. In much of breeding zone it is the only breeding species, requiring separation only from vagrant Mediterranean and eastern species. Pallid Swift is the main identification pitfall, especially in the Mediterranean and Middle East where the breeding ranges overlap. Common appears structurally more rakish, less broad-winged (especially on the primaries), with more

pointed wing tips and a longer more deeply forked tail. The body appears thinner and more rounded, especially on the head and across the rump. Perhaps as a result of structural differences the flight of Common Swift can appear more dashing and agile than Pallid. Plumage details are more important than rather subjective structural and flight differences. The plumage of Common is darker black-brown than the olive-brown Pallid Swift. The forehead and lores of Common rarely contrast as strongly with the crown and eye patch as Pallid except in some *pekinensis*, exceedingly worn adults of either race (which in these circumstances are inseparable) and juveniles. The throat patch of Common is typically small, indistinct and rounded (except in the most extreme individuals of *pekinensis* and juveniles) as opposed to the large triangular or flared patch of Pallid that usually appears contiguous with the pale lores and forehead (in Common it is only contiguous in juveniles). The darker *illyricus* shows the most striking white patch of Pallid races due to the surrounding darker feathers. Usually the throat patch of Pallid appears rather ill-defined around its border. In Common the saddle is less striking than in Pallid, although still present and in strong sunlight surprisingly prominent, and in particular it lacks the strong contrast between nape and crown (and ear-coverts) giving Pallid a pale-headed appearance especially in *brehmorum* and in some *pallidus* and *illyricus*. In the saddle of Common the upperpart contrast is between the whole of the back and the rump whereas in most Pallid (except some *illyricus*) the contrast is between the saddle and the lower back and rump creating a smaller more pronounced saddle. *Pallidus* shows the least prominent patch of the Pallid races but its pale grey plumage should render identification from Common straightforward. Pallid typically shows a more scaly feather pattern than Common leading to a more 'moth-eaten' or lightly barred appearance. In Common Swift fresh adults show light scaling caused by narrow white fringes to the body feathers. This is still apparent on their return to the breeding quarters but the feathers are mainly abraded during the breeding season. When worn the paler feather bases may be exposed. Juvenile Common Swifts show even more marked scaling on the body and wings, but the effect is less barred than in Pallid. Wing patterns of the two species differ. In Common the wing is more uniform with inner/outer wing contrast being most marked in worn birds and even in these individuals the whole of the outerwing appears uniform (also uniform with the mantle). In Pallid the outer primaries appear blacker than the rest of the upperwing and often blacker than the mantle. In this way the broad black wedge of the outer primaries can appear as an isolated patch. In Common the outer primary coverts appear uniform with their respective coverts whereas in Pallid the paler, uniform primary coverts appear as a pale nick into the outerwing. Importantly, in Pallid the secondaries, inner primaries and greater coverts (and to lesser extent, though still greater than in Common, the median coverts) contrast markedly with the outer wing and lesser coverts. On the underwing the contrast is greatest in Pallid and the median coverts appear a little paler than the lesser coverts causing the darkest area on the underwing to appear smaller. On the upperwing Pallid shows a greater tendency towards fringing than in Common Swift (except the rather black-winged juvenile).

Figure 53. Undersides of adult nominate Common Swift (left) and adult *brehmorum* race of Pallid Swift (right). Note broader area of dark underwing-coverts in Common and the generally more broken appearance of the wing pattern in Pallid.

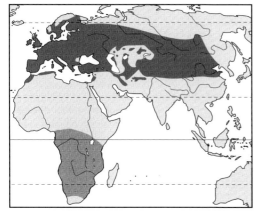

DISTRIBUTION Breeds throughout an extensive Palearctic range, and winters in sub-Saharan Africa (small numbers winter in northern India and Arabia). Breeds throughout Europe and the Mediterranean islands, except the far north (Iceland, N Scandinavia, and the tundra zones of European Russia). Also absent as a breeder from far northern Scotland, the Faroes, the southern mountains of Scandinavia, and parts of the Alps and Balkans. Also found on the Mediterranean coast of NW Africa and the adjoining Atlas mountains. In the Middle East breeds on the Mediterranean coast, at scattered locations in Syria, and throughout Asia Minor. Breeds in Asia (south of the 60° latitude except in the far west of Asia) south to Iran and east to Mongolia and N China. Migration to Africa on a broad front south through Europe, but ringing recoveries indicate that birds move primarily south to south-west in autumn (Glutz and Bauer 1980), with a more easterly trend on the return movement suggested (Cramp 1985). This theory is supported by Fry *et al.* (1988), stating that the species is commonest in Senegambia in autumn, but from Mali eastwards to Nigeria it is commonest in spring. An absence of West African ringing recoveries suggests a diagonal movement across the Sahara by birds moving through western Europe to NW Africa. *Pekinensis* moves south through E Africa via the Middle East (and the southern coast of the Red Sea in particular). Although not averse to sea crossings the relative scarcity of migrant records in Libya and abundance in NW and NE Africa is significant. Winters primarily south of the equator with the western nominate race found primarily from Zaïre and Tanzania in the north, to Zimbabwe and Mozambique in the south. *Pekinensis* winters primarily in Namibia and Botswana. Complex intra-African movements occur, and are

believed to be mainly weather provoked in attempts to avoid rainfall. These movements could account for winter records away from main wintering grounds e.g. West Africa. Nominate birds leave S Africa generally by late January or early February and *pekinensis* leaves by early March. Arrival in the Western Palearctic occurs in strength from mid March in the northern Mediterranean with arrivals further north continuing until early June. Autumn migration from Europe occurs from late July (late individuals can be seen in November, rarely) with arrival in wintering grounds from September. Small numbers winter, perhaps only irregularly, in N Africa, Arabia and N India. Ali and Ripley (1970) state that those birds wintering in India are *pekinensis*. Has been recorded, and no doubt could be recorded, anywhere south and west of the breeding range. Accidental records from Spitsbergen, Iceland, Faroes, Azores, Seychelles, Assumption Island, Aldabra Island and, most spectacularly, from Bermuda, where a single was recorded in November 1986. It is an uncommon visitor to the Cape Verde Islands which has been recorded mostly in December and January.

The only commonly observed swift throughout most of the northern range (north of the Mediterranean and east of Lake Baikal). Fry *et al.* (1988) states that the west European population is 2.5 million birds, with a world population of perhaps 25 million. Figures collated in Cramp (1985) suggested little recent population change across Europe and perceived increases in some areas.

HABITATS In the breeding range occurs throughout a broad geographic zone, inhabiting desert and steppe areas in the extreme east and south, the full range of European Mediterranean and temperate habitats, to boreal conditions above the Arctic Circle in the far north. In much of the breeding range it is found primarily in areas of human habitation. In the wintering range regularly recorded from all African sub-Saharan habitats. Throughout wintering and breeding range can be found at any altitude.

DESCRIPTION *A. a. apus*
Adult Sexes similar. **Head** Small rounded dirty-white throat patch, often poorly defined, especially when worn (when can be mottled brown). Patch size is variable, not reaching the gape at the eye, but usually extending onto the lower throat. Grey-brown forehead, lores (sometimes slightly darker) and narrow line over eye, contrast slightly with black eye-patch. Head gradually darkens, through dark grey-brown on crown and sides of neck, to black-brown on nape. Lower ear-coverts and feathers along edge of throat patch dark grey-brown with pale grey fringes when plumage fresh. **Body** Underparts black-brown from below throat patch on breast to vent; undertail-coverts paler grey-brown. When fresh off-white fringes to feathers; when worn, fringes abraded and some paler grey feather bases can be seen (these bases are visible to some extent even before abrasion on undertail-coverts). On the upperparts, an oily-black saddle extends onto lower back, contrasting with slightly paler black-brown rump and uppertail-coverts, but appearing very uniform with head (some individuals show slight contrast, especially in bright light). Feather pattern similar on underparts, though fringes narrower and pale bases less apparent (except uppertail-coverts). Plumage appears browner when worn. **Upperwing** Darkest on outerwing, with outer four or five primaries black (outerweb) or black-brown (innerweb), outer primary coverts as primaries and median primary

195

coverts, alula and lesser coverts black. Inner primaries, secondaries and tertials appear uniformly black-brown (secondaries can appear blacker when plumage fresh and viewed in the hand, but this is seldom seen in the field). Greater coverts and inner primary coverts appear black-brown, showing little contrast with rectrices. Median coverts appear blacker than greater coverts though slightly paler than lesser coverts. When fresh wing pattern is at its darkest and most uniform, but wear makes innerwing paler and browner, increasing contrast between leading edge and innerwing, especially between lesser coverts and other coverts. Wing feathers, especially coverts and secondaries, narrowly fringed paler brown; these fringes are reduced by abrasion. Leading-edge coverts black with distinct white fringes. **Underwing** Similar to upperwing though remiges appear paler (dark grey) and degree of contrast greater. Greater primary coverts and greater coverts appear uniform with corresponding remiges. Median primary coverts and median coverts blacker though slightly paler than black-brown lesser coverts and axillaries. Greater primary coverts, greater coverts, median primary coverts and median coverts narrowly tipped white, with greater coverts showing a slightly darker subterminal crescent. Inner primaries and secondaries can appear translucent when viewed from below. **Tail** Black-brown, slightly paler-fringed, appearing paler below. **Body/Wing Contrast** Upperbody contrasts with wings especially tertials, greater and median wing coverts. Underbody usually appears uniform with darker underwing-coverts and clearly darker than rest of underwing.

Juvenile Similar to fresh adult though a little blacker (and therefore considerably blacker than autumn adult) and more extensively white-fringed, most notably forehead and lores which are white and concolorous with white throat patch which is broader and more distinct than in adult. Fringing becomes less dense and narrower away from forehead and least distinct on mantle, becoming more marked again on rump. Underbody is more extensively 'scaled' than on fresh adult. Wing pattern is like that of fresh adult though remiges and wing-coverts are extensively white-fringed. Some individuals will abrade white body fringing before winter and will appear more like adults although fringing on wings remains longer.

First Year Plumage resembles adult, but more heavily worn remiges and more rounded outer rectrices (figure 54).

Figure 54. Outer tail feathers of Common Swift: adult (left) with well-defined emargination and juvenile (right). (Based on Cramp 1985.)

Measurements Wing: adult male (83) 167-179 (173); adult female (81) 164-180 (173); juvenile male (14) 157-174 (166); juvenile female (13) 162-173 (168). Tail: adult male (50) 71-82 (75.8); adult female (35) 69-79 (74.3); juve-

nile male (7) 65-73 (69.1); juvenile female (11) 64-70 (67.2). Tail-fork: adult male (48) 28-36 (31.6); adult female (34) 25-34 (29.7); juvenile male (7) 24-32 (27.6); juvenile female (11) 22-27 (24.5) (Cramp 1985). Delta-length: male (17) 5-11.5 (7.8); female (9) 5-11.5 (8.2) (Brooke 1969a). Weight. Gibraltar: (24) average 44.9. East Africa: (November-April) (13) 35-44 (39.0) (Moreau 1933). Namibia: male (8) 31-40 (34.9); female (5) 31-42 (36.2) (Fry *et al.* 1988).

GEOGRAPHICAL VARIATION Two races recognised. *Marwitzi*, a race blacker than *apus* and lacking bronze gloss to the plumage, has also been described. It breeds in SW Asia and winters from Zimbabwe to Transvaal and Natal.

A. a. apus (West of the range, eastwards towards Lake Baikal, and south eastwards towards Iran) Described above.

A. a. pekinensis (Iran (possibly eastern Iraq), eastwards through the W Himalayas to Mongolia and N China. The two forms are clinal, with intergrades appearing where they meet. *Pekinensis* is essentially slightly paler than the nominate *apus*. This is most apparent on the slightly larger, whiter and more pronounced throat patch. Forehead also on average slightly paler. The upperparts are a little browner and the underparts do not appear so deeply black. On the wing the coverts are more clearly fringed and the innerwing appears slightly paler. Wing: male (6) 164-178 (168.3); female (6) 164-174 (170). Tail: male (6) 72-79 (75.2); female (6) 72-77 (74.2). Tail-fork: male (6) 27-33 (28.8); female (6) 28-31 (29.5). Delta-length: male (6) 7-10 (8.3); female (7) 7.5-9 (8.3) (Brooke 1969a). Weight: (7) 30-46 (37.3) (Fry *et al.* 1988).

Worn adults of both subspecies are probably inseparable in the field and, although *pekinensis* is slightly browner and paler, the areas of contrast are the same. Juvenile *pekinensis* is very similar to juvenile *apus*, but it may average slightly paler in the wing.

VOICE Most commonly heard call is a wheezy, high-pitched, screaming *shree*. Variation in the duration and pitch can be heard, and the call may be delivered in several short bursts. Disyllabic calls are produced by a drop in tone of the standard call, *shree-eee*. Amongst 4 main pitch and duration variations distinguished by Stadler and Schmitt (1917) the abrupt and sibilant '*i*' represents the greatest diversion from the norm. Greatest call variety is heard in the breeding season. Migrants and wintering birds are less frequently heard.

HABITS Highly gregarious regularly occurring with a large number of different swift species and also hirundines. Mass flocking occurs particularly on migration, or when inclement weather causes local movements. Particularly active in lower airspace in the late evening. Normally feeds from 2-100m but exceptionally has been noted at 3600m above ground (Fry *et al.* 1988). Cramp (1984) notes that it tends to feed in higher airspace than resident swifts when on wintering grounds.

BREEDING Common Swift is a colonial breeder nesting primarily in buildings, but also in rock crevices and, in the far north of its range, in hollows in trees; in Algeria it has been recorded nesting in cedars. The nest is a cup, measuring 125 x 110mm with an internal diameter of 45mm, made of small pieces of vegetation and feathers, stuck together with saliva. Clutches have between 1 and 4 eggs. In North Africa the average is 2 whilst in Europe it is somewhat higher. The eggs average (100) 25 x 16mm (Cramp 1985 and Lack 1956c).

REFERENCES Ali and Ripley (1970), Brooke (1969a and 1972b), Chantler (1990 and 1993), Cramp (1985), Finlayson (1979), Fry *et al.* (1988), Glutz and Bauer (1980), Lack (1956a and 1956c), Moreau (1933), Stadler and Schmitt (1977).

80   PLAIN SWIFT
Apus unicolor   **Plate 21**

IDENTIFICATION Length 14-15cm. Medium-sized dark *Apus*, similar in plumage to the closely related Common Swift. Within known range requires separation from Common and Pallid Swifts. Structural differences important, with Plain appearing more streamlined and rakish than either, but Pallid in particular (as Common is in turn more rakish than Pallid). This is particularly apparent on the body, especially the rump (slimmer and more streamlined with tail, lacking the more 'lumpy' feel of the larger species). The wings and tail also appear proportionally slightly longer and the tail-fork is heavier. This coupled with the species' more dashing frantic flight action, can draw comparison with the Palm Swifts *Cypsiurus* spp. Differs from Pallid in generally darker black-brown or dark grey-brown plumage. This is most marked on the head with Plain's darker forehead and lores, showing little or no contrast with the eye patch, and the typically restricted, indistinct, mottled pale grey-brown (at palest off-white) throat patch. Unlike Common Swift there is probably no overlap in throat patch size between Pallid and Plain. The upperparts show less saddle contrast (as in Common) with little or no contrast on head and less contrast on the rump than in Pallid. The underparts are similar in pattern to Pallid though the general shade is darker. The wing pattern is as Common and therefore darker with, especially in fresh plumage, less contrast between the inner- and outerwing. This is best seen on the outer primary coverts and the lesser coverts. Plumage separation from Common is far more hazardous and requires great care. Juvenile Common with its highly contrasting face pattern can be excluded easily. Fresh adult Common Swifts will appear blacker than the essentially dark grey-brown Plain, especially on the underparts, which however, will show some indistinct light scaling, though not to the extent of fresh Plain. Once worn, however, the plumage tone will be very similar. Fresh adult Common Swifts show distinct, though small, off-white throat patches, although worn birds can show indistinct mottled-brown patches (figure 55).

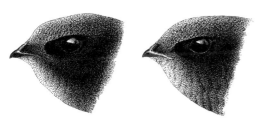

Figure 55. Heads of adult nominate Common Swift (left) and Plain Swift (right). The adult Common Swift is a worn individual with a very ill-defined throat patch. Note slightly paler head of Plain Swift and paler more barred lower throat.

In these cases particular care should be paid to the underparts: slightly barred pattern in Plain, but uniform in worn Common, though some paler feather bases may be seen. The importance of structural, and to a lesser extent flight action, differences in the separation of Common from Plain cannot be overstated.

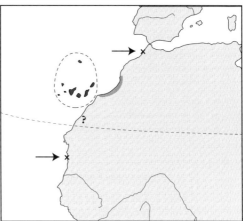

DISTRIBUTION Restricted as a breeding species to Madeira and Canary Islands. Present all year round on these archipelagos, but much reduced on both in winter. Recently this species has been recorded visiting holes and crevices in sea cliffs in Morocco. Wintering grounds unknown, presumably in Africa. Recent winter records from SW Morocco, and previous records of dark *Apus* swifts from this area suggest that wintering grounds are in NW Africa rather than further south. Has been recorded in the winter from Mauritania (Meininger *et al.* 1990) and from N Morocco (N. J. Redman verbally).

Common breeding species on Madeira and Canary Islands.

HABITAT On both archipelagos, present in all habitats, including towns, from sea-level to around 2500m. Especially common in deep gullies (often coastal). Migrants depart September to mid October, returning January-March (mainly February). When encountered in the winter in Morocco occurs over sea cliffs.

DESCRIPTION *A. unicolor*
Adult Sexes similar. **Head** Small indistinct throat patch typically grey-brown, occasionally grey-white; mottled when worn. Smallest examples restricted to chin, largest extending just beyond level of eye, but still ill-defined and tending to merge with surrounding progressively darker grey-brown tracts. Forehead and lores dark grey-brown, slightly paler than crown, with pale grey fringes. Often, when viewed head-on areas around bill clearly paler than rest of head. Black eye-patch. Ear-coverts and crown slightly paler than, or uniform with, black-brown saddle. Light fringing on head when fresh. **Body** Underparts progressively dark grey-brown away from throat, then uniform to vent, becoming paler grey-brown on undertail-coverts. Black-brown feathers distinctly pale grey-tipped and grey-based. Fringes abrade and some bases visible when worn. Feathers of upperparts similar, though blacker, with less distinct fringing and less extensive pale bases. Blackest on saddle which extends to lower back, contrasting with dark grey-brown rump and uppertail-coverts. **Upperwing** Outer primaries, alula, outer primary coverts, median pri-

mary coverts and lesser coverts black-brown. Inner primaries and secondaries black-brown, uniform with or slightly paler than corresponding greater coverts, with median coverts darker, closer to lesser coverts. On remiges, innerwebs (and tips to secondaries) palest. Wear in wing increases contrast between inner and outerwing. Coverts indistinctly pale-fringed, especially greater coverts. **Underwing** Remiges paler grey, uniform with or slightly paler than corresponding greater coverts which are white-fringed, as are median coverts which are clearly darker (though slightly paler than the lesser coverts). Greater coverts have darker subterminal crescents. Inner primaries and secondaries can appear translucent. **Tail** Similar to remiges. **Body/Wing Contrast** As Common Swift though darkest underwing-coverts often appearing darker than body as can be seen in Pallid Swift.

Measurements Wing: male (26) 150-158 (153); female (12) 152-159 (155). Tail: male (22) 67-74 (70.3); female (9) 68-72 (69.6). Tail-fork: male (22) 27-34 (29.4); female (9) 28-30 (28.7) (Cramp 1985).

GEOGRAPHICAL VARIATION None. Monotypic.

VOICE Calls of adult very similar to *shreee* scream call of Common Swift, perhaps more wheezing and a little higher pitched. Rapid trill can be recorded. Calls heard throughout year on Canaries.

HABITS Often seen in very large flocks, and associates with Pallid and Common Swifts.

BREEDING Plain Swift is a colonial breeder that nests either in caves or fissures on the sheer faces of cliffs or gullies (coastal or interior), under bridges, in holes on houses or under tiles. Saucer-shaped nest, 10-12cm wide, with a shallow depression 1-2cm deep is composed largely of the downy seed cases of Compositae flowers, (various other plant matter and small pieces of man-made items are also sometimes added) glued together with saliva with some feathers on the surface. The clutch is of two eggs and double-brooding is frequent. Eggs average (14) 22 x 15 (Cramp 1985).

REFERENCES Bannerman and Bannerman (1965), Chantler (1993), Cramp (1985), Meininger *et al.* (1980).

---

81   NYANZA SWIFT
Apus niansae   Plate 22

Other name: Brown Swift

IDENTIFICATION Length 15cm. Typical uniform *Apus*, with rather uniformly brown body plumage and the strongest contrast between inner and outerwing of any of the genus. Throughout the species' range the main confusion species are Forbes-Watson's Swift, in northern and coastal East Africa, African Swift in interior East Africa and, during the northern winter and on passage, Common Swift, potentially anywhere within its range. The striking contrast between the pale grey-brown, almost straw-coloured greater coverts and secondaries (and inner primaries) and the remainder of the wing is the best feature to exclude all of these other species, although this pattern is apparent, but less pronounced in all of the uniform *Apus* species. In Nyanza it appears as a pale rectangle on the innerwing, with the greater coverts contrasting markedly with the greater primary coverts and the medi-

an coverts and the secondaries contrasting markedly with the inner primaries. This pattern is most closely matched in African Swift although the degree of contrast is less. Otherwise the wing pattern of Nyanza is closest to a worn Common Swift with the lesser coverts only slightly darker than the median coverts and more or less uniform with the alula, outer primary coverts, outer primaries and the saddle. In African, as in Pallid, the lesser coverts typically appear slightly paler than both the saddle and the outer primaries and coverts. The upperbody of Nyanza is more uniform than in either Forbes-Watson's or African Swifts, with the upper head and the rump appearing uniform with or only slightly paler than the brown saddle. However considerable contrast is shown between the pale rectangle and the saddle. On the underparts the off-white throat patch is small and ill-defined. The underbody appears mid brown, typically somewhat 'moth-eaten' in the style of Pallid due to indistinct pale fringes, and the undertail-coverts appear clearly paler. Both Common and in particular African appear far blacker in plumage. The shape of Nyanza is closest to Common, with typically scythe-shaped wings and a small rounded head. However, the tail appears rather shorter and fuller, a little reminiscent of Pallid in consequence, and significantly shallower-forked with less sharply pointed outer rectrices. It is unlikely that Pallid is often recorded with Nyanza, but identification would be possible by paying attention to the prominent saddle of Pallid (especially the contrast between the nape and the head), the less extreme wing contrast and the larger paler throat and forehead of Pallid.

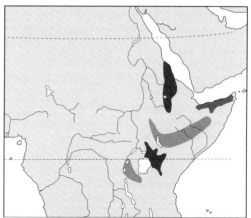

DISTRIBUTION Sub-Saharan East African endemic. Ethiopia from central coastal Eritrea south to Addis Ababa. Further east occurs in the coastal mountains of northern Somalia and over the border in adjacent areas of northeast Ethiopia. Occurs from the mountains of E Uganda in Kidepo Valley National Park and Mount Moroto, eastwards through WC Kenya and just into extreme N Tanzania in Arusha National Park. Scattered records outside the breeding season have occurred in N Zaïre at Oka, NE Zaïre at Semliki river and N Tanzania south of Lake Victoria. There is a scattering of these winter records away from the breeding grounds in Somalia, Ethiopia and Kenya.

Common and occasionally abundant. A huge and very well known colony occurs at Hell's Gate Gorge, near Lake Naivasha in Kenya.

MOVEMENTS The Addis Ababa population is present in

the rainy seasons of February to April and June to September. Outside these periods it is absent or at least greatly reduced. The northern Somalia breeding population is present from late March to late September, although it may be to some extent resident. Presumed wintering birds in Kenya are present from November to March. Outside these areas it is resident.

HABITAT Occurs in rather arid mountainous areas, preferring gorges, cliff faces, high crags and highland towns. Also in lowlands.

DESCRIPTION *A. n. niansae*
Adult Sexes similar. Head Poorly defined round throat patch, usually grey-white and showing some variation in size (averages slightly larger than nominate Common). Worn examples are very indistinct. Forehead, lores and narrow line of feathers over eye pale grey-brown. Black eye-patch. Crown and ear-coverts uniformly grey-brown and clearly darker than forehead. Fresh plumage shows some pale grey fringes. Body Underparts appear distinctly brownish with a grey tinge and are fairly uniform below the throat with the undertail-coverts appearing distinctly paler and greyer. The feathers are distinctly grey fringed with a broad brown subterminal band and a grey base. In reasonable views the impression is rather 'moth-eaten' as with Pallid Swift. The upperpart pattern is similar to Common Swift with the nape uniform with the crown and the mantle only slightly darker brown or uniform. The lower back, rump and uppertail-coverts are similar or slightly paler than the nape and therefore the saddle effect is not very distinct. Upperwing The outerwing is black-brown appearing very uniform through the lesser coverts and into the outer primary coverts and outer primaries. The inner primaries and their coverts are slightly and progressively paler than the outer primaries and coverts. The secondaries and in particular the greater coverts are pale grey-brown and contrast strongly with the surrounding tracts, showing as a pale rectangle. The median coverts are darker brown than the greater coverts and a little paler than the lesser coverts. Underwing The remiges are paler and greyer than above and contrast most with the dark-brown lesser coverts. The greater coverts are fairly uniform with the remiges and the median coverts appear paler than the lessers. On the underwing the pale rectangle of the secondaries and greater coverts is mirrored by increased translucency in the secondaries. Tail Mid brown, with paler fringes and innerwebs. Paler below.

Juvenile As fresh adult, but paler-fringed with a darker, more mottled throat.

Measurements Wing: male (36) 143-162 (154); female (23) 146-161 (154). Tail: (13) male 59-65 (61.2); female (12) 53-64.5 (60.6). Tail-fork: male (22) 19-25 (22.20); female (21) 17-23 (20.6). Weight: male (20) 24-39 (32); female (17) 25-37.5 (33) (Fry *et al.* 1988). Delta-length: male (38) 1.5-6.5 (3.3); female (27) 1.5-5.0 (3.0) (Brooke 1969b).

GEOGRAPHICAL VARIATION Two races.
A. n. niansae (W Ethiopia, E Uganda, W Kenya and across border in N Tanzania) Described above.
A. n. somalicus (North-east of range in N Somalia and in the adjacent parts of Ethiopia. Some birds winter in Kenya) Paler than *niansae*, with a well-defined throat patch; also somewhat smaller. Wing: male (10) 144-160 (150.8); female 147-156 (154). Tail: male 59-64; female 62-64. Delta-length: male 3-5; female 3-4. Tail-fork: male 18-24; female 20-23 (Brooke 1969b).

VOICE Slightly different from the typical scream call of Common Swift. The scream is shorter, more twangy and more disyllabic; *zee-u* or when longer and more rasping *zzzziiiiiirrrrr*. In pre-roost circusing the call has a harsh, almost Starling-like tone *bizeer-zee-zee-zee-za* and solitary birds can make a staccato *pip pip pip* call.

HABITS Very gregarious, occurring in large flocks and with other swift species. Most visible in morning and late afternoon/evening. However, around breeding colonies it is active throughout the day. Like most swift species low level feeding is associated with bad weather and in particular stormy weather. Outside the breeding season, at the huge colony on the cliffs at Hell's Gate National Park in Kenya, the birds leave en masse in the early morning to feed at height over the nearby Lake Naivasha, returning at dusk in vast flocks to roost.

BREEDING Nyanza Swift nests on buildings and crevices in gorges and cliffs. The nest is a shallow cup consisting of feathers and grasses (up to 12cm wide) and is glued together with saliva. The base is made of grass and possibly wet mud, whilst loose fragments of concrete, up to 30mm in diameter, may be incorporated. Mud is apparently used as a technique to close a small space, after which feathers and even bat bones may be added. The internal cup may have a diameter of 85-110 and a depth of 25-30. Clutch size is 1-3 eggs that average (4) 23.5 x 14.5 (Fry *et al.* 1988).

REFERENCES Brooke (1969b), Fry *et al.* (1988).

82   PALLID SWIFT
Apus pallidus   Plate 21

Other name: Mouse-coloured Swift

IDENTIFICATION Length 16cm. A rather large bulky uniform *Apus* species with a measured and somewhat laborious flight and rather blunt wing tips. Records in sub-Saharan Africa from Kenya, Uganda, Zambia and South Africa suggest, however, that the species can occur anywhere within Africa although the main wintering grounds are in West Africa and the Sahel. In the two darker northern races *brehmorum* and *illyricus* identification is best achieved by examination of the striking pale throats, foreheads and lores. These are particularly striking, especially the throat patch, in *illyricus* as the plumage is the darkest, heightening contrast. In most *brehmorum* and the pale *pallidus* the dark eye and eye patch stand out prominently against the pale head. In *brehmorum* in particular, and in the other two races to a lesser degree, the head appears significantly paler than the saddle as does the lower back. In *illyricus* the saddle is very black and larger extending in some individuals onto the back of the crown and the lower back, creating a less obviously pale head. In the very pale *pallidus* the saddle is smallest and least prominent. The bodies of all three races although variable in darkness, all show the typical 'moth-eaten' or slightly barred effect caused by the pale-fringed brown subterminal band and grey base to each feather. The wing pattern in all races is similar and a useful tool for identification. The innerwing contrasts strongly with the outerwing (though less so than in Nyanza) with the outer primaries appearing as a blackish wedge, blacker than the rest of the wing and either equal with or darker than the saddle. In *illyricus* the effect is less contrasting with the lesser coverts closer in colour to both the mantle and outer primaries. A useful feature occurs due to the uniformity of the greater primary coverts which on the outer coverts appear paler than their respective primaries causing a pale nick into the outerwing. To summarise, on the upperwing the pattern is typically rather broken, a feature that will separate it on good views from Common or Plain Swifts. On the underwing the pattern differs from Common Swift in that the lesser coverts are the only truly darker feathers and therefore the area of contrast appears smaller than in Common or Plain Swifts. In this way it resembles most other plain African species.

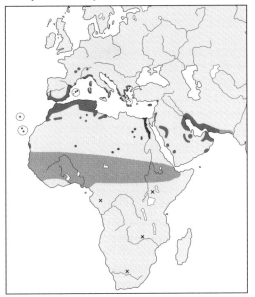

DISTRIBUTION Predominantly Western Palearctic breeding range, also extends to W Asia and the Middle East. Winters mainly in Africa. In Europe has a rather broken, largely Mediterranean, distribution. In Portugal it occurs in the west coastal plain from Leiria and south along the southern coast into Spain, and also northwards through the interior along the Spanish border to the Provinces of Salamanca in Spain, and Guarda in Portugal. The species may be regular further north along the coast in Portugal as it has been recorded from Porto. In N Iberia it occupies the coastal strip and hinterland from Portugal to Barcelona and further north in the Costa Brava at Cap de Creus. In France occurs in the Mediterranean coastal foothills of the Pyrenees, and inland around the southern Massif Central; further east in Bouches de Rhône, S Provence, Alpes Maritimes on the Italian Border and Switzerland; in N Italy on the coast of the Riviera and inland at Turin, and further south in the Tuscan archipelago and the adjacent mainland coastline. The distribution in S Italy is a little more continuous with breeding occurring from the south of the Gulf of Salerno to Sinni in the Gulf of Taranto. Further east in Italy on the large peninsula in E Apulia around Cape Santa Maria di Leuca, and the Adriatic around the peninsula of Monte Gargano. In the eastern Adriatic the species occurs in Slovenia on the coast around Rijeka in the north, and in Croatia on some islands off the Dalmatian coast, and inland close to the

Albanian border further south. The Greek distribution is a little confused. Occurs on the Adriatic from Corfu southwards and then eastwards in the southern foothills of the Taiyetos Oros and the adjacent peninsula, on Cerigo and on several islands in the Cyclades group. Northwards occurs on the eastern and southern shores of Korinthiakos Kolpos and on the northern and western shores of the Aegean as far as the Turkish border. Occurs on the Balearics, the Sardinian coast, except the north, around the west and south Corsican coast, the NW Sicilian coast, Malta and the adjacent islands, C and E Crete and C and E Cyprus. In N Africa continuous on the coast and the coastal Atlas mountains from Agadir in Morocco to the Libyan border of Tunisia. Isolated populations south of the High Atlas in Morocco and the Grand Erg Occidental, Algeria. In Libya on the coast around Tripoli, the desert south of Tripoli and further south on the coast and hills of Cyrenaica east to Tobruk. In Egypt occurs in the Nile Delta and south along the Nile to Lake Nasser. Isolated oasis populations occur east of the Nile. Further south in N Africa it occurs on the Atlantic coast at Banc d'Arguin in Mauritania, and in the Aouderas mountains of Niger, and probably the Hoggar mountains of S Algeria, the Bandiagara escarpment, SE Mali, W Darfur in Sudan and the Ennedi mountains, Chad. In Turkey only in the mountains around Mustafa Kemalpasa in the west and around Iskenderun on the extreme south-east coast. Isolated populations in E Lebanon and Syria. In Israel in the northern Rift Valley and just south of the Dead Sea. Occurs in E Jordan from Amman in the north, south towards Aqaba. Eastwards is found in the central Euphrates valley, Iraq and from the Euphrates delta east along the Gulf of Arabia through Iran to Pakistan, where it has been recorded east to Karachi and Hyderabad. South of the Arabian Gulf, in S Iraq through Kuwait as far as, but not actually reaching, Qatar, and in N Oman from the Strait of Hormuz to Sur. A population occurs in the Red Sea mountains from Asir in southern Saudi Arabia into Yemen. The N African wintering range is quite extensive, but it is most frequent in the Sahel from Gambia to Sudan. It is, however, present in the winter from S Egypt and S Morocco in the north to N Sierra Leone in a band across W Africa to C Sudan in the south. Some birds winter in coastal Pakistan. On migration can occur in all of the intervening countries and eastwards in Africa as far as Djbouti where it is a common passage migrant. This suggests that the species may occur in Ethiopia and N Somalia although it has never been officially recorded there. There are claimed sightings of the species from Kenya but no specimens exist to support the records. Vagrants have been recorded north of the range on a handful of occasions in the British Isles in both spring and late autumn, from the Cape Verdes ,and south of the wintering range in Zambia, Uganda and South Africa.

Varies throughout range from locally uncommon, such as on the Canary Islands, France and Turkey, to abundant, as in parts of N Africa and the Mediterranean basin. In June 1974 between 8000 and 12000 individuals were present along 125km of cliffs in SE Mali. In France an estimate of between 100-1000 pairs was made in 1976 (Yeatman 1976); the range has apparently expanded this century with the first mainland breeding recorded in 1950, following the first Corsican breeding in 1932. The first colony was not located until 1966. It is possible that identification difficulties cloud the picture in many countries.

MOVEMENTS Migratory in all but the more southerly breeding areas. As a result of double brooding it is present for much longer in the Western Palearctic than Common Swift. In the north of the range it arrives in France in early April leaving by mid November. Present further south on the Canaries from January to September, and Gibraltar and Morocco from late February to October. In the winter present in Gambia from October to December with lesser numbers through to April. In more central North African areas such as around Lake Chad and in Mali present to some extent for most of the year, with less occurring in July but some present, in Mali at least, throughout the year. Locally resident in the Middle East. Strong migration through Eilat in spring with an early March peak. In NW Africa migration from late February (or mid March) to early May, and from August to November. Migration dates are confused by the return of some birds to Moroccan breeding sites as early as mid December and the not infrequent incidence of wintering as far north as S France.

HABITAT Rocky gorges and cliff faces with a marked preference, in most of the range, for coastal sites. However, not restricted to coastal habitats and can be seen in continental zones, particularly in the south of the range where it occurs in the Saharan oases and highlands. In the Mediterranean occurs commonly in towns on the coast and in the hinterlands. Recorded to 1750m on Cyprus and 1200m in the Canaries. Feeds over a variety of habitats, including the sea.

DESCRIPTION *A. p. brehmorum*

Adult Sexes similar. **Head** Broad-based, very pale grey or off-white throat patch, varying in size but typically spreading to gape, lower throat and upper breast. Narrower patches show distinct line of grey-brown feathers around gape. Throat patch often ill-defined as a result of pale grey-brown surrounding tracts. Shaft streaks grey-brown and, when worn, darker grey-brown feather bases can be seen. Forehead, lores and line over eye pale grey, slightly darker than throat though this is often hard to see and areas around bill often appear uniformly pale; this is especially striking when viewed head-on. Pale on forehead variable in extent, often passing eye, but appearing more uniform when pale fringes have been abraded. Forehead and lores contrast clearly with black eye-patch. Crown, ear-coverts, side of throat and feathers around gape grey-brown. Pale fringes across head when fresh are lost through abrasion. **Body** Feathers have pale grey bases, dark brown subterminal crescents and thin grey-white fringes. On upperparts, black-brown or dark olive-brown saddle (pale fringes narrowest, brown crescents broadest) extends onto upper back, contrasting with paler grey-brown lower back, rump and uppertail-coverts (pale bases more extensive). Dark saddle extends slightly onto nape, contrasting with crown and ear-coverts, emphasising pale-headed appearance. Underparts, from below throat patch to vent, olive-brown or dark grey-brown, lightly scaled or (when pale fringes abraded and bases more apparent) lightly barred, with undertail-coverts slightly paler. **Upperwing** Darkest on outerwing with outer two to four primaries black with dark brown innerwebs, black-brown alula and median primary coverts and dark olive-brown lesser coverts. Inner primaries progressively paler grey-brown, with secondaries and tertials uniformly grey-brown. Greater primary coverts with pale fringes and innerwebs, appear quite uniform, causing slight indentation into dark outer primaries. This is less apparent when worn. Greater coverts and median coverts similarly grey-brown, as

secondaries and inner primaries or slightly darker, especially median coverts. In fresh plumage secondaries can appear darker than greater coverts though this is rarely the case in the field. Leading-edge coverts pale brown, with broad, though indistinct, pale fringes. All wing feathers narrowly fringed paler, less so on darker feathers and most prominently on secondaries and innerwing coverts. **Underwing** Similar to upperwing but remiges paler grey-brown and degree of contrast greater. Greater primary coverts and greater coverts appear uniform with remiges (upperwing greater primary coverts can appear paler than outer primaries). Median primary coverts and median coverts appear darker than greater coverts though slightly paler than lesser coverts and axillaries. Greater primary coverts, median primary coverts, greater coverts and median coverts narrowly fringed white, and greaters show slightly darker subterminal crescents. **Tail** Dark olive-brown, appearing paler below and with paler fringes and innerwebs. **Body/Wing Contrast** Outer primaries often appear as darkest area in wing or indeed upperparts, with saddle appearing uniform with or slightly darker than lesser coverts and clearly darker than other wing-coverts and tertials. On underwing, darkest wing-coverts often appear darker than body although this is light-affected.

Juvenile Similar to adult though pale fringes more extensive and subterminal crescents narrower on feathers and hence plumage paler, especially forehead and throat. Remiges distinctly pale-fringed.

Measurements Variety of sites from Canaries east to Italy. Wing: male (18) 168-178 (171); female (20) 167-176 (171). Tail: male (19) 68-75 (71); female (20) 67-74 (70). Tail-fork: male (19) 23-30 (26.3); female (20) 23-27 (25.1) (Cramp 1985). Delta-length: male (6) 4-5.5 (5.1); female (6) 3.5-5.5 (4.7) (Brooke 1969a). Weight: (100) 41.3 (Finlayson 1979).

GEOGRAPHICAL VARIATION Three races.

A. p. brehmorum (Widespread European form from Portugal through to Turkey, except on the east coast of the Adriatic where *illyricus* occurs. Also in coastal N Africa from Morocco to NW Egypt, and the Canary Islands and Madeira) Described above. Within *brehmorum* there is considerable variation in the darkness of plumage. Generally speaking the birds of coastal North Africa are paler than those of Europe and the Canary Islands. In all ranges there is a slight variation in size that is hard to correlate with geographical variables.

A. p. pallidus (Banc d'Arguin through the Saharan highlands in Egypt and through the Middle East to Pakistan) Typically the palest race. Mantle less distinct and throat patch tends to be slightly larger than in *brehmorum*. Smallest race. Wing: male (6) 160-175 (166.2); female (6) 161-168 (163.0). Tail: male (6) 62-74 (68.0); female (6) 64-74 (68.0). Delta-length: male (6) 4.5-5.5 (4.9); female (6) 3-5.5 (4.4). Tail-fork: male (6) 22-28 (25.4); female (6) 22-28 (23.8) (Brooke 1969a).

A. p. illyricus (The former Yugoslavian coast and possibly the east coast of Italy) Some authors have also considered those birds occurring in Cyrenaica, NW Egypt, the Aegean and Cyprus to be *illyricus*. In the E Adriatic this is the largest and darkest race. The crown is dark and usually uniform with the large blackish saddle. Rump darker than other races, but still paler than the saddle. Broader dark subterminal bands on feathers of the underparts cause the underparts to appear darker than other races, but the large white throat patch is consequently more striking. Outerwing darker than in *brehmorum*. Wing: male 166-179; female 165-169. Tail: male 63-73; female 67-74. Tail-fork: male 26-29; female 21-28. Delta-length: male 4.5-5; female 4 (Brooke 1969a). Fry *et al.* (1988) state that this race is smaller than *brehmorum*, but include birds from SE Europe and the Egyptian coast in the range.

VOICE The call is subject to great variation. The most distinctive call is a rather grating disyllabic *shree-er* that is not as shrill as Common Swift. A single *cheeoik* call is often heard after a series of screams (Fry *et al.* 1988). All calls are rather shorter than the typical *shree* of Common Swift.

HABITS Gregarious species usually encountered in small groups but occasionally in hundreds or even thousands. Regularly mixes with other swifts including *Tachymarptis*. Like other *Apus* often feeds at low level.

BREEDING Pallid Swift is a colonial species building a nest of straw and feathers bound together and to the surface by saliva. The nest is 8-10 x 10-12cm in diameter and is 4-5cm deep; the centre of the cup is 2.5cm deep. A variety of nest sites have been noted including caves, cliff face fissures, under the eaves of buildings, in holes in palms and even in freshly made House Martin *Delichon urbica* nests, after evicting the owners! (Kennedy 1986). The clutch is usually 2-3 eggs, but 1 and 4 have been recorded. In Gibraltar it has been noted that first clutches average (19) 2.9 eggs whilst second clutches average (22) 1.9 eggs. Eggs average (11) 25 x 16 (Cramp 1985).

REFERENCES Brooke (1969a), Chantler (1990 and 1993), Cramp (1985), Finlayson (1979), Fry *et al.* (1988), Kennedy (1986), Yeatman (1976).

---

83 AFRICAN SWIFT
Apus barbatus   Plate 22

Other names: Black Swift; African Black Swift; Madagascar Swift (race *balstoni*); Fernando Po Swift (race *sladeniae*)

IDENTIFICATION Length 16cm. Typical large *Apus* within the *pallidus* superspecies which has the blackest body plumage of any of the confusion species. From an identification perspective it is important to note that this species is only sympatric, as a breeding species, with Nyanza Swift and, during the northern winter and on passage, with Common Swift. Separation from Nyanza in East Africa is often quite straightforward. The body of African appears almost jet-black on the mantle and black on the underparts with a notable degree of saddle contrast (both with the head and the rump) and, in the East African race *roehli*, a more obvious white throat patch than Nyanza. This is quite different from the rather uniformly mid brown-bodied Nyanza which shows an indistinct throat patch and little saddle contrast. A further difference occurs in wing pattern which is similar though far less contrasting in African, which may also show the lesser coverts slightly paler than the black saddle and outer primaries (less apparent in the darkest races). However, the rectangular block of the paler secondaries and greater coverts contrasting with the outerwing and upperbody serves as a useful feature to separate Common Swift. In addition the body of Com-

mon, although similar in shade (slightly paler especially prior to the winter moult), shows less saddle contrast and a smaller less distinct throat patch. The shape of African is also close to Pallid appearing bulkier than Common, especially in the broader less pointed wings, the flatter broader head, the bulkier and deeper body (especially behind the wings) and the shorter more rounded tail (which appears rather full and heavy when closed). On the rare occasions that Pallid comes into contact with African the significantly paler-bodied Pallid, with its broad white throat patch and pale head, should prove little problem. The exceptionally pale Bradfield's Swift which might occasionally occur with this species is very uniform on the upperparts and much greyer in general plumage tone.

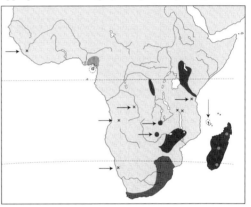

DISTRIBUTION Scattered sub-Saharan range, mainly East African. Also on Madagascar. Accurate assessment of the range is complicated by the presence of many records away from areas of certain occurrence and difficulties in identification. It is highly possible that the actual breeding range is more extensive than indicated here. The species has been recorded in W Africa from Sierra Leone, S Nigeria, SW Cameroon, NE Angola and central coastal Namibia, but these are possibly no more than migrants (see Geographical Variation). Occurs on Bioko (Fernando Po) and in nearby SW Cameroon. In E Africa it is found from E Uganda (Mount Moroto) into W Kenya and south into NE Tanzania. Further west it has a small range in E Zaïre at Rutshuru in the Ruzizi valley, Kivu and into SW Uganda. Continuous from W Mozambique, through S Malawi and into E Zimbabwe, and south into Transvaal, Natal and S Cape Province west to Cape Town. Between these two main ranges there are many isolated records believed to represent migrants. However isolated breeding populations also occur in S Zambia (Zambezi escarpment), W Zimbabwe (around Victoria Falls), NE Botswana and NE Namibia. Found throughout the Madagascan lowlands and the Comoro Islands.

The status of this species varies considerably. In South Africa it can be locally abundant, whereas throughout its East African range it is only locally common. The three endemic West African races are much scarcer and known from only a handful of records. The *sladeniae* race is given specific status by Collar *et al.* (1994) and included in the category of Data Deficient Species on the World List of Threatened Birds.

MOVEMENTS Occurs only as a breeding bird in South Africa where it is present in Cape Province between August and mid May. The wintering grounds of this population are not known, but migration through Kariba, Zimbabwe in May and August is presumed to be of these birds. A bird of this race, *barbatus*, was collected in northern Mozambique. Langrand (1990) notes apparent population decline of the endemic Malagasy race, *balstoni*, from April to July in the south-west of the island and between April and September on the High Plateau. This coupled with sightings of large flocks apparently arriving from the Mozambique Channel might lead one to think that *balstoni* migrates to Africa but this remains hypothetical.

HABITAT Primarily a species of damp mountainous regions. However, also over adjoining habitats including lowlands. In some areas it occurs on coastal cliffs, such as close to Cape Town, South Africa. It breeds mainly in the altitudinal range of 1600-2400m, and less often from below 1000m to sea-level.

DESCRIPTION *A. b. barbatus*
Adult Sexes similar. **Head** Rather triangular whitish throat patch varies in size and is not always striking. Rather black surrounding feathers increase contrast. Pale grey-brown forehead, lores and narrow line over eye, black eye-patch. Crown slightly darker than forehead with the nape and ear-coverts appearing blacker especially when pale grey fringes of the fresh plumage have abraded. **Body** Blackish underparts with prominent white fringes and paler bases (only seen when worn). Undertail-coverts browner. Upperparts show a prominent black saddle contrasting with the head, and slightly less with the lower back, rump and uppertail-coverts. Plumage appears browner in strong sunlight or when worn. **Upperwing** Blackest on outer two primaries with the others becoming paler inwards. The median primary coverts and alula are also blackish. The primary and lesser coverts are black-brown, contrasting both with the primaries and mantle. Both sets of primary coverts become progressively (though slightly) paler inwards and are narrowly pale-fringed. The secondaries are grey-brown with the greater coverts appearing slightly paler (and broadly pale-fringed, as are the slightly darker median coverts) and contrasting with both the saddle, the primaries and their coverts. The lesser coverts are black-brown. **Underwing** Paler remiges rather uniform with greater coverts and only slightly paler than the median coverts, though clearly paler than the black-brown lesser coverts. **Tail** Black-brown, appearing greyer below, with slightly paler fringes and innerwebs.

Juvenile As fresh adult with considerable grey-white fringing and more noticeable throat patch.

Measurements Wing: male (5) 172-186 (180); female (6) 172-185 (179). Tail: (5) 71-77 (74). Weight (19) 35-50 (42.1) (Fry *et al.* 1988). Delta-length: 4-6 (Brooke 1969a). Delta-length (10 BMNH): 1.5-4.5 (3.15).

GEOGRAPHICAL VARIATION Nine races recognised here, but Fry *et al.* (1988) questioned the validity of the three West African races due to scarcity of specimens. Two forms, *sladeniae* and *balstoni*, are sometimes considered separate species.
   A. b. barbatus (S Africa; migratory) Described above.
   A. b. glanveillei (Only two specimens from Rokupr, Sierra Leone) Darker than, but otherwise resembling, *barbatus*.
   A. b. sladeniae (This form has a rather dispersed W African range with specimens collected from SE Nigeria, Mount Kupe in W Cameroon, Bioko (Fernando

Po) and Mount Moco in Angola) This is a very dark race with blacker plumage than either *barbatus* or *glanveillei*, and a very dark throat often lacking any grey-white feathering. Considerable variations in the measurements of this race add to questions regarding its validity, with the wing of the four specimens varying from 168-184 (174) (Fry *et al.* 1988).
   A. b. serlei (Only recorded from Bamenda, Cameroon) From the specimens obtained this race is apparently darker than *barbatus*, lacks the blue gloss to the mantle feathers and has underpart fringing restricted to the lower belly. Wing: (2) 165 + 170. Tail: (1) 70. Tail-fork: (2) 19 + 25.5. Delta-length: (2) 1 + 2 (De Roo 1970).
   A. b. roehli (Widespread E African race, found from N Uganda and Kenya to Malawi and E Zaïre and also Sombo in Angola) Smaller than *barbatus* and darker in plumage. Wing: male (13) 167-176 (171); female (14) 164-176 (171). Weight: male (8) 40-46 (41.8); female (11) 35-50 (42.4) (Fry *et al.* 1988). Tail: male (6) 67-71 (69.3); female 68-71. Tail-fork: male (6) (21-27) (23.2); female (6) 22-27 (24.0) (Brooke 1969a). Delta-length: (BMNH) 1.7-4.0 (3.14).
   A. b. hollidayi (Restricted to the Victoria Falls area on the Zimbabwe/Zambia border) Paler than *barbatus*. Mantle uniform with the wing-coverts. Wing: (4) 177-178 (177.5) (Fry *et al.* 1988). Tail: male (1) 76 (Brooke 1969a).
   A. b. oreobates (Zimbabwe from Melsettwe to Mount Inyangani, in Mashonaland and on Mount Gorongoza in Mozambique) Similar to *roehli* but with longer wings. Wing: (14) 169-182 (176) (Fry *et al.* 1988).
   A. b. balstoni (Madagascar) Similar plumage to *roehli*, though perhaps a little darker-plumaged. This is most notable in the lesser coverts which are closer in darkness to the primaries and mantle than in *roehli*. Wing: (6 BMNH) 162-164 (163). Delta-length: (8 BMNH) 3.0-5.5 (3.8).
   A. b. mayottensis (Comoro Islands) Similar to *balstoni* but a little browner in plumage. Wing (4 BMNH); 159-165 (161.8). Delta-length: (2 BMNH) 3.8-4.0 (3.9).

VOICE Typical *Apus* scream but at a much greater frequency than in Common Swift and consequently sounds rather hissing.

HABITS Like Common Swift this is a gregarious species that can be seen in flocks of many hundreds as well as singly. Freely mixes with other swift species. Tends to feed at lower levels than White-rumped or Alpine Swifts, and like all swifts it will feed at low level, especially in bad weather.

BREEDING African Swift is a colonial breeder that has been recorded in a mixed colony with Alpine Swifts. In East Africa it tends to nest in hollow trees especially large cedars *Juniperus procera*, whilst in southern Africa it breeds on cliff faces, but not those composed of granite. Some colonies around Cape Town are coastal and one is close enough to the sea for the nest cracks occasionally to become covered in sea spray. The nest is stuck with saliva to the surface and is a shallow pad composed of grass, and sometimes thistle-down, with feathers (of this species and occasionally others). The base of the nest appears untidy, but the top is rather neat, and sturdy but thin. One or two eggs are laid, averaging (4) 25.9 x 16.8 (Fry *et al.* 1988).

REFERENCES Brooke (1969a, 1970a and 1972b), Collar *et al.* (1994), De Roo (1970), Fry *et al.* (1988), Langrand (1990).

84   FORBES-WATSON'S SWIFT
Apus berliozi   Plate 22

Other names: Berlioz's Swift; Watson's Swift

IDENTIFICATION Length 16cm. Very typical *Apus* most closely resembling one of the darker populations of Pallid. Although this species is probably only very rarely sympatric with Pallid Swift the two would be best separated using the following features. The throat of this species is more clearly defined than in Pallid. The forehead and lores are darker than the throat unlike Pallid in which the eye patch and eye stand out clearly; the forehead is also paler than in Common. The saddle contrasts only slightly with the hind neck, which is unlike *brehmorum* Pallid Swift but similar to the darker-backed *illyricus* and the much paler *pallidus*. Clearly any Pallid-like bird observed within the range of Forbes-Watson's Swift should be scrutinised with these features in mind, but such an individual would prove very hard to separate. Nyanza Swift can be separated by its smaller less bulky appearance, pronounced pale innerwing panel, less heavily marked and somewhat browner (and hence paler) body plumage, less distinct throat patch and darker rump (hence more uniform upperparts). This last point is particularly noteworthy with regards to *berliozi*, as *bensoni* is darker rumped than the nominate race. African Swift is clearly blacker than the olive-brown Forbes-Watson's. As a member of the Pallid superspecies this species appears slightly larger and bulkier, with blunter wing tips and a flatter more triangular head than the Common Swift superspecies, and it typically has a more leisurely, heavier flight.

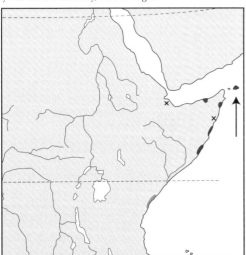

DISTRIBUTION Small coastal East African range. Occurs on Socotra Island and coastal E Somalia (though also up to 90km inland). Possibly also in coastal mountains of N Somalia. Winter range includes coastal Kenya where it has been recorded from Sokoke, Gazi, Kilifi, Ribe, Tiwi and Shimba Hills.

The total population is probably not large, but it is locally common in the Somalian breeding grounds and common on Socotra.

MOVEMENTS Partially migratory, with movement only from the N Somalian breeding grounds, south as far as coastal Kenya.

HABITAT Rather arid country species. When at the coast it breeds in sea caves and cliffs and can be seen feeding over vegetated permanent dunes. On Socotra it occurs over most habitats, including towns. In more mountainous areas of the range it occurs between 700-1200m in rather craggy areas.

DESCRIPTION *A. b. berliozi*
Sexes similar. **Head** Well-defined whitish triangular throat patch lacking visible dark shaft-streaking or bases. Feathers bordering patch on the breast are quite uniform creating an abrupt division between the whitish throat and its grey-brown border. Brown lores slightly darker than the forehead which is clearly darker than the throat patch, being a light grey-brown that is lightly fringed, with the fringing extending to the narrow line of feathers over the eye (where it is slightly whiter). Black eye-patch. The olive-brown ear-coverts are tipped buffish-white and merge with the ear-coverts to a greater extent than the breast feathers. From the forehead backwards the head becomes progressively darker, becoming blackish-brown on the hind crown. **Body** Immediately below the throat patch the feathers have extensive blackish-brown subterminal bands and due to overlap the pale bases are hard to detect. These feathers lack any pale tips. Below the breast the subterminal bands become smaller, revealing paler brown bases, and the feathers have whitish tips. In worn examples these tips are lacking, and in others the subterminal bands are less blackish (with resultant paler underparts). The undertail-coverts are slightly paler, due to the more visible pale bases. On the upperparts the mantle and upper back are blackish-brown, causing a saddle that contrasts with the lower back and rump. **Upperwing** Innerwing coverts darker towards leading edge. Longest lesser coverts blackish-brown, paler brown distally. Median coverts dark brown with buffish fringes, somewhat paler than lesser coverts and clearly darker than the greater coverts. Greater coverts mid brown, fringed pale whitish, especially on the outerwebs. Tertials olive-brown with darker, dull brown subterminal bands and narrow whitish fringes, which are broadest at the tip. Furthermore, the longest tertials have broader fringes than the shorter ones. Secondaries, with dark brown centres and sharply defined white fringes, are marginally darker than (centres of) greater coverts. The pale darker coverts, tertials and secondaries together form a distinct buffish-brown wing panel (similar to but less pronounced than Nyanza Swift). The shorter leading edge coverts have complete fringes. Lesser and median primary coverts dark blackish-brown with bluish-green gloss (hardly visible). Alula black-brown (darker on the outerweb) tipped buffish (especially innerweb). Median primary coverts form darkest part on the closed (and probably the opened) upperwing. Greater primary coverts have innerwebs blackish-brown, glossed bluish-green, outerwebs brown, paler towards sides and tips. Greater coverts correspond, more or less, with respective primaries. Primary outerwebs blackish-brown, glossed bluish-green. Innerwebs browner, but darker subterminally. Whole of primary fringed buffish very narrowly, whiter towards base of innerweb. Primaries are black-brown on the outers,

through to dull dark-brown on the inners. **Underwing** Coverts heavily fringed, especially on median coverts. Marginal primary coverts warm mid brown, fringed buffish-white, especially on the outerwebs. Lesser primary coverts dark brown with narrow buffish-white edges and broader, whiter tips. Median primary coverts centred mid brown and with broad, whitish outer edges and tips, merging in brown centre; overlap results in more tips and less bases being visible, so creating a very pale impression. Greater primary coverts brown-grey with sharply demarcated white fringes. Primaries brown-grey, outer three or four darkening considerably. Inner primaries pale brown-grey. **Tail** Rectrices centred olive-brown, glossed greenish. Subterminally darker and very narrow (easily worn-off) paler fringes. Fringes even narrower (or lacking) on outer feathers. Undertail paler, brownish-grey. **Body/Wing Contrast** Tertials paler than mantle, and lesser and median coverts. Very pale-fringed underwing-coverts may appear paler than body, but this may be rather different in the field.

Measurements Wing: male (18) 164-173 (168); female (11) 156-170 (165). Tail: male (18) 65-76 (69.1), female (11) 63-69 (66.2). Tail-fork: male 25-32.5 (27.6); female 23.5-30 (25.3). Delta-length: male (18) 4.5-8.0 (5.9); female (11) 4-6 (4.9). Weight: male (18) 34-42 (37.6); female 37-46 (40.1) (Brooke 1969b).

GEOGRAPHICAL VARIATION Two races.
A. b. berliozi (Socotra Island) Described above.
A. b. bensoni (Somalia; partial migrant to the Kenyan coast in winter) A little browner and darker than *berliozi*, with a slightly darker rump showing less saddle contrast. Less feather patterning on upperparts, darker forehead and more rounded throat patch. Wing: male 168-177; female 170-179. Tail: (6) 68-77 73.2; female 76-80. Delta-length: (6) 6-11 (7.8); female 6-9.5. Tail-fork: male 26-32; female 28-35. Weight: male (7) 38.5-46 (42.4); female 41.1-51 (47.4) (Brooke 1969b).
The population occurring at Hal Hambo is considered intermediate between the two forms in plumage characters and in measurements. Wing: (146) 161-178 (168.5). Tail: (94) 65-77 (71.1). Tail-fork: (86) 21-30 (25.1). Outer rectrix is (95) 3-8 (5.2), longer than the penultimate one. Weight: (146) 37-54 (43.5) (Fry *et al.* 1988).

VOICE Typical scream call *shreee* is less piercing and somewhat shorter than that of Pallid Swift, and in addition a highly audible *chip* call can be given rapidly in a sequence (Fry *et al.* 1988).

HABITS A gregarious species usually seen in small groups of around 10 but on occasions up to 200 have been seen together.

BREEDING Forbes-Watson's Swift is a colonial species that is known to nest in holes in the roofs of sea caves. Nests are saucer-shaped or pad-like with an external diameter of 110-130mm and a thickness of 30-60mm. A cup may be present (i.e. a depression in the pad or saucer). All recorded nests have included a dried seaweed *Cymodocea*, bound together with saliva to a number of different materials, including sand, feathers, fishing line, plant down, plant stems and seed heads. The clutch consists of two eggs averaging (4) 26.6 x 16.35 (Fry *et al.* 1988).

REFERENCES Brooke (1969b), Fry *et al.* (1988).

85 BRADFIELD'S SWIFT
Apus bradfieldi
Plate 22

IDENTIFICATION Length 18cm. This very geographically restricted species is the palest of the uniform *Apus*. Most similar to the pale *pallidus* race of Pallid Swift. Pallid, however, has only been recorded with certainty on two occasions south of the equator, although one of these specimens was recorded in the eastern part of Bradfield's range. It is likely that Common is the only confusion species that regularly occurs within the range of Bradfield's Swift and then only during the northern winter. African Swift is the only other confusion species that is likely to be recorded within the range of Bradfield's Swift. Bradfield's may be distinguished by its pale grey-brown body feathers, appearing very uniform on the upperparts with only a very slight saddle (indeed some individuals have a darker crown than the mantle). Some but not most *pallidus* have similarly indistinct saddles, but it is important to consider that vagrant Pallid Swifts reaching southern Africa are more likely to be the larger, darker, more migratory races. Bradfield's Swift further differs from Pallid in that both the blackish tail and sometimes the lesser coverts (and outerwing generally) can appear strikingly darker than the adjacent body. The wing pattern of Bradfield's is also quite distinct. The greater coverts are clearly paler than all surrounding feather tracts, including the secondaries. Other swifts occasionally show this but never such a striking panel. Like Pallid, Bradfield's shows a distinct off-white throat patch and forehead. The very dark-plumaged African and Common Swifts with their small, though often distinct, rounded throat patches should be easy to separate. They further differ from Bradfield's in both wing pattern and the contrast between the tail/wing and body.

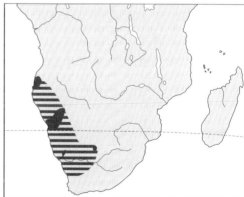

DISTRIBUTION Restricted SW African range. Occurs in Namibia: in the central highlands (including the coastline around Skeleton Coast Park), in the Great Fish Canyon (S Namibia), in the north-west and in adjacent areas in S Angola. It also occurs in South Africa around Kimberley, North Cape.

Generally rather common in suitable habitat throughout range. Indeed this is the commonest breeding swift in C Namibia.

MOVEMENTS Believed to be resident throughout range, although after breeding it is largely nomadic and searches for suitable feeding areas.

HABITAT Essentially a species of arid landscapes in both true desert and open savanna, where it occurs over rocky hills.

DESCRIPTION *A. b. bradfieldi*
Adult Sexes similar. **Head** Broad off-white indistinct throat patch which shows some variation in extent. Whole of head pale grey-brown reaching its palest on the forehead, lores and in the line of feathers over eye. The head darkens slightly towards the nape and the ear-coverts appear slightly darker. Like Pallid there is considerable contrast with the black eye-patch. **Body** The underparts are very uniformly pale grey-brown below the throat. The feathers have narrow, but distinct dark brown subterminal crescents, white fringes and visible pale grey bases. This creates a slightly barred impression which is most marked on the chest and belly. The grey-brown upperparts are the most uniform of the larger *Apus* with the head or the mantle appearing slightly darker in some individuals. The individual feather pattern is similar to the underparts but the subterminal band is slightly broader making the feather bases less apparent. **Upperwing** Blackest on the outer primaries, but the tendency for the innerprimaries and secondaries to become progressively pale is less apparent than in Pallid; indeed the secondaries appear clearly darker than the greater coverts which form a clearly visible panel in the wing. This panel varies individually. The lesser coverts, alula and primary coverts are blackish like the outer primaries. The median coverts are slightly less blackish than the lesser coverts. In many individuals the outerwing and lesser coverts appear strikingly darker than the mantle. **Underwing** Typical *Apus* underwing. Remiges paler and greyer than above and fairly uniform with the greater coverts. The pattern, however, is closer to Pallid than Common in that the median coverts are only slightly darker than the greater coverts and therefore the extent of the darkest coverts is restricted to the lesser coverts. **Tail** Black-brown and appearing distinctly darker than the tail-coverts.

Juvenile Much as fresh adult, with more marked feather fringing especially the remige tips.

Measurements Wing: male (10) 162-176 (170); female (8) 163-178 (172). Tail: 70-77. Tail-fork: male (8) 21-27 (22.6); female (8) 20-24 (21.5). Delta-length: male (11) 3.0-5.5 (4.3); female (9) 1.5-5.0 (3.3). Weight: (35) 33-50 (42.4) (Fry *et al.* 1988).

GEOGRAPHICAL VARIATION Two races.
A. b. bradfieldi (Angola and Namibia) Described above. Namibian birds average larger. Wing: (30) 167-182 (175) (Fry *et al.* 1988).
A. b. deserticola (This race occupies the most southerly areas of the range in S Africa) This form has a slightly darker body than *bradfieldi* and the general colour tone of the plumage is grey-brown as opposed to brown.

VOICE Not known.

HABITS A colonial species that is very gregarious. Feeding habits generally very typical of the genus, but it has been noted landing on the ground whilst catching bees near hives in Namibia. Bees are eaten with apparently no ill-effect and hives can attract large numbers of the species.

BREEDING Bradfield's Swift is a colonial nester breeding primarily in fissures within granite and basalt rocks but is also thought to nest in dead palm fronds (Ryan and

Rose 1985). A nest described by Dean and Jensen (1974) was constructed from vegetable matter, in the form of twigs, straw, grasses, down, the inflorescences of *Stripagrostris*, and small feathers. The straw was probably taken from the nests of Pale-winged Starlings *Onychognathus nabouroup*. The nest was a flat cup-like structure measuring 100mm long and 105mm wide, with a central depression 87mm in diameter and 24mm deep. The thickness of the nest varied from 35-55mm and the nest was attached to the angle between the floor and wall of a fissure by saliva. Two recorded clutches had two eggs measuring 26.5-27.4 x 16.8-17.0 (Fry *et al.* 1988).

REFERENCES Brooke (1970a), Fry *et al.* (1988), Loutit (1979).

86 PACIFIC SWIFT
Apus pacificus Plate 20

Other names: Large White-rumped Swift; Fork-tailed Swift

IDENTIFICATION 17-18cm. The largest *Apus*. Slightly longer-winged than Common, appearing more rakish, though similar wing shape. Tail-fork deeper, very thickset (outer tail of fork appearing broader than other fork-tailed species), tapering less from rump. When tail is held tightly closed fork cannot easily be seen, but tail appears heavy and parallel-edged. Head larger and more protruding than Common Swift. Body shape as Common. White rump and heavily marked underparts make identification straightforward. Distinguished from all other white-rumped *Apus* by structure; most similar to White-rumped Swift which is perhaps the biggest potential pitfall to an inexperienced observer. White-rumped's slim build, especially its tail (thin and spike-like when closed, deeply forked with thin pointed outer tail when open), is quite different from the powerful Pacific Swift. Examination of the underparts of White-rumped reveals uniform black-blue or black-brown feathers only very narrowly edged paler when plumage fresh (hard to discern in field). A semi-albino individual of any of the larger *Apus* could be troublesome, and close examination of the underparts is needed to clinch a record of a vagrant (figure 56).

Figure 56. Partially albino Common Swift (left) and Pacific Swift (right).

DISTRIBUTION Eastern Palearctic range. Breeds in north from Siberia east to Kamchatka and Japan, and then southwards to C Annam, Vietnam, and to Thailand and Burma. Also on Taiwan, Hainan and Lanyu Islands. A geographically separated population breeds in the outer Himalayas and the hills of Assam. The northern populations migrate south to winter in Malaysia, the Sundas, New Guinea and Australia. Some wintering apparently occurs throughout peninsular India, and there are several recent records on the Seychelles. A lack of records of the northern *pacificus* suggests that the north-south migration occurs in the east of the species' range through China and the E Indies (Cramp 1985). Vagrants recorded from western Aleutians and Pribilofs in North America (5 records, 13 June - 24 September), the North Sea off Norfolk and north Norfolk in England, and New Zealand and the subantarctic Macquarie Island.

Common in breeding range, but little detailed data.

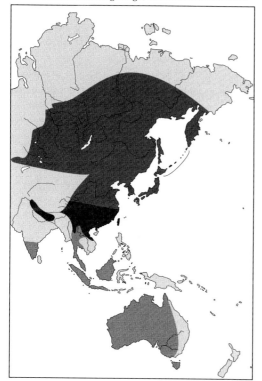

MOVEMENTS The presence of up to three subspecies in parts of SE Asia clouds the timing of movements and location of wintering grounds. *Pacificus* is a long distance migrant breeding in northern east Asia and wintering in Indonesia, Melanesia and Australia. Siberian breeders depart from August to mid September, with the majority returning in May. Common throughout Australia from October-April. Main movements to wintering grounds occur through E Asia, avoiding large mountain ranges, with the movements occurring on a broader front through SE Asia, the Greater and Lesser Sundas and the Philippines. Occurrence throughout the Indian subcontinent is debatable, but specimens have been collected from N Indian states. Movement through the Malay Peninsula occurs from mid September to mid November, and from late February to the end of May. Vast numbers cross the Malacca Straits and migrate along the west coastal plain of the peninsula. Surprisingly few records from Wallacea, Sumatra and Borneo, though a regular passage migrant in Java. Occurs in Papua New Guinea with regularity only in the south, often in large flocks, where it has been recorded from mid October to late December and in mid

February. The movements of *leuconyx* are poorly understood. Appears to wander throughout the Indian subcontinent and is perhaps best regarded as a dispersive resident. *Kanoi* is a short-distance migrant that has been recorded wintering in the Philippines, Indonesia and Malaysia. Mees (1973), however, studied 37 Javan specimens all of which agreed with *pacificus* and he considered the earlier comments of Vaurie (1965) that *kanoi* probably wintered in the S Malay Peninsula, Sumatra and Java were based on supposition. *Cooki* the resident SE Asian subspecies, has been recorded in the Malay Peninsula. In some winters, at least, birds resembling this race occur in large numbers. Deignan (1956) considered that no specimens of *cooki* occurred south of Surat Thani in the N Thai Peninsula and that those birds claimed in the Malay Peninsula as *cooki* where *kanoi*. Some migration occurs at night with 10 individuals, that were not subspecifically identified, captured at Fraser's Hill from 11 October to 3 November (1966-69).

HABITAT As with Common Swift, found through a great range of geographical and climatic zones, both continental and oceanic: from low arctic in the north, south to the tropics. Found at both low and high altitudes. In summer can be found to 3800m in Nepal, but mainly descends to the lowlands in winter. Regularly found around human habitation. Migrants winter mainly in lowlands.

DESCRIPTION *A. p. pacificus*
Adult Sexes similar. **Head** Large, broad, white triangular throat patch, not contrasting sharply with underbody, especially when worn when it appears greyer. Upper head black-brown fringed white, most extensively on lores, forehead and on line over eye, less so on darker crown. Black eye-patch. Lightly fringed black-brown ear-coverts, and side of neck slightly paler than crown. **Body** Heavily marked feathers of underparts black with broad white fringes, giving marked scaly appearance. When worn some pale bases can be seen and underparts appear very mottled. Black saddle very lightly fringed, extending across back. Rump white, extending slightly onto rear flanks and in front of rump, giving U-shape. Breadth of rump averages c. 20mm (Deignan 1956). Uppertail-coverts black, only fringed immediately below rump. **Upperwing** Very dark. Outer primaries, alula, outer greater primary coverts, median primary coverts, median coverts and lesser coverts black, with remainder of wing black-brown. Innerwebs of remiges palest. When worn contrast in wing increases. Secondaries lightly fringed paler. No marked fringing on coverts. **Underwing** Remiges slightly paler, can appear translucent in strong light. Greater primary coverts and greater coverts slightly darker than remiges, though paler than browner median coverts, and blacker lesser coverts. All coverts extensively white fringed. Axillaries uniform with lesser coverts. **Tail** Black, uniform with uppertail-coverts, slightly paler below.

Juvenile Similar to fresh adult, but narrowly tipped white on secondaries and inner primaries.

GEOGRAPHICAL VARIATION Four races.
    A. p. pacificus (North of range south to N China and S Japan. Its occurrence, as a migrant, in all other areas of the species' breeding range makes subspecific field identification problematic) Described above. Little variation occurs within the large range of *pacificus*, although it has been suggested that those in Japan are slightly larger; this comes from only a small sam-

ple of specimens.

A. p. kanoi (Taiwan west to SE Tibet) Blacker than *pacificus*, especially underparts and upperparts, the crown, with smaller throat and rump patches. Breadth of rump averages c.15mm (Deignan 1956). Throat is greyer than *pacificus* and more heavily streaked (as is rump); the scaling on the underparts is sometimes less obvious. Less deeply forked tail. Wing measurements intermediate between the Japanese and Russian populations of *pacificus*. Wing: male (5) 173-179 (176) (Lack 1956a). Deignan (1956) gives 170-186. Weight: (2) 50 + 56 (Cramp 1985).

A. p. leuconyx (Outer Himalayas and Assam Hills) Similar in plumage to *kanoi* but is smaller. Wing: 147-160 (Baker 1927). This subspecies is thought to account for records from peninsular India.

A. p. cooki (SE Asia south of *kanoi*) The rump is narrowest and the body fringing most restricted. Rump and throat streaking most pronounced in *cooki*. Breadth of rump averages c.10mm (Deignan 1956). Wing: 163-172 (Baker 1927).

MEASUREMENTS Breeding and wintering grounds. Wing: male (23) 176-186 (180); female (13) 173-182 (177). Tail: male (26) 75-83 (79.4); female (16) 76-88 (79.1). Tail-fork: male (26) 30-38 (33.3); female (16) 29-40 (32.7) (Cramp 1985). Weight (Mongolia): male (7) 38-54 (48.1); female (2) 38-47 (42.5).

VOICE Typical scream call less wheezy and softer than Common Swift, *sreee*.

HABITS Gregarious. Tends to occupy higher airspace than many sympatric swifts, especially in the winter when is often only seen at low levels in periods of poor weather. On migration often seen with needletails.

BREEDING Pacific Swift is a colonial species that makes a half-cup of grass and other vegetable matter stuck together with saliva against the sloping face of a cliff fissure. In Nepal it has been recorded nesting in a Nepal House Martin *Delichon nipalensis* nest. Two or three eggs are typical in the clutch and these average (11) 22.7 x 15.0 (Ali and Ripley 1970).

REFERENCES Ali and Ripley (1970), Chantler (1993), Coates (1985), Cramp (1985), Deignan (1956), Dementiev and Gladkov (1951), Inskipp and Inskipp (1985), Lack (1956a), Medway and Wells (1976), Mees (197?), Parker (1990), Roberson (1980), Smythies (1981), Vaurie (1965), White and Bruce (1986).

---

87 DARK-RUMPED SWIFT
Apus acuticauda   Plate 20

Other names: Khasi Hills Swift; Dark-backed Swift

IDENTIFICATION Length 17cm. Within range this is the only regularly occurring *Apus* species with a dark rump. Closely related to Pacific Swift and distinguished from that species on plumage mainly by the dark rump. Darker on upper head than Pacific. Underparts very similar with broad white tips and a broad but indistinct pale throat; the throat patch may average more heavily streaked. Smaller than Pacific Swift and differing slightly in tail shape. The outer rectrix of Dark-rumped is very sharply pointed producing a finer outertail than Pacific which has the

heaviest of all *Apus* outertails. Identification from Pacific based purely on a view from below is inadvisable. House Swift is easily distinguished by virtue of its tail shape. In the event of confusion with more westerly *Apus* swifts that could occasionally appear within this species' range the underpart pattern and the very black upperparts should be noted. All other sympatric species can be excluded by a combination of the very black upperparts, the underpart pattern and the tail shape.

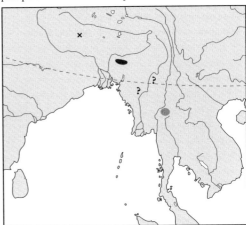

DISTRIBUTION Poorly understood S Asian range. Known to breed with certainty only in the Khasi Hills near Cherrapunji in Meghalaya, NE India (25°N 91°E). The type specimen, however, was described from Nepal in the last century (where it has not been recorded since), but the specimen may have originated from India. Ali and Ripley considered that it also breeds (or at least once bred) in Mizoram. It is quite possible that its breeding range was formerly more extensive. The population is apparently migratory, being absent from the breeding range outside the breeding season. It is interesting to note that unidentified swifts were observed in Myitkyina district, Burma by Stanford and were thought to be either Common or Dark-rumped Swifts. Outside the breeding season there are winter records from Thailand. A single nineteenth century record on the Andaman Islands was a misidentified specimen of *Apus apus*.

Believed to be one of the world's most threatened swifts due to its apparently small range and population. The lack of research into the species obviously clouds the picture. Listed by BirdLife International as Vulnerable (Collar *et al.* 1994).

MOVEMENTS Apparently some short-distance movements to NW Thailand in winter, but there are too few records to establish its precise status and distribution outside the breeding season.

HABITAT Occurs around the cliffs and deep gorges of the Khasi Hills above the Sylhet plains.

DESCRIPTION *A. acuticauda*
Adult Sexes similar. **Head** Indistinct pale grey throat patch. Feathers very pale grey with darker shaft streaking. Upper head uniform, not paler on the lores or forehead, deeply black with a greenish gloss. Ear-coverts dark black-grey becoming paler and a little mottled towards throat. Black eye-patch. **Body** Mantle uniform with head and nape, or perhaps even blacker as a result of a darker, more

purple gloss. Rump and uppertail-coverts similarly uniform though, like the head, more green rather purple-glossed. Feather bases browner and therefore some lightening and mottling to plumage when worn. Below the palest area of the throat (which extends to the lower throat) the feather pattern changes a little as the upper breast feathers are brown-based with broad white fringes. The lower breast/upper belly sees a marked change in the pattern to deep glossy-black bases with broad white fringes. This pattern covers the remainder of the underparts apart from the black undertail-coverts. **Upperwing** Black outer primaries, alula, outer greater primary coverts, median primary coverts, median coverts and lesser coverts. The remainder of the wing is black-brown. The innerwebs of the rectrices are paler and browner than the black outerwebs. The contrast between the inner and outerwing increases when worn. Some lighter fringing can be seen on the secondaries when fresh, though not markedly on the coverts. **Underwing** The remiges are paler and greyer than above, uniform or slightly paler than the greater coverts and markedly paler than the blackish median and lesser coverts. The outer primaries are the darkest remiges. The coverts are all broadly white-fringed. The leading edge coverts are blackish with pale grey fringes. **Tail** Black as the remiges and uniform with the uppertail-coverts, whilst below it is paler and greyer.

Juvenile This plumage has not been studied but is likely to closely resemble the fresh adult, and in addition should have some narrow white tips to the secondaries and inner primaries (compare Pacific Swift).

Measurements Wing: 167-174 (one 177). Tail: 70-74. Tail-fork: 21-26 (Baker 1927).

GEOGRAPHICAL VARIATION None. Monotypic.

VOICE Presumably similar to Pacific Swift.

HABITS General behaviour presumably similar to other cliff-dwelling species.

BREEDING Dark-rumped Swift breeds colonially on ledges in cliff fissures. The nest is a shallow cup composed mainly of grass and feathers bound together with saliva. Clutches are usually of 2 or 3 eggs (4 has been recorded) and these measure, on average, (50) 26.0 x 16.3 (Ali and Ripley 1970).

REFERENCES Ali and Ripley (1970), Baker (1927), Collar *et al.* (1994), Inskipp and Inskipp (1991), Lack (1956a), Smythies (1953).

---

88 LITTLE SWIFT
Apus affinis   Plate 23

Other name: House Swift

IDENTIFICATION Length 12cm. Throughout much of its range Little Swift can be recognised with ease by virtue of its distinctive tail shape. The tail is rather square-ended when closed and rounded when open although it sometimes seems rather uneven and often shows a slight cleft. Never shows the relatively thin-tailed and deeply-forked shape shown by all other *Apus* species except for the related House and Horus Swifts. In these species, when closed, the tail can appear broader than the body, unlike any other species. Furthermore Little, House and Horus

are rather different from other *Apus* appearing rather compact and broad-bodied with quite blunt-ended, relatively short wings. Less graceful flight than other *Apus*, being rather weak with a great deal of gliding. Within the Western Palearctic the white rump is shared only by White-rumped Swift and the vagrant Pacific Swift. White-rumped is blacker than the pale Western Palearctic *galilejensis* which further differs, like all Little Swifts, in its broader rump patch which extends considerably onto the rear flanks. Pacific is clearly a large *Apus* with a very different jizz and a long, strongly forked tail (which when closed appears broad and pointed) and long wings. Pacific has blacker wings, narrower rump patch, broader throat patch and heavily marked underparts. In sub-Saharan Africa Horus shares very similar plumage characters to Little but has a very different tail shape, being longer and somewhat forked. Also in tropical Africa several spinetails have white rumps and squarish tails, and in particular Mottled Spinetail can be troublesome. Little Swift is best separated by its different jizz and wing structure. All spinetails have rather short secondaries and bulging inner primaries (beware moulting Little Swift), with generally more protruding heads and usually quite different flight action. Furthermore all spinetails show slight plumage differences (see relevant species descriptions). In the Indian subcontinent nominate *affinis* comes close to, but shows no sympatry with, the most widespread race of House Swift *nipalensis*, which occurs eastwards from the Himalayas through Assam and Bangladesh to the China Sea and south to the Malay Peninsula. House Swift appears very similar to Little in both plumage and structure. It differs in all forms in having a slight tail-fork (5-8mm) and a longer tail generally. The plumage is also generally blacker above and below. This is particularly noticeable in *subfurcatus*, which shows a blue-black crown and brown forehead, blue-black mantle and uppertail-coverts, with a slightly narrower rump band. *Nipalensis*, which comes closest to *affinis*, is darker than that form especially on the upper head (not paling to grey even on the forehead) and black-brown tail-coverts. Some observers have stated that House Swifts are more barred on the underparts but this feature is not obvious even in the hand.

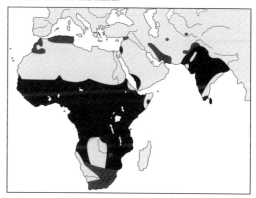

DISTRIBUTION Extensive sub-Saharan African and southern Asian range with more scattered populations locally in the southern Western Palearctic. The Western Palearctic range has shown recent signs of expansion with the first Algerian record being in 1924, expansion from only two breeding sites in Morocco at the end of the 19th century to many in the present day and the first Turkish breeding in 1970 (although Kirwan and Martins pers. comm. cite reliable earlier records from as far back as 1881). In Tunisia the population is apparently going through a period of contraction after early 20th century expansion. The Western Palearctic range is rather scattered on the Banc d'Arguin, in NW Africa in Morocco along the Atlantic coast as far south as the 30th parallel (Oued Massa), and inland in the High Atlas and Middle Atlas. Occurs in highly scattered and isolated populations in Algeria and Tunisia. In the E Mediterranean found in scattered populations in the extreme SE and E of Turkey with the largest colonies at Birecik, Halfeti and Kilis (Kirwan and Martins pers. comm.), northern and south-central Israel and western Jordan. Found in all areas south of the Sahara, with the exception of the most arid regions: fareastern Ethiopia and the Somali Peninsula, SE Angola, SW Zambia, W, C and S Tanzania, most of Botswana (except the eastern and southern borders), and most of Namibia east and west of the central highlands. Elsewhere ubiquitous. Occurs in south-west Arabia in the mountains on the edge of the Red Sea in Asir, Saudi Arabia, and throughout Yemen (except parts of the far east). Eastwards, continuous through Iran, Afghanistan, Pakistan and into India south of the Himalayas from the Punjab to W Bengal and Sri Lanka.

Abundant in its optimum habitat of tropical African and Indian towns and cities. Throughout sub-Saharan African range considered common although rather localised away from urban centres. Throughout Africa the population has increased enormously during the 20th century as a result of adaptation to man-made habitats. In the Western Palearctic it is locally common to scarce.

MOVEMENTS Throughout tropical range the species is a resident, but in the extreme north and south of the range either wholly or partially migratory. In the Western Palearctic the small Turkish population is the only one that appears to be entirely migratory with the birds being present from March-September. Elsewhere the populations appear to be only partial migrants with absence reported from some Middle Eastern breeding sites in mid winter and reduced numbers on other sites. The Turkish population's wintering grounds are unknown. In N Africa numbers are reduced in the winter; migrant *galilejensis* has been recorded from S Saharan countries and an October passage through Chad has been noted. In Asia the species is a summer visitor to Tadzhikistan, Uzbekistan and Turkmenistan, but the wintering grounds are unknown. In the Indian subcontinent the populations in the north leave during the winter months. The wintering grounds are unknown but it is unlikely that the Pakistan population, at least, winters in the peninsula as *galilejensis* has not been recorded there. It is possible that the population migrates westwards through Arabia into Africa, as it has been noted that there are a great many records in Arabia away from known breeding sites. In South Africa numbers apparently reduced during June-July and the population of South Africa as far north as the Transvaal and into southern Namibia, Botswana and Mozambique is thought to be partially migratory. The central plateau population of Zimbabwe is also reduced during the winter months. In Freetown, Liberia only small numbers remain in January-March. Despite these suggestions of migration in the sub-Saharan regions Brooke (1993 pers. comm.) is not convinced that these populations are even partially migratory.

HABITAT Found in lower middle, subtropical and equatorial latitudes. Generally avoids the most arid areas and although *galilejensis* breeds sparsely in areas such as the Dead Sea this is clearly not optimum habitat. Throughout much of range breeds exclusively in man-made sites and seldom seen far from them. A familiar sight and sound in many African and Indian cities. Natural sites such as gorges, around rocky crags in mountains and hills and cliff faces are still used in some areas. Travels up to 15-20km from breeding sites in daily pursuit of food if conditions dictate, and therefore found over a variety of habitats, often at great height.

DESCRIPTION *A. a. galilejensis*
Adult Sexes similar. **Head** Large white rounded throat patch, typically extending to gape in width and beyond eye onto lower throat in length. Forehead, lores and line over eye pale grey (lores often slightly darker), darkening to grey-brown on crown and black-brown on nape. Black eye-patch. Ear-coverts grey-brown darkening on nape and sides of neck. Worn individuals show indistinct mottled grey throat patch. **Body** Underparts from breast to vent deeply black, slightly paler on edge of throat, some greywhite fringes when fresh, paler grey bases can be seen when worn. Undertail-coverts pale grey, especially outer coverts. Glossy black-blue saddle extends onto back, less black just above rump. Saddle feathers deeply black at tip with paler grey bases only visible when worn. Broad white rump band extends onto rear flanks and ensures that some white can be seen at almost any angle. Uppertail-coverts grey-brown, paler than saddle, with outerwebs of outer coverts palest. **Upperwing** Primaries and secondaries dark grey-brown, with blacker outerwebs, and slightly darker outer primaries (less contrast than larger *Apus*). Secondaries indistinctly paler-tipped. Greater primary coverts and greater coverts uniform with or slightly darker than remiges, and pale-fringed. Other coverts becoming progressively darker towards leading edge. Leading edge coverts broadly fringed white. **Underwing** Remiges paler and greyer than upperwing, appearing translucent in strong light. Greater primary and greater coverts slightly darker or uniform with remiges and white-fringed (also slightly darker subterminal marks), with other coverts becoming darker towards leading edge. Axillaries blue-black. **Tail** Pale grey, slightly darker at centre, and when spread appearing translucent, with outer tail contrasting with centre and body.

Juvenile Plumage less glossy than adult, lightly fringed like fresh adult, but more extensively on wing-coverts.

Measurements NW Africa. Wing: male 133-140 (137); female 134-141 (136). Tail: male 40-43 (41.4); female 39-43 (40.9). Levant. Wing: male 137-138 (137); female 132-137 (135). Tail: male 40-44 (42); female 39-44 (41.8). Weights, based on a variety of 64 specimens (not only *galilejensis*): 18-30 (25) (Cramp 1985).

GEOGRAPHICAL VARIATION Six races. The four races of House Swift are sometimes included in this species.

A. a. galilejensis (Occurs in a broad band through N Africa westwards through the Middle East into Pakistan, and south of the Sahara in E Sudan, Ethiopia and NW Somalia) Described above. This is the palest race, and is uniform in colour throughout the range. Varies in size with populations south of the Sahara, S Arabia, SE Iran, Baluchistan and Pakistan being notably smaller with an average wing length of 129 as

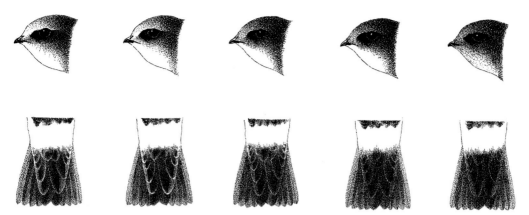

Figure 57. Heads and tails of Little Swift races. Left to right: *galilejensis, theresae, affinis, aerobates* and *bannermani*.

opposed to 136-137. These small populations within *galilejensis*, have sometimes been separated as *abessynicus*.

A. a. affinis (India, east of Pakistan and south of Himalayas, and E Africa from S Somalia to N Mozambique. Also the Pemba and Zanzibar Islands) Generally slightly blacker in plumage than *galilejensis*, less contrast between the body and the head, darker forehead, darker tail-coverts, slightly narrower rump and throat patch. E Africa: wing (10) 126.5-131.5 (128) (Fry *et al.* 1988). N India: wing (6) 130-135 (132.5). S India. Wing: (18) 123-132 (127.5). Tail throughout India: 37-44 (Abdulali 1966). Weight E Africa: male (18) 18-28 (24.75); female (2) 25 + 25 (Fry *et al.* 1988).

A. a. aerobates (Extensive African range from Mauritania through W Africa, eastwards to Somalia and south to Transkei, S Africa) Darker even than *affinis*, especially on wings and tail. Population on the coast between the Limpopo and Transkei still darker and also larger and has been considered a separate race *gyratus*. Zaïre. Wing: (10) 129-138 (133) (Clancey 1980). Angola. Wing: (26) 118-134 (126) (Brooke 1971b). Mali to Cameroon. Wing: (14) 124-135 (129) (de Naurois 1972). Weight: (19) average 23.3 (Fry *et al.* 1988).

A. a. theresae (S Africa from W and S Angola, to S Zambia, and southwards through S Africa) Similar to *galilejensis* and similarly paler than *aerobates*. Differs from *galilejensis* in darker undertail-coverts. Populations south of the Cunene River larger than those to the north. South. Wing: male (26) 123-139 (133); female (17) 128-140 (135). North. Wing: male (7) 119-131 (128); female (14) 123-135 (129). Weight (Angola): male (10) 21-30 (24.5); female (16) 19-30 (25) (Fry *et al.* 1988).

A. a. bannermani (Islands in Gulf of Guinea: Bioko (Fernando Po), Príncipe, São Tomé) The darkest race. Also streaked on the throat. Wing: male (19) 128-139 (133.9); female (15) 126-138 (133.0). Weight: (5) 23-25 (23.90) (Brooke 1971b).

A. a. singalensis (S India and Sri Lanka) Blacker than the more widespread Indian subspecies, *affinis*, especially the head and uppertail-coverts. Can show slight tail-fork. Wing: (6) 127-130 (129.3). Tail: 43-45. Weight: (2) 23 + 24 (Abdulali 1966).

VOICE Very vocal. Most often heard call is a harsh, rapid,

rippling trill *der-der-der-dit-derdiddidoo*. A great variety of thin screams or shrieks may be heard especially around the nest site.

HABITS Extremely gregarious and can be seen in flocks of several hundred in areas of abundance. Often joins groups of other *Apus* swifts. Less frequently seen feeding at very low level. In winter feeds at altitude being seen only for a few minutes around colony after waking and for slightly longer before going to roost. As Brooke (pers. comm.) points out this may explain the supposed winter decline in abundance in some sub-Saharan areas.

BREEDING Primarily a colonial nester although solitary nesting is not unknown. Nesting most often occurs on buildings wherever suitable eaves or overhangs provide an angle between wall and roof. Cliff sites are also used and abandoned swallow *Hirundo* spp. nests may be used. On occasions swallows may even be evicted from their nests. The nests are sturdy, but externally rather scruffy, hemispherical bags. They are composed largely of grass, small twigs and vegetable down with a variety of feathers. The interior is notably more tidy and smoother than the exterior. Nests are built together in clumps, with a great degree of overlap, and have 1-3 entrances which may rarely be communal (to more than one nest). Clutches of 1-3 eggs vary with range. Eggs average (96) 22.6 x 14.6 (Fry *et al.* 1988) (figure 58).

Figure 58. Little Swift nests.

211

REFERENCES Abdulali (1966), Brooke (1971b), Clancey (1980), Cramp (1985), Chantler (1993), Fry *et al.* (1988), de Naurois (1972).

89   HOUSE SWIFT
Apus nipalensis   Plate 23

Other name: Little Swift

IDENTIFICATION Length 15cm. Within its range the combination of white rump and shallowly forked tail are diagnostic. The much larger, longer-winged, longer-tailed and more powerful Pacific Swift has a very different jizz, heavily forked tail, which even when closed, appears long and pointed (and unforked for short periods only), is quite different in shape to the rather short broad tail of House Swift. The two small white-rumped spinetails within its range, Silver-rumped and White-rumped, have square tails and unmistakable spinetail jizz, most notably in the typical butterknife wing shape, cigar-shaped bodies with more protruding heads and rapid shimmering flight. Several swiftlet species are sympatric and have pale rumps, though none are as striking as that of House Swift. Their plumages, in particular the underparts, are paler and more grey-brown. Furthermore the swiftlets are far weaker in flight than *Apus* species and typically show a curious rocking motion on down-tilted wings.

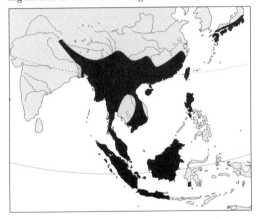

DISTRIBUTION Range mainly east of Little Swift, predominantly SE Asian. In the Himalayas, from Nepal eastwards into Assam and southwards into Bangladesh, Mizoram, Manipur and Nagaland. Eastwards as far as SE China in S Yunnan, Guangxi, Guangdong, Fujian and Hainan, and S Japan below 36° N. South of these northern points the species is found throughout SE Asia in Burma, Thailand, Laos, Vietnam, Cambodia and Malaysia. Widespread on island groups of region, through Sumatra (including Belitung, Riau and Lingga groups), Java, Bali, Borneo (including Anambas, N Natunas and Tambelan), S Sulawesi and the N Philippines on Camiguin Norte, Luzon, Mindoro and Negros.

Abundant in suitable habitat throughout range.

MOVEMENTS The northern populations of *nipalensis* are believed to be migratory, although little work has been undertaken into this subject. Winter specimens of *nipalensis* have been collected from Luzon, Philippines and the Indian subcontinent. Elsewhere, a resident species including most of the range of *nipalensis*.

HABITAT Similar to Little Swift, although slightly less colonial. Like that species occurs most commonly in an urban setting. Away from human habitation is encountered far less frequently, even a short distance from a town. However, can be observed over a great variety of habitats in the course of diurnal food sorties and often at great height, particularly over forested areas. Found from the lowlands up to 2000m.

DESCRIPTION *A. n. subfurcatus*
Adult Sexes similar. **Head** Throat patch white and rounded reaching to lower throat, extending to the gape and base of the ear-coverts. Forehead, lores and line over eye brownish contrasting slightly with black crown. Black eye-patch. Ear-coverts black-brown darkening on nape and sides of neck. In fresh plumage some narrow pale grey fringing on the head and the throat patch can appear grey and mottled. **Body** Underparts from below throat patch to vent deeply black, with the border of the throat patch usually appearing clear-cut. In fresh plumage the underparts are narrowly fringed grey-white. These fringes abrade but some paler bases can be seen. The undertail-coverts are uniform with the rest of the underparts. Nape black uniform with the crown, mantle and back. Some very narrow fringes can be seen when fresh though they are less apparent than on the underparts. The white rump band

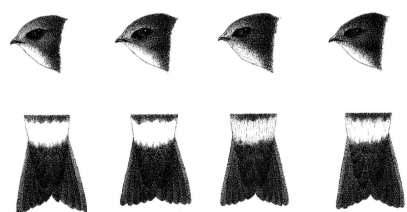

Figure 59. Heads and tails of House Swift races. Left to right: *furcatus*, *nipalensis*, *kuntzi*, and **subfurcatus**.

contrasts strongly with the surrounding tracts and extends slightly onto the rear flanks to create a rather U-shaped patch. This patch often has visible dark shaft-streaking. The uppertail-coverts are black, uniform with the saddle. **Upperwing** Primaries and secondaries black-brown, with blacker outerwebs, and slightly darker outer primaries (less contrast than larger *Apus*). The secondaries are paler-tipped, sometimes notably visible in the field. Greater primary coverts and greater coverts are uniform with remiges or slightly darker. The median and lesser coverts appear blacker being uniform with the saddle. The coverts are at most very indistinctly paler-fringed. Leading edge coverts are black-brown with pale grey fringing. **Underwing** The remiges appear paler and greyer than above and uniform with the greater coverts. The median coverts and lesser coverts appear black, fairly uniform with the underbody and creating great contrast in the wing. The outer primaries usually appear clearly darker than the other remiges, an effect at odds with the upperwing. The coverts, particularly the greater and median, often have white fringes and the appearance of darker subterminal bands. **Tail** Similarly black-brown as the remiges though when spread the outer rectrices can appear paler than the centre of the tail. The undertail is greyer and less black than above.

Juvenile Similar to the fresh adult but with a greater tendency to show fringing on the remiges.

Measurements Wing: 130-140. Tail: 47-55. Tail-fork: 5-8. Weight (3 N Borneo): 22-25 (Brooke 1971b).

GEOGRAPHICAL VARIATION Ali and Ripley (1970) state that *nipalensis* does not show a forked tail and does not occur to the east of the Kamrup district of western Assam. This is at odds with Brooke (1971b) whose studies indicate that *nipalensis* does show tail furcation as in *subfurcatus* and that there are clinal increases in tail and wing length and in the darkness of the underside of the rectrices eastwards to the China Sea. Four races.

A. n. subfurcatus (Malay Peninsula south from the Straits of Tenasserim, Borneo, Sumatra, Anambas, Biliton, N Natunas and Rhio) Described above.

A. n. nipalensis (Nepal eastwards to China in Fujian province, southwards through Assam, Burma, Thailand, Laos, Vietnam and Cambodia) Not as deeply blue-black as *subfurcatus*, especially the crown. Wing central Nepal: (22) 128-138 (132.8). Wing (coastal China): (4) 134-142. Tail (C Nepal): (22) 42-49. Tail (coastal China): (4) 48-55 (Abdulali 1966).

A. n. furcatus (Java and Bali) Similar in plumage to *nipalensis*, but underparts a little browner. Deepest tail-fork of any race. The two outermost rectrices are staggered in length, unlike *subfurcatus* in which they are roughly equal in length. Wing: male (9) 135-144 (139.2); female (6) 134-145 (138.0). Tail: (4) 55-59 (57.3). Tail-fork: (16) 8.0-13.5 (11.0) (Brooke 1971b).

A. n. kuntzi (Taiwan) Somewhat intermediate in darkness between *subfurcatus* and *nipalensis*, with a rather grey-white rump that shows a greater tendency towards dark shaft-streaking than the other races. Wing: (51) 124-141 (133). Weight: (5) 20-32 (21.10); female (9) 22-35 (26.13) (Brooke 1971b).

VOICE Very similar to Little Swift a harsh, urgent, rippling trill *der-der-der-dit-derdiddidoo*. Also similarly vocal.

HABITS Gregarious, feeding readily with other swift species, most frequently Pacific Swift, the main sympatric *Apus*

species. However, will loosely associate with other genera and hirundines. Like Little Swift rarely feeds at a genuinely low level.

BREEDING House Swift has very similar nesting behaviour to Little Swift. It is similarly colonial with perhaps, on average, even larger clusters. Nest construction is similar, but some man-made materials such as rags may be incorporated. Clutch size is normally 2 or 3 and averages (50) 22.7 x 14.9 in *nipalensis* (Smythies 1981, Ali and Ripley 1970).

REFERENCES Ali and Ripley (1970), Abdulali (1966), Chantler (1993), Brooke (1971b), Clancey (1980), Cramp (1985), Smythies (1981).

---

90   HORUS SWIFT
Apus horus   Plate 23

Other names: Loanda Swift (race *fuscobrunneus*); Toulson's Swift (morph '*toulsoni*')

IDENTIFICATION Length 15cm. Small fork-tailed *Apus* with rather black plumage broken by prominent white rump and throat patches. The normal phase can be distinguished from all sympatric species, with the exception of White-rumped, by the combination of white rump and forked tail. The very similarly plumaged Little Swift never shows such a deeply forked or long tail as Horus. However, it must be remembered that when banking the effect of a tail-fork will be reduced or even eradicated and indeed Horus appears closer in general jizz to Little than White-rumped. Flight, like Little Swift, is very fluttery lacking the almost hirundine-like grace of White-rumped Swift. Best told from White-rumped by slightly larger size, chunkier, long-winged build, and by its tail shape; the tail of Horus is not as deeply forked as White-rumped and the outer rectrices are not as streamer-like as that species. When closed the tail of Horus is wider and slightly shorter than the sharp spike of White-rumped. Plumage features are much the same as Little and White-rumped, but Horus has a deep rectangular rump patch that wraps around the body, a paler forehead and lacks the white trailing edge to the wing so often prominent in White-rumped Swift. Additionally, the underwing-coverts of Horus are paler than in White-rumped, though this is hard to see in the field. The race *fuscobrunneus* and morph *toulsoni* have been considered by some authors as a separate species, Loanda Swift, and can be recognised by their dark brown rumps and restricted white throat patches with obvious dark streaks. *Toulsoni* is dark-bodied as *horus*, but *fuscobrunneus* has a generally paler body, especially the head and undertail-coverts (figure 60).

Figure 60. Horus Swifts of *toulsoni* morph (right) and individual intermediate between nominate and *fuscobrunneus* (left).

These forms, however, are identical in shape to the nominate race and retain its jizz, easing separation from the larger, uniform *Apus* swifts. Furthermore, they are often accompanied by normal individuals.

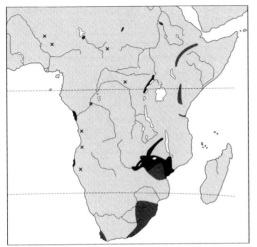

DISTRIBUTION Endemic to sub-Saharan Africa, with a mainly E and S African range. Scattered populations further west. Identification problems confuse the picture and away from the most continuous parts of the range evidence of breeding is often hard to come by. Possibly far more widespread than currently thought. Occurs in the C Ethiopian highlands and C Kenya (south to just across Tanzanian border) although the two populations are not continuous as the species does not occur in N Kenya. NW Uganda to Burundi, from the Rusizi River to Gihanga, and into E Zaïre (Kasenyi, Lake Kivu, Kitendwe and Lukolela). In S Africa in S Zambia (Zambezi basin to Balovale, Barotse province and the middle Zambezi and Luangwa valleys), Zimbabwe, N Botswana, C Mozambique and the Republic of South Africa, from SW Cape in a narrow strip along the southern coast and more extensively in the east from Port Elizabeth north to the Transvaal (including Lesotho, the Orange Free State and Swaziland). Away from these more continuous ranges it is found in: Nigeria, on the Mambilla Plateau; Cameroon on the Adamawa Plateau; W Sudan at Darfur; E Sudan at Sennar and Wad Medani; and on the River Chari between N Cameroon and W Chad. In C and S Africa there are records from N and W Zaïre and W Angola, especially from the northwest, Cabinda and the south-west. Possible vagrant to Agalega, Seychelles in the Indian Ocean.

Locally common throughout much of the range, though locally abundant in the C Ethiopian highlands.

MOVEMENTS Mainly resident, with some intra-African migration in S Africa, and possibly E Africa. Essentially a breeding visitor to the highveld of South Africa and Zimbabwe (possibly resident in the South African lowveld and definitely resident in the Zimbabwean lowveld), breeding from October to May. These populations move in winter, possibly only to the Zimbabwean lowveld where it is also a winter breeder. Dispersive throughout E African range being present at 3000m at Mau Narok, Kenya from late April-early July. Such movements are thought to be only local, although there are few East African records away from the empty colonies. The situation is similar on

the Zimbabwean highveld where the species is present from October to April (sometimes May), when it is believed to move to lower altitudes.

HABITAT Habitat linked to that of bee-eaters, kingfishers and martins as Horus normally utilises disused nest sites of these species. Therefore often found, in the vicinity of sandy banks, both natural and man-made. Forages over most adjacent habitats, but seldom over sizeable human settlements.

DESCRIPTION *A. h. horus*
Adult Sexes similar. **Head** Large white, highly contrasting, rounded throat patch, extending to gape in width and onto lower throat in length. When worn darker feather bases can be seen giving the throat a slightly mottled appearance. Forehead, lores and line over eye pale grey-brown, darkening to grey-brown on crown and bluish black-brown on nape. Black eye-patch. Grey-brown ear-coverts uniform with nape/rear crown. **Body** Deep black underparts from below throat patch to vent with under-tail-coverts a little greyer. When plumage fresh some narrow white fringes can be seen. Deep blue-black mantle and back uniform with nape. Feathers broadly and deeply blue-black tipped with paler grey bases. Broad white rump band highly contrasting against the very black surrounding tracts. The rump band extends onto rear flanks, giving wrapped-around impression like Little Swift, and similarly some white can be seen at almost any angle. Uppertail-coverts brown, very slightly paler than saddle. **Upperwing** Primaries and secondaries dark grey-brown, with blacker, glossy outerwebs, and slightly darker outer primaries (little contrast as with White-rumped and Little Swifts). Secondaries lightly paler-tipped, not easily viewed in the field. Greater primary and greater coverts uniform with or slightly darker than remiges, and only very indistinctly fringed. Other coverts darker, becoming progressively darker to leading edge. Leading edge coverts broadly fringed white. **Underwing** Remiges paler grey than upperwing, and can appear translucent in strong light. Greater primary coverts and greater coverts fairly uniform with remiges and broadly white-fringed (also slightly darker subterminal marks), with other coverts becoming darker to leading edge though still broadly fringed. **Tail** Black-brown as remiges, paler and greyer below.

Juvenile Plumage similar to fresh adult, but with more pronounced body-feather fringing.

Measurements E Africa. Wing: male (32) 143-161 (153); female (18) 147-158 (152). Tail: (19) 49-59 (55.4). Delta length (31) 3-6 (4.55). Tail-fork: 9-13. S Africa. Wing: (12) 153-161 (157). Tail: (12) 55-61.5 (57.2). Weight. S Africa: (214) 17-31.3 (26.2). E Africa: male (5) 23-28 (25.5); female (6) 24-30 (26.3) (Fry *et al.* 1988).

GEOGRAPHICAL VARIATION Due to the extreme abrasion that this species suffers as a result of its tunnelling habits, museum specimens can show considerable tonal differences. It is therefore considered safest not to subdivide the nominate-type populations (i.e. those with white rumps) and to consider *toulsoni* as a morph and *fuscobrunneus* as a subspecies. This follows Brooke (pers. comm.) and Fry *et al.* (1988). It has been noted that in Zimbabwe, at Esigodini, intermediates between *toulsoni* and 'normal' birds can be observed. Prigogine (1985), however, considers that these two dark types are a paraspecies within a *horus* superspecies.

A. h horus (range of species except SW Angola) Described above.

A. h fuscobrunneus (SW Angola) Greyish throat patch smaller than in nominate race and with dark shaft streaks. Rump dark brown. Head and undertail-coverts paler than nominate form. Lacks strong gloss to plumage. Wing: (10) 149-157 (153.6). Tail: (10) 49-56 (53.0). Tail-fork: (10) 12-18 (14.95). Delta-length: (10) 3-5 (3.9). Weight: (10) (Brooke 1971a).

Toulsoni morph (NW Angola and Zimbabwe) Darker than *fuscobrunneus*, but similarly has smaller throat patch and a darker rump than the nominate race. Wing: (5) 149-151 (149.8). Delta-length: (5) 3-5 (3.9). Weight: (3) 25-28 (26.7) (Brooke 1971a).

VOICE Scream call is lower-pitched than either Little or White-rumped Swift and is rather buzzing in tone *preeeooo preeeooo* and, whilst breeding, calls more frequently than either of these species. Generally rather quiet outside the breeding season.

HABITS Typically gregarious, though usually seen in small groups, up to 30 having been recorded, often with other swifts and hirundines. Feeding believed to occur mainly between foraging altitudes of African Palm Swift (lower) and White-rumped Swift (higher), but this is very dependent on feeding conditions as White-rumped can often be observed undertaking very low-level feeding.

BREEDING Horus Swift is a solitary breeder, although up to 12 pairs may be present within a colony. Uses deserted nests of burrow-nesting species and burrows tend to be situated in the banks or rivers. There is also one record of nesting in a bank on a beach. Although the occupancy of these burrows may be disputed eviction is not known to occur. Hopcraft (1974) recorded a 45cm burrow with a rather slit-like entrance that was 10cm wide and 6cm high terminating in a disc-shaped cavern 30cm by 45cm with a height of 6cm. The nest is sited at the end of the burrow and is a thin hard platform with a diameter of 10-14cm and constructed of vegetable matter including grass, down, maize leaves and moss, bound together with hair, feathers and even fragments of rag by saliva. Clutches consist of 1-4 eggs, but usually 2 in Zimbabwe and 3 in South Africa (Fry *et al.* 1988).

REFERENCES Brooke (1971a), Brooke and Steyn (1979), Fry *et al.* (1988), Hopcraft (1974), Prigogine (1985).

---

91   WHITE-RUMPED SWIFT
Apus caffer   **Plate 23**

IDENTIFICATION Length 14cm. The very black plumage and contrasting white rump and throat patch exclude all sympatric species except Little and Horus Swifts. The jizz is very distinctive due to the slim body, and long tail and wings. The tail, which is very deeply forked when opened, is characteristically held closed over long periods when it appears as a needle-thin spike; it is remarkable amongst the *Apus* by its very thin outer rectrices caused by a heavy emargination on the innerweb (in the juvenile the outer rectrix is more gently rounded, the tail slightly shorter and the fork slightly shallower). The wings appear long, rather narrow and very pointed. The very slim body tapers gently into the tail without the 'lumpiness' of the larger *Apus*. This structure coupled with a graceful flight

action, which can be rapid and rather fluttering, contribute to an elegant appearance. Horus can be distinguished by its larger more bulky appearance, its broader body, longer wings and proportionally shorter, less deeply forked tail; its tail shows a broader less pointed outer rectrix and is less frequently held tightly closed. Little Swift is excluded on structure mainly as a result of its diagnostic tail shape which is square-ended when closed and rounded when spread, never showing more than a light cleft in the tail. Its tail always appears broad, often wider than the adjoining body, quite unlike White-rumped Swift. Further structural differences are in Little's particularly chunky appearance with its rounded body and rather blunt-tipped wings. The plumage of Little and Horus Swifts are particularly similar, with both differing from White-rumped in their broader white rumps which extend round onto the rear flanks and are less U-shaped. They also lack a noticeable white trailing edge to the secondaries and inner primaries (although this can be lacking in a worn White-rumped) and have comparatively paler heads which contrast more markedly with the saddle than in White-rumped. In the Western Palearctic the race of Little Swift, *galilejensis* further differs through its noticeably paler tail coverts.

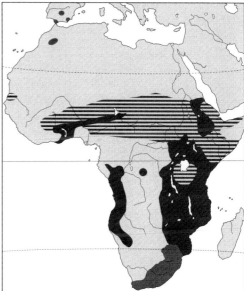

DISTRIBUTION Extensive though widely scattered sub-Saharan African range, which has apparently recently expanded to the Western Palearctic, where it is found in C Morocco and S Spain. In Spain, first recorded in the extreme south in 1962 at the Laguna de la Handa, and in 1964 in the Sierra de la Plata, near to the town of Tarifa, Cadiz Province, where breeding was first recorded in 1966. Since this time the range in Spain has slowly spread east to Almeria, and north to Cordoba. In Morocco it was first recorded in 1968 in the Imlil Valley, High Atlas, and subsequently at Ijoukak, Ouarzazate and Immouzer. The Moroccan range does not appear to be expanding. In sub-Saharan Africa occurs at its most westerly location in the Niokolo Koba National Park, Senegal. It was also recorded from Nèma, Mauritania in 1971. Also recorded from Mali. Further east the range becomes continuous from C Ivory Coast, central and coastal Ghana, Togo, Benin, Nigeria, mainly through the central region of the country north of the Niger delta and the south-central lowlands to Chad. There are a number of records both north, south and east of this broad West African range: in N and S Nigeria, Cameroon, E Chad and W Sudan (Darfur). The status of these records is uncertain, but it is possible that they are the result of dispersal from the breeding range, migrants or, perhaps, overlooked breeders in these regions. Further south in W Africa the range is continuous from S Gabon, W Congo (though not to the coast), Cabinda, W and coastal Zaïre, the whole length of coastal (except the far south) Angola and into C Namibia south to Windhoek. Found in E Africa at its most northerly point on the western and south-western slopes of the Ethiopian Highlands as far north as the Red Sea coast, and westwards into WC Sudan. The range is then disjunct until southernmost Sudan and NE Zaïre where it is continuous in E Africa in the west through E Zaïre, Uganda, Rwanda and Burundi southwards into Zambia (except the west), Zimbabwe (except the far west), extreme E Botswana and into S Africa except for western Transvaal and N Cape. An isolated population occurs in SC Zaïre at Luluaburg. In the east the range is continuous from S Kenya, coastal Tanzania as far west as the central plateau, and throughout Mozambique and Malawi and into S Africa. In E Africa it has bred in S Somalia in the past and there are two records from C Tanzania, where it may be present at low density.

In the Western Palearctic the species is still highly localised and scarce particularly in Morocco, although in Spain an increase in population does seem to be occurring. Up to 50 birds have been seen together at one site near Tarifa, and it is no longer unusual to observe the species away from traditional sites. Most abundant in S Africa, particularly Mozambique. A common breeder in S Africa. Throughout E Africa it is believed to be increasing and is thought to be the commonest swift in rural areas. In W Africa, although widespread, it is far more localised.

MOVEMENTS The Western Palearctic (at least in southern Spain) and the southern African populations (south of the Zambezi) are largely migratory, with the remainder of the sub-Saharan populations thought to be resident with some post-breeding dispersal. In S Spain breeding birds arrive in late May and leave from August-October, with records over the Straits of Gibraltar in autumn from 11 August - 13 October. The wintering range is not known and it is noteworthy that there are two Spanish records for the period December-January. Nothing is known of the smaller Moroccan population's wintering behaviour and there is a possibility that the Spanish population may be wintering close to home in the West Palearctic. The southern African migrant breeding population is present between August-May and is virtually absent from S Cape Province during June-July with numbers much reduced in the northern part of this migrant range during the same period. The wintering grounds of this population have not been discovered, although it has been postulated that they may be north of the equator. In the north of the sub-Saharan range it is considered that there may be some wet season movement north into the Sahel. These localised migrations or dispersals are poorly understood, and probably account for the appearance of the species away from known breeding areas.

HABITAT This is largely dictated by the preferences of the retort-nestbuilding swallows and Little Swifts whose nests are parasitised by the White-rumped Swift. In sub-Saharan Africa, natural nest sites are rarely used because of its dependence upon these species: Lesser Striped Swallow *Hirundo abysisinica*, Greater Striped Swallow *H. cucullata*, Rufous-chested Swallow *H. semirufa*, Wire-tailed Swallow *H. smithii*, Rock Martin *H. fuligula*, Angolan Cliff Swallow *H. rufigula*, and Red-rumped Swallows *H. daurica* Brooke (pers. comm.) points out that the colonisation of new areas by these swallow species especially of man-made structures has led to much expansion of the White-rumped Swift. White-rumped Swifts do not normally venture far from their nesting and roosting sites and are commonly found in towns and cities in Africa, whereas in the Western Palearctic they have yet to be recorded breeding in truly urban areas. Within its range the species may be recorded over a wide variety of habitats from fairly arid savanna to equatorial forest, and at altitudes from sea-level to 2500m (in the High Atlas). In S Spain the species' traditional breeding site is a large unfinished beach hotel and the birds spend much of their time over the beach area or climbing to great height over adjacent sierras.

DESCRIPTION *A. caffer*
Adult Sexes similar. **Head** Broad, white throat patch, contrasting sharply with black underparts. When worn can show darker streaking. Forehead and line over eye pale grey-brown, lores usually darker, uniform with feathers around gape. Black eye-patch. Forecrown and ear-coverts grey-brown progressively darker towards nape and sides of neck. When worn, head becomes paler contrasting more with saddle. **Body** Saddle uniformly blue-black, feathers of lower back, immediately above rump band, slightly paler. Narrow but striking white rump band, extending slightly onto rear flanks and forward slightly under wing, causing U-shape. White visible from side, but not from below. When worn darker streaks can be seen on rump. Lower rump slightly paler than blue-black uppertail-coverts. Underbody uniformly blue-black from breast to undertail-coverts, with very narrow paler fringes when plumage fresh. When worn grey-brown feather bases can be seen. **Upperwing** Primaries and secondaries black-brown (innerwebs paler), with outer primaries slightly darker, though less so than in larger *Apus*. Narrow white tips to secondaries, and often some inner primaries, show as white trailing edge though this is prone to wear. Greater primary and greater coverts, uniform with or slightly darker than remiges, with smaller coverts becoming increasingly blue-black towards leading edge. Leading edge coverts narrowly tipped white. **Underwing** Remiges paler grey than upperwing, can appear translucent in strong light. Greater primary and greater coverts uniform with or slightly darker than remiges, and white-tipped. Smaller coverts increasingly darker to leading edge, with median primary and median coverts white-tipped. **Tail** Black-brown; little contrast with uppertail-coverts, more so with undertail-coverts. **Wing/Body Contrast** Median and lesser coverts appear fairly uniform with under and upperbody, with remiges and greater coverts paler. The very black underbody can appear darker than the underwing-coverts.

Juvenile Plumage very similar to fresh adult, but white trailing edge perhaps more prominent and has narrow white fringes on greater and median coverts (these are soon lost through abrasion).

Measurements E Africa, north to Sudan and Ethiopia. Wing: male (13) 137-145 (141); female (10) 135-143 (139).

Tail: male (11) 66-75 (70.3); female (10) 68-74 (69.9). Tail-fork: male (10) 30-36 (32.6); female (10) 27-32 (30). Weight: 18-30 (22.1) (Cramp 1985). Africa south of the Cunene and Zambezi rivers. Wing: (61) 143-157 (150). Weight (170) 18-30 (22.1) (Fry *et al.* (1988).

GEOGRAPHICAL VARIATION Currently considered monotypic, although previously two subspecies were recognised: *caffer* and *streubelii*. Within this monotypic species variation occurs only in size with any consistency, with those populations bordering the Gulf of Guinea averaging smallest. In S Africa the longest winged birds occur from Pretoria southwards, whilst those in N Transvaal and Zimbabwe are noticeably shorter-winged.

VOICE Generally rather silent, especially compared to Little Swift. The most commonly heard call is a twittering trill based around *sip* notes *sip-sip-sip-seep-sip-sip*, which is rather bat-like in character and deeper than Little Swift. A further call described as a chattering crescendo *pi-pi-pi-pi-pi-pi-pee-pee*, has been described although it is possibly a different transcription of the *sip* trill.

HABITS Highly gregarious, mixing with other swift species and swallows. Within the Western Palearctic it can be seen with Red-rumped Swallows, though more commonly with Pallid and Common Swifts. Known to migrate in flocks of up to 100 birds.

BREEDING White-rumped Swift nests either in loosely associating colonies or solitarily. Although it will nest in crevices or ledges within rock fissures or on buildings, it more commonly uses disused Little Swift nests or those of certain swallows *Hirundo* spp. Nests are lined with feathers and down adhered on the edges by saliva. When rock sites are used for nesting, a shallow cup is made from feathers, grass, roots and down from a variety of species and bound together using saliva. When it nests within the eaves of buildings it always builds against the masonry, never wood. Usually has 2 eggs in a clutch, but 1 and 3 are not rare. In some parts of the range it is double-brooded (Cramp 1985, Fry *et al.* 1988).

REFERENCES Alström *et al.* (1991), Brooke (1971a), Chantler (1993), Cramp (1985), Fry *et al.* (1988).

92   BATES'S SWIFT
Apus batesi   Plate 11

Other name: Black Swift

IDENTIFICATION Length 14cm. An aberrant species not falling conveniently into either the white-rumped or uniform *Apus* groups. It is a tiny swift perhaps closest in shape to White-rumped with its slim build and long deeply forked tail. The plumage is blacker and more uniform than any other member of the genus and the indistinctness of the throat patch is shared only by Plain Swift. A continuous rapid flight, described as rapid winnowing wing beats interspersed with occasional short glides (Iain Robertson pers. comm.), add to this species' uniqueness. The similarly small Scarce Swift appears far greyer, indeed quite silvery-grey in sunlight, and is considerably paler on the throat. The enigmatic Schouteden's Swift is closer in darkness to Bates's Swift but probably shares the distinctive jizz of Scarce Swift.

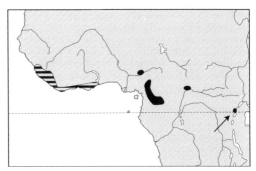

DISTRIBUTION Primarily West African range. W Cameroon and neighbouring areas of N Gabon and Rio Muni. Occurs in C Africa in C African Republic, and Kivu Province and in the Semliki Valley in E Zaïre. Scattered sight records have been recorded from many West African countries from Nigeria to Sierra Leone.

Throughout most of the range this is believed to be a rather rare species although it may be common in the Cameroon highlands.

MOVEMENTS Considered to be entirely resident throughout its range.

HABITAT Primarily a species of hilly rainforest, especially in which rocky crags. Believed to be restricted to areas in which cliff swallows occur, particularly the Dusky Cliff Swallow *Hirundo fuliginosa*, in whose vacated nests it has been recorded breeding. There is a possibility that it will sometimes occupy nests of Mottled Spinetail as it has been recorded examining the nests of this species.

DESCRIPTION *A. batesi*
Adult Sexes similar. **Head** Upper head glossy blue-black with slightly duller forehead. Indistinct grey-brown throat patch appears rather mottled. Ear-coverts similar to crown. As plumage becomes worn the gloss is lost and it becomes a little browner. Black eye-patch indistinct due to darkness of surrounding feathers. **Body** From crown to uppertail-coverts uniformly blue-black. Paler brown feather bases can be apparent when plumage worn. Below the throat patch the plumage is deeply blue-black through to the undertail-coverts. When very fresh some narrow pale fringing can be present on the body or head. **Upperwing** Wholly blackish-brown with the outerwebs appearing most blackish and glossy. Doubtless in the field the closer spacing in the outer primaries of the outerwebs produces the typical *Apus* effect of a darker outerwing. **Underwing** The remiges are slightly paler but even more glossy below. The median and lesser coverts appear typically darker than the greater coverts and remiges. **Tail** As remiges and showing no contrast with the very black tail-coverts.

Juvenile Narrow white fringes to body feathers and remiges. Shorter tail.

Measurements Wing: male (2) 125 + 132; female (2) 122 + 13. Male and female: (6) 122-127 (124). Tail: male (2) 54-71; female (2) 60-68. Tail-fork: (3) 20-27 (23.3) (Fry *et al.* 1988).

GEOGRAPHICAL VARIATION None. Monotypic.

VOICE Not recorded.

HABITS Often seen alone or in small groups, although up to 30 have been recorded together.

BREEDING Bates's Swift nests in the abandoned nests of Dusky Cliff Swallow *Hirundo fuliginosa* and is not colonial. The interior of the mud nest, even the spout, is covered with feathers, vegetable down and saliva. The clutches in the nests recorded were two eggs. Two eggs measured 22 x 14.5 and 22 x 14.3 (Fry *et al.* 1988).

REFERENCES Dyer *et al.* (1986), Fry *et al.* (1988).

HEMIPROCNE

The family Hemiprocnidae has only one genus of four small to large species. The family ranges from the Indian subcontinent through SE Asia, the Philippines, Indonesia, New Guinea, the Bismarck Archipelago and the Solomon Islands.

Separation from sympatric genera

The plumage of the family is quite different from any other members of the Apodiformes having bright sheens on green and blue plumages and elaborate head markings with red or chestnut patches and loose moustachial markings.

Not confusable with other swifts whilst perched or indeed, as a result of the very long wings and tail and the very upright stance with head held high and chest puffed out, with any other family. In flight the larger species, with their deep wing beats and long periods of gliding, can recall bee-eaters *Merops* spp, but equally they can at times recall some of the larger swifts when they move rapidly, especially when feeding.

93 CRESTED TREESWIFT
Hemiprocne coronata   Plate 24

IDENTIFICATION Length 23cm. Very distinctive species and the only treeswift in its range (but some overlap in SW Thailand at Kaeng Krachan). Instantly identified when perched by its upright stance, usually with crest erect, and very long wings crossing considerably with the very long tail-streamers held between them. In flight the species looks very much like one of the true swifts, with long, thin scythe-shaped wings and long tail which is usually tightly closed, appearing needle-thin, but occasionally opened when it can be seen to be deeply forked. The blue-grey plumage, and in particular the reddish face of the male, give the species a rather swallow-like appearance. Best distinguished from the closely related Grey-rumped Treeswift by the following features: the plumage is generally darker in Grey-rumped apart from the pale grey rump which contrasts strongly with the dark grey remainder of the upperparts; the darker breast of Grey-rumped contrasts more strongly with the whitish remainder of the underparts than in Crested; the male's face patch is far brighter and more extensive in Crested extending from the ear-coverts to the bill; the dark ear-coverts of the females are far more contrasting in Crested by virtue of its paler upper head; and the gloss on the plumage of Crested is bluish as opposed to greenish in Grey-rumped. When perched, the tail of Crested extends well beyond the wing tips, whereas in Grey-rumped the wings extend beyond the tail (figure 61).

Other distinguishing features have previously been recognised but are rather subjective. The underwing-coverts of Crested have been said to be concolorous with the rest of the underwing whereas they are darker in Grey-rumped, but this feature is very hard to discern in either species. The tertials of Grey-rumped are slightly whiter than those of Crested, but this can only reliably be seen in close perched views. Juvenile Crested Treeswifts have remarkable scaly plumage that is not unlike a cuckoo, but they are seldom seen far from the nest and soon moult their scaly body feathers after fledging. The flight of this species is not unlike that of the typical swifts when gliding and soaring in rather graceful, leisurely broad arcs interspersed with flicking wing beats, and some rocking from side to side. Often glides for long periods without flapping wings. More active flight is very powerful with deep wing beats, prompting comparison with larger swifts such as Alpine especially when this mode of flight is used to fly to a higher altitude.

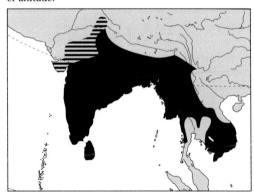

Figure 61. Perched adult male Grey-rumped Treeswift (left) and adult male Crested Treeswift (right). Note difference in comparative position of tail to wing tips.

DISTRIBUTION Extensive S and SE Asian range. Indian subcontinent from the Himalayan foothills southwards and from the Punjab, Rajasthan and Gujarat eastwards, wherever suitable habitat occurs. Also Sri Lanka. In the north it occurs in the lowlands of Nepal where it is least common in the west and most frequent at Chitwan. Eastwards the species is found throughout Bangladesh and into Burma as far south as C Tenasserim. In China resident in SW Yunnan. In Thailand throughout the north-west, avoiding the central plain with the range extending southwards in the mountainous region along the Burmese border to Kaeng Krachan, and to the east of the central plain southwards in Thiu Khao Phetchabun range as far south as Khao Yai at the western end of the Phnom Dangrek range. In Indochina it occurs in C and S Laos, throughout S Vietnam as far north as Central Annam and in Cambodia.

Common in suitable habitat throughout most of range.

MOVEMENTS Largely resident but in the Indian subcontinent believed to undergo some seasonal movements.

HABITAT Primarily a species of forested areas, both secondary and primary, where it is mostly found feeding over the canopy or in clearings. Also in more cultivated areas including gardens in small towns. In Nepal it has been recorded to 1280m, but most commonly below 365m, in Sri Lanka from the lowlands to 1200m, and in Thailand to 1400m. Typically found a around favourite perch site, usually a large tree with leafless outer branches, from which it flies in broad circles looking for prey.

DESCRIPTION *H. coronata*

Adult Male. Head Upper head bright blue-grey, with notably darker green-blue crest from forehead. Crest laid flat in flight, but frequently raised when perched. Crest appears rather untidy and is 1-1.5cm long. Black eye-patch. Feathers over eye pale-fringed. Ear-coverts dull orange, extending along the side of the throat and across the chin though less intense than on ear-coverts. Central throat pale blue-grey. Lower throat slightly darker. **Body** Upperparts uniformly blue-grey from head to rump, with uppertail-coverts slight darker grey. Underparts whiter away from the darkest point on the lower throat/upper breast. **Upperwing** Remiges uniformly black-brown, appearing darkest on outer primaries. Greater coverts and greater primary coverts a little darker and glossed slightly blue. Median coverts, lesser coverts and alula deep blue and heavily glossed, appearing darker than remiges and greater coverts. Tertials pale grey, especially inner ones, clearly paler than surrounding tracts though often hard to see. At rest wing appears much blacker than body. **Underwing** Remiges appear paler and greyer, with greater coverts uniform or slightly darker. Median coverts slightly darker still, with lesser coverts notably sooty-grey. Leading edge coverts uniform with lesser coverts. **Tail** Dark blue-grey above, paler grey below.

Adult Female Differs from male in lacking orange facial feathering. Ear-coverts and lores dark grey forming striking facial mask. This effect is further heightened by presence of white moustache extending from behind the ear-coverts, along the base of the ear-coverts to the chin.

Juvenile Strikingly scalloped plumage. Upperparts green-grey or blue grey (feathers perhaps darkest on the head), with extensive white fringes. This fringing is less marked on the mantle. The lower back and rump are the palest areas on the upperparts. Feathers of underparts are pale grey-white with grey-brown subterminal bands and white tips. The remiges are broadly fringed white at the tips as are the tertials.

First winter Post-juvenile moult replaces juvenile body feathers with adult-type feathers. Wing and tail feathers will be juvenile until the first complete moult in second autumn.

Measurements Wing: male 141-156; female 148-160. Tail: male 110-135; female 124-127. Tail (centre): male 40-47, female 40-46 (Ali and Ripley 1970). Wing: (53 BMNH) 142-165 (155.5).

GEOGRAPHICAL VARIATION None. Monotypic.

VOICE Typical harsh call of the genus. More disyllabic than Grey-rumped a *kee-kyew*, uttered on the wing, with the second note lower (Lekagul and Round 1991). A tri-syllabic call heard from the perch, with emphasis on the middle note, has been rendered as *kip-kee-kep* (Lowther, in Ali and Ripley 1970). These calls have been described as reminiscent of screams of parakeets and of Shikra *Accipiter badius*.

HABITS Not particularly gregarious. Found mainly in small groups of 6-12. Does not associate any more than loosely with other species. Particularly active late evening.

BREEDING The nest is a tiny bracket-shaped structure placed on the side of a thin horizontal branch 4-18m above the ground. It is composed of papery bark scales and small feathers bound together with saliva. Nests measure 50x30mm and 10-12mm deep. The clutch is always just one egg and this is adhered to the nest surface with saliva as with the Palm Swifts (Ali and Ripley 1970) (figure 62).

Figure 62. Treeswift nest.

REFERENCES Ali and Ripley (1970), King and Dickinson (1975), Lekagul and Round (1991), Robson *et al.* (1989), Wildash (1968).

94   GREY-RUMPED TREESWIFT
Hemiprocne longipennis   Plate 24

IDENTIFICATION Length 21cm. Easily recognised species that rarely overlaps with the confusable Crested Treeswift (see under that species for full discussion of differences). Like Crested Treeswift, males can be identified from the females by their orangey ear-coverts as opposed to the female's dark-grey ear-coverts. The much smaller Whiskered Treeswift is very different in plumage, and in jizz where its awkward rocking flight and flycatching sorties are very different from the graceful more typically swift-like flight of the larger treeswifts. The very cryptic juvenile plumage is reminiscent of a cuckoo, but it is seldom seen away from the adults or the vicinity of the nest and soon after fledging undergoes a body moult.

DISTRIBUTION Limited SE Asian range, but occurs extensively in the Greater Sundas and also in Wallacea. Occurs in the Thai peninsula from Phetchaburi province in the north, mainly on the west coast, avoiding the eastern coastal strip, but also in the mountainous areas surrounding the peak of Khao Laung in Nakhon Si Thammarat province. Also in S Tenasserim, in Burma. In peninsular Malaysia throughout in suitable habitat including Singapore and the islands of Penang and Tinggi. In Sumatra throughout the mainland and on the Riau and Lingga archipelagos, Bangka, Belitung, Simeulue, Nias, the Batu group, the Mentawi islands and Enggano. On

Java and Bali the species is widespread. It also occurs on neighbouring Lombok at Pusuk, on the west coast, and Senaru on the north side of Gunung Rinjani (RWRJ Dekker pers. comm.). On Borneo occurs throughout the lowlands, and also in the Kelabit highlands. Occurs on the small islands off the north coast, and on the Anambas and Natuna Islands. In Wallacea the species is found throughout Sulawesi, and on the islands of Lembeh, Butung (Buton), Kabaena, Salayar, Banggai and Sula.

Locally common throughout range in suitable habitat, and very common on the Indonesian island of Butung.

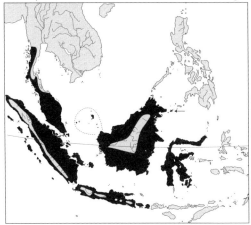

MOVEMENTS Believed to be resident throughout the range. Vagrant to the Tawitawi, Philippines (two records).

HABITAT Essentially a species of forest edges and clearings, although a variety of forest types, including primary and secondary, mangrove, interior, dry and moist, are utilised. Gardens are sometimes frequented. Often found over water where it will hawk after insects. Like all other treeswifts tends to feed close to a favoured perch, often in dead or leafless branches of a large tree. Mainly a lowland bird although it has been recorded to 600m on Sulawesi, 1000m at Benkulu on Sumatra, to 1550m on Java and to 970m in peninsular Malaysia.

DESCRIPTION *H. l. harterti.*
Adult Male. Head Upper head dark glossy green, with untidy crest from forehead that is flattened in flight and often raised repeatedly when perched. Ear-coverts dull brick-orange, not extending to side of throat though cut-off indistinct on lower edge of ear-coverts. Throat uniformly dark green-grey. **Body** Nape and mantle uniform with head, back paler grey and rump strikingly pale grey. Uppertail-coverts dark grey. Underparts uniform with throat to upper breast then rapid progression to paler grey becoming off-white on belly. Flanks and undertail-coverts slightly darker, more uniform with breast. **Upperwing** Blackish-blue remiges glossed blue (secondaries appearing palest) fairly uniform with greater coverts and appearing slightly darker than glossy dark-green alula, and median and lesser coverts. Tertials conspicuously paler than surrounding tracts, but this can be hard to discern in the field. At rest the wings appear darker than the underbody but uniform with the mantle and head. **Underwing** Remiges paler and greyer than above, and paler than dark green-brown underwing-coverts. This pattern can often be hard to see. **Tail** Similar to remiges above and below.

Adult Female Differs from male in having dark blackish-green ear-coverts.
Juvenile Extensive rusty-brown fringing on upperparts, least marked on rump which appears clearly paler than rest of the upperparts. The underpart feathers are off-white, with untidy brown subterminal bands and white fringes. Remiges and upper rectrices have broad white-fringed tips and the tertials and scapulars are broadly white-tipped.

First winter The juvenile body feathers are replaced with adult feathers after the post-juvenile moult, but remiges and rectrices are retained until the post-breeding moult.

Measurements Sumatra, Belitung and Borneo. Wing: male (5) 154-169 (159.2); female (5) 157-165 (161.8). Tail: male tail 72-99 (88.4); female 81-101 (95.2) (Hoogerwerf 1965). Malay Peninsula (north to south Burma). Wing: male (16 BMNH) 152-165 (160.1); female (9 BMNH) 154-173 (161.6). Sumatra and Borneo. Wing: male (9 BMNH) 155-173 (167.6); female (5 BMNH) 158-173 (165.6).

GEOGRAPHICAL VARIATION There has been some considerable debate in the past regarding the races of this species. Hoogerwerf (1965) came to the conclusion that *harterti* is a dubious race and only *wallacii* and *perlonga* are valid races together with the nominate form. Populations of this species from Nias (*ocyptera*) and from the Batu, Pagai and Enggano islands (*thoa*) have at times been considered distinct subspecies but many authors suggest they are synonymous with *perlonga*. Somadikarta (1975) recognised two more subspecies: *mendeni* and *dehaani*. *Mendeni*, from Peleng, was recognised as distinct on the basis of its broad bill. However, Eck (1976) considered a larger sample and concluded that the bills of Somadikarta's specimens cannot have been normal and rejected the race. *Dehaani*, from Sula, is considered by Somadikarta to have the throat and breast more silvery-grey than *wallacii* and to be smaller. Four races are recognised here.

H. l. harterti (S Burma and peninsular Thailand, Malaysia, Sumatra and Borneo) Usually darker than *longipennis*, although some overlap occurs, and with less distinct borders between areas of dark and pale plumage than *longipennis* or *wallacii*.

H. l. longipennis (Java and Bali) The lightest race, although some *harterti* are as light as some darker individuals of *longipennis*. Java. Wing: male (4) 167-168 (167.5); female (5) 155-168 (162.8). Tail: male 94-107 (101); female 90-106 (96.2). Ujung Kulon. Wing: male (1) 162; female (1) 165. Tail: male (1) 105; female (1) 106. Panaitan. Wing: Male (1) 164; female (1) 156. Tail: male (1) 86. female (1) 85. Kangean. Wing: male (1) 171; female (1) 168. Tail: male (1) 106; female (1) 103 (Hoogerwerf 1965). Java and Bali. Wing: (5 both sexes BMNH) 160-167 (162.8).

H. l. perlonga (Simeulue Island, Sumatra) Shows less contrast between dark and pale plumage tracts than any other race. Larger than either *harterti* or *longipennis*, but smaller than *wallacii*. Wing: male (2) 175 + 178, female (1) 180. Tail: male 110 + 112; female (1) 99 (Hoogerwerf 1965).

H. l. wallacii (Sula Islands and Sulawesi) Largest race. Shows the greatest contrast between light and pale areas in the plumage. Sulawesi. Wing: male (5) 179-183 (181.2); female (3) 177-184 (181). Tail: male (5) 98-119 (111.8): female (3) 114-119 (116.7) (Hoogerwerf 1965). Wing: (7 BMNH) 174-193 (183).

VOICE Highly vocal with very shrill wader-like or tern-like calls. Much variation noted, but the commonest call is a harsh, piercing *ki* often repeated as part of a series *ki-ki-ki-kew*. Like Crested Treeswift, will also call disyllabically *too-eit*, with the second note being very metallic. Calls are made both from the perch and on the wing. They are often made most excitedly when just coming into land.

HABITS Usually in small groups, although up to 50 have been recorded together. Very active at dusk, when it will loosely associate with other aerial feeders. In areas of dense forest without larger trees will drink regularly from favoured small pools. In these situations it is a spectacular sight drinking at high speed by diving in graceful arcs whilst dipping its lower mandible into the water, often just feet away from bemused swimmers! It has also been observed flying in small groups on the edges of bee swarms, flying with the swarm whilst picking off individual bees (Smythies 1981).

BREEDING Nest and nest site similar to Crested Treeswift. Other aspects of breeding biology probably also similar.

REFERENCES Eck (1976), Hoogerwerf (1965), Smythies (1981), Somadikarta (1975), White and Bruce (1984).

---

95   MOUSTACHED TREESWIFT
Hemiprocne mystacea   Plate 24

Other name: Whiskered Treeswift

IDENTIFICATION Length 28-30cm. Quite unmistakable. Range does not overlap with any other treeswifts. A large species presenting few identification problems. In flight shows very long scythe-shaped wings and tail; the latter is either held tightly closed appearing very thin and spike-like or open when it can be seen to be very deeply forked with thin streamer-like outer tail feathers. Head rather bulbous and protruding. When perched, usually on a prominent perch, it has a typically upright stance with the breast puffed-out and the long wings crossed in an open-scissor fashion, with the long tail held closed (or opened) between the wing tips. Flight often rapid on powerful wing beats especially when feeding.

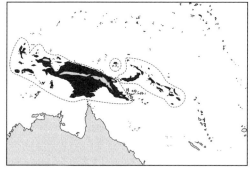

DISTRIBUTION Restricted to E Indonesia and New Guinea. Found throughout Papua New Guinea except the highlands. Occurs on the Moluccas (as far as Buru, Ambon and Seram in the south), Aru Islands, West Papuan Islands, the islands of Geelvink Bay, Bismarck Archipelago (Long, Umboi, New Britain, Watom, Duke of York, New Ireland, New Hanover, Tabar, Lihir, Mahur, Tanga, Feni,

Mussau, Emira), Admiralty Islands (Manus, Rambutyo, Lou), Bougainville, Buka, and Solomon Islands (Vella Lavella, Kolombangara, Guadalcanal, Malaita, Rennell and San Cristobal).

Fairly common to common throughout range.

MOVEMENTS Essentially resident though its sudden disappearance from areas of frequent occurrence suggests that the species is to some extent nomadic.

HABITAT A species of forest edge and open country with some remaining trees which are used as perching sites. Although primarily in the lowlands one unusual record is of a bird rescued from ice at 4400m in the Carstensz Massif, Irian Jaya (Schodde *et al.* 1975). However, not usually encountered over 1580m.

DESCRIPTION *H. m. mystacea*
Adult Male. Head Dark black-blue upper head with a violet gloss. Long white supercilium from bill to nape where it meets, though feathers rather wispy and not tight to head. Black eye-patch and lores. Ear-coverts blue-grey with some rusty feathering distally. White moustache runs from chin, where it joins and is at its broadest, to the rear nape and, like the supercilium, is rather wispy. Throat grey. **Body.** Upperparts uniformly blue-grey from nape to back, becoming darker grey on uppertail-coverts. Underparts are uniform from throat to mid breast becoming pale grey or off-white on undertail-coverts. **Upperwing** Remiges uniformly blackish, slightly paler on innerwebs. Coverts, particularly the median and lesser coverts, are glossed blue-violet and can appear darker than remiges. The tertials and scapulars are extensively white and appear as large patches in flight and when perched. When perched the wings look darker than the body. **Underwing** Remiges paler and greyer than above, fairly uniform with the greater coverts but contrasting with the dark black-blue median and lesser coverts. Leading edge coverts uniform with lesser coverts. **Tail** Blackish-blue above, greyer beneath.

Adult Female Lacks rusty feathering on the ear-coverts.

Juvenile Differs from adult in its spotted and barred plumage. The lores are white and the upper head is extensively buff-spotted with some light buff fringing across the upperparts. Throat feathers are dark grey with extensive buff fringes, considerably darker than the rest of the underparts whose feathers are broadly straw-fringed with buff subterminal bands and grey bases.

Measurements Wing: (12) 221-237 (229.5). Tail: 164-197 (178.8). Tail as percentage of wing: 72-85 (77.8) (Salomonsen 1983). New Britain: Wing: female (8 BMNH) 210-233 (222); female (5 BMNH) 214-229 (219.8). Weight: (5) 69-74 (72.2) (Diamond 1972).

GEOGRAPHICAL VARIATION Six races.
H. m. mystacea (New Guinea and the West Papuan Islands) Described above.
H. m. confirmata (Moluccas and Aru Islands) Similar in plumage to nominate race but consistently smaller. Moluccas: Wing: (6) 208-236 (220.3) (Salomonsen 1983).
H. m. aeroplanes (Bismarck Archipelago, except Admiralty Islands) Distinctly paler above than *mystacea* or *confirmata*, with less contrasting white on abdomen which has more grey intermixed. Variations occur within the range with birds from New Britain most distinct. Small subspecies. New Britain: Wing: (35) 209-230 (216.6). Tail: (35) 149-195 (179.1). Tail as per-

centage of wing: (35) 67-87 (82.4). Weight: (11) 56-79 (70.3) (Salomonsen 1983).

H. m. macrura (Admiralty Islands) Longer-tailed and slightly paler grey upperparts than *aeroplanes*. Rambutyo population rather intermediate between the two forms. Six adults from Manus: Wing 216-226 (222.5); tail 188-203 (198.8); tail as percentage of wing 83-94 (89.4) (Salomonsen 1983).

H. m. woodfordiana (Solomon Islands, except San Cristobal. Also on the Feni Islands, Bismarck Archipelago) Dark race. Underparts wholly dark grey. Some have small amount of white on undertail-coverts. Upperparts as *mystacea*. Wing: (98) 195-217 (204.3). Tail: (98) 142-201 (177.4). Most have tail 170-185 with tail percentage of wing: 81-92 (86.7) (Salomonsen 1983).

H. m. carbonaria (San Cristobal, Solomon Islands) Much darker than *woodfordiana*, being lead-grey below and having darker grey upperparts with bluish gloss. Measurements as *woodfordiana*.

VOICE Typical call is a shrill, downwards inflected *kiiee, whiiee* or *siiee*, described as almost hawk-like in quality (Coates 1985). Coates also describes an ascending high-pitched upslurred *owi-wi-wi-wi* similar in quality to the usual note.

HABITS Rather crepuscular in main feeding habits. Observed mainly at dawn and dusk, and even in darkness, feeding in flocks over open areas. These feeding flocks usually number from 10-20 up to several hundred. An immense evening flock of around 20000 was noted in the Jimmi Valley, Western Highlands Province in Papua New Guinea. During the day feeding is usually restricted to isolated sorties after insects from tree perches. Insects are pursued and caught in the air. Rather aggressive species and frequently harries raptors as a flock.

BREEDING Clutch consists of a single egg, 29-33x20-21, laid in a tiny cup composed of feathers and some plant material bound together with saliva. The nest is sited on a bare horizontal limb high in a large tree (Coates 1985).

REFERENCES Coates (1985), Diamond (1972), Salomonsen (1983), Schodde *et al.* (1975).

---

96   WHISKERED TREESWIFT
Hemiprocne comata   Plate 24

Other names: Lesser Treeswift

IDENTIFICATION Length 15cm. Highly distinctive species, easily identified whilst perched or flying. When perched appears obviously darker and smaller than Grey-rumped Treeswift, which is the only sympatric treeswift. When seen well the bronzy plumage and white facial stripes are diagnostic. Both facial stripes are somewhat loosely held against the head. The lower one in particular extends beyond the back of the head. The crest is less marked than in the other treeswifts and appears more as a ruffling of the crown feathers. Like other treeswifts the perched posture is very upright and puff-chested, rather more so than other species. In flight jizz is unmistakable. Tends to fly less than other species of treeswifts and often makes short rapid fly-catching sorties, employing a rather mechanical rapid flickering flight during which it rocks

rapidly from side to side. This curious rocking motion is not seen when gliding.

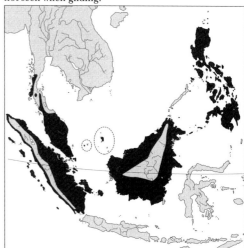

DISTRIBUTION Limited SE Asian, Sumatran, Bornean and Philippine range. In Thailand only south of the Isthmus of Kra and also the south of Mergui in Tenasserim, Burma. The Thai range is not continuous; it extends down the west coast of the peninsula to Krabi and further east an isolated population occurs in the hills on the borders of the Trang and Krabi provinces. A third population occurs from the inlet of Ao Ban Don on the Gulf of Thailand through the central highlands and then along the west coast from Trang southwards to the southern provinces where it occurs along the whole of the border with Malaysia. In peninsula Malaysia and Singapore it occurs throughout in suitable habitat. On Sumatra occurs throughout the mainland, and on the island groups of Riau, Lingga, Bangka, Nias, Batu and Mentawai. On Borneo found throughout the lowlands and on the North Natunas and the Anambas. Also occurs locally on most major islands in the Philippines except Palawan.

Generally a common resident throughout the range, in suitable habitat.

MOVEMENTS Believed to be entirely sedentary.

HABITAT Primarily a bird of the forest edge and clearings. In S Thailand found in high density in clearings formed by power cables through second-growth forest. As well as large trees with exposed branches, overhead cables are often utilised as perch sites. In areas of dense, primary forest often found close to rivers. Tends to be far more loyal to one site than the larger treeswifts and will often be seen day after day in exactly the same location. Mainly a bird of the lowlands occurring to 800m in Thailand, to 1000m in Sumatra and on the Philippines and to 770m in Malaysia.

DESCRIPTION *H. c. comata*
Adult Male. Head Glossy blackish-blue upper head. Rear lores and eye-patch black. Front of lores white at start of long white supercilium which runs along top of ear-coverts and onto nape. Supercilium meets at back of nape but feathers loose and wispy in appearance. Ear-coverts chestnut-olive with bronze gloss. White whisker starting from chin (meets under bill, where at its broadest) extends along gape under ear-coverts and onto rear nape.

223

Like supercilium, loose and wispy and when seen in pro-file often stands out beyond nape. The side of head rear of the ear-coverts, between the two facial streaks, is blackish-blue, uniform with the upper head. Throat ol-ive-black below the white chin, becoming paler and dark olive-bronze from the lower throat onto the breast. **Body** Blackish-blue on upper head not extending beyond nape. Rest of upperparts are olive-bronze, palest on rump. Un-derparts uniform bronze from upper breast to lower belly; vent and undertail-coverts are white. **Upperwing** The whole wing is very uniformly blackish-blue with a slight gloss that is most marked on the coverts. Remiges appear slightly paler than the coverts in flight. The tertials ap-pear very white. When perched the wings look considerably blacker than the body. **Underwing** Remiges slightly paler than on the upperwing and quite uniform with the greater coverts, but clearly paler than the very black median and lesser coverts. The leading edge cov-erts are uniform with the lesser coverts. **Tail** Glossy black-blue like remiges, and slightly paler below.

Adult Female The female has blackish-blue ear-coverts.

Juvenile Not known.

Measurements Wing: 125-127. Tail: 74 (Hachisuka 1934). Wing (Tenasserim and Malay Peninsula): (26 BMNH) 118-133 (124.27). Wing (Sumatra and Borneo): (15 BMNH) 116-130 (124.73).

GEOGRAPHICAL VARIATION Four races.

H. c. comata (Peninsular Thailand and Malaysia, Su-matra and Borneo) Described above.

H. c. stresemanni (Pagai Island, Sumatra) Similar to *comata*, but darker with olive-green wash, and with the gloss on head and upperwing-coverts greener than rather blue-glossed. Wing: 123-127 (Neumann 1937).

H. c. major (Philippines: Cebu, Guimaras, Luzon, Marinduque, Masbate, Mindoro, Negros, Panay, Sa-mar, Tablas) Significantly larger than *comata*. Wing: male 140; female 138-144. Tail: male 80; female 81-85 (Hachisuka 1934). Wing: (17 BMNH) 126-144 (134.8).

H. c. nakamurai (Mindanao and Basilan, Philippines) Larger than *comata*. Differs from *major* in having the upperparts and underparts more strongly tinged green. Wing: male 128-134; female 128-132. Tail: male 79-81; female 78-80 (Hachisuka 1934).

VOICE Vocal species with call not unlike Grey-rumped Treeswift. A shrill chatter, rather higher pitched and less piercing than that species, *she-she-she-she-SHOO-she* becom-ing rather higher in pitch on the *SHOO* note. In addition to this chatter, and variations on it, a Greenfinch-like *chew* is often heard from perched birds. Another call tran-scribed as *squeawk* has been recorded from the perch (M M Norman in Smythies 1981).

HABITS Does not readily associate with other species, al-though tolerant of Grey-rumped Treeswifts which can be seen sitting in the same tree as this species, although usu-ally on higher perches. Usually seen alone or in pairs, although can be seen in groups of up to six. When set-tling on a branch it has a rather endearing habit of shuffling its plumage as if it is getting comfortable or as Smythies (1981) puts it: 'it has a curious trick of closing its long wings "by numbers" in three or four quick jerks'.

BREEDING Breeding biology similar to other species of treeswift, but nest dimensions undoubtedly smaller.

REFERENCES Hachisuka (1934), Lekagul and Round (1991), Medway and Wells (1976), Neumann (1937), Smythies (1981).

BIBLIOGRAPHY

Alström, P., Colston, P., & Lewington, I. 1991. *The Rare Birds of Britain and Europe*. HarperCollins, London.

Abdulali, H. 1966. Notes on Indian Birds 5 - The races of *Apus affinis* (J E Gray) in the Indian region. *J. Bombay Nat. Hist. Soc.* 62 (3): 521-528.

Alcorn, J. R. 1988. *Birds of Nevada*. Fairview West, Fallun, Nevada.

Ali, S., & Ripley S.D. 1970. *Handbook of the Birds of India and Pakistan*, Vol. 4. Oxford Univ. Press, Bombay.

Anon. 1988. *American Birds*. 42 (2): 289.

Amos, E. J. R. 1991. *A Guide to the Birds of Bermuda*. Amos, Warwick, Bermuda.

Arn-Willi, H. 1959. Photographic studies of some less familiar birds 99: Alpine Swift. *Brit. Birds* 52: 221-225.

— 1967. *Tierwelt* 77: 1119.

Avelado, R., & Pons, A. R .1952. Aves nuevas y extensiones de distribucion a Venezuela. *Nov. Cient. Contrib. Ocas. Mus. Nat. La Salle*. 7: 7-8.

Baker E. C. S. 1927. *The Fauna of British India*. Vol. 4. Taylor and Francis, London.

Bannerman, D. A. 1953. *The Birds of West and Equatorial Africa*. Oliver & Boyd, Edinburgh and London.

— & Bannerman, W. M. 1965. *Birds of the Atlantic Islands*. Vol. 2. Oliver & Boyd, Edinburgh and London.

— & — 1966. *Birds of the Atlantic Islands*. Vol 3. Oliver & Boyd, Edinburgh and London.

— & — 1968. *Birds of the Atlantic Islands*. Vol. 4. Oliver & Boyd, Edinburgh and London.

Beaman, M. 1994. *Palearctic Birds: A Checklist of the Birds of Europe, North Africa and Asia north of the foothills of the Himalayas*. Harrier, Stonyhurst.

Becking, J. H. 1971. Breeding of *Collocalia gigas*. *Ibis* 113: 330-333.

Beebe, W. 1947. Avian migration at Rancho Grande in North-central Venezuela. *Zoologica* 32 (4): 164-167.

Beehler B. M., Pratt, T. K. & Zimmerman, D.A. 1986. *Birds of New Guinea*. Princeton Univ. Press, Princeton.

Beklova, M. 1976. Contribution to the characteristics of population dynamics of certain hemisyananthropic species of bird. *Zoo. Listy* 25: 147-155.

Belton, W. 1984. Birds of the Rio Grande do Sul. *Bull. Amer. Mus. Nat. Hist.* 178: 564-570.

Benson, C. W., & Benson, F. M. 1977. *The Birds of Malawi*. Mountfort Press, Limbe.

Blincow, J. (1992). 'Little swifts' with unusual plumage. *Birding World* 5: 160.

Bond, J. 1956. Nesting of the Pygmy Palm Swift. *Auk* 73: 457.

— 1985. *Birds of the West Indies*. 5th edition, Houghton Mifflin, Boston.

Bowen, P. St. J. 1977. European Swift mortality on migration. *Bull. Zamb. Orn. Soc*. 9: 61.

Brandt J. H. 1966. Nesting notes on the *Collocalia* of Micronesia and peninsular (sic) Thailand. *Oologist's Record* 40: 61-68.

Bravery, J. A. 1973. Sight records of swiftlets. *Emu* 73: 29.

Brazil, M. A. 1991. *The Birds of Japan*. Christopher Helm, London.

Bregulla, H. L. 1992. *Birds of Vanuatu*. Nelson, Oswestry.

Bromhall, D. 1980. *Devil Birds: The Life of the Swift*. Hutchinson, London.

Brooke, R. K. 1966. The Bat-like Spinetail *Chaetura boehmi* Schalow (Aves). *Arnoldia* (*Rhod*.) 2(19):1-18.

— 1967. *Apus aequatorialis* (Von Müller) (Aves) in Rhodesia and adjacent areas with description of a new race. *Arnoldia* (*Rhod*.) 3(7):1-7.

— 1969a. Notes on the Identification of Swifts in Southern Africa. *Bokmakierie* 21:39-40.

— 1969b. Age characters in swifts. *Bull. Brit. Orn. Cl.* 89: 78-82.

— 1969c. *Hemiprocne coronata* is a good species. *Bull. Brit. Orn. Cl.* 89: 168-169.

— 1970a. Geographical variation and distribution in *Apus barbatus*, *A. bradfieldi* and *A. niansae*. (Aves: Apodidae). *Dur-*

ban *Mus. Novitates* 8 (19): 363-374.

— 1970b. Taxonomic and evolutionary notes on the subfamilies, tribes, genera and subgenera of the Swifts (Aves: Apodidae). *Durban Mus. Novitates* 9 (2): 13-24.

— 1971a. Geographical variation in the Swifts *Apus horus* and *Apus caffer* (Aves: Apodidae). *Durban Mus. Novitates* 9 (4): 29-38.

— 1971b. Geographical variation in the Little Swift *Apus affinis* (Aves: Apodidae). *Durban Mus. Novitates* 9 (7): 93-103.

— 1971c. Geographical variation in the Alpine Swift *Apus (Tachymarptis) melba* (Aves: Apodidae). *Durban Mus. Novitates* 9 (10): 131-143.

— 1971d. Taxonomic notes on some lesser known *Apus* swifts. *Bull. Brit. Orn. Cl.* 91 (12): 33-36.

— 1972a. Geographical variation in Palm Swifts *Cypsiurus* spp. *Durban Mus. Novitates* 9 (15): 217-229.

— 1972b. Swift migrations in Southern Africa. *Bokmakierie* 24: 31-32.

— 1972c. Generic limits in the Old World Apodidae and Hirundinidae. *Bull. Brit. Orn. Cl.* 92: 53-57.

— 1993. Review of Chantler 1993. *Ostrich*.

— & Steyn, P. 1979. The white-rumped swift seen at the Agalegas and migrations of the Horus Swift *Apus horus*. *Bull. Brit. Orn. Cl.* 87: 124-125.

Brooks, T. M. *et al.* 1992. *Birds surveys and conservation in the Paraguayan Atlantic forest*. Study Report No. 57. BirdLife International, Cambridge.

Browning, M. R. 1993. Species limits of the cave swiftlets (*Collocalia*) in Micronesia. *Avocetta* 17:101-106.

Buden D. W. 1987. *The birds of the Southern Bahamas*. BOU Check-list No.8. BOU, Tring, Herts.

Bull, J. 1976. *Birds of New York State*. Cornell Univ. Press, Ithaca and London.

Burke, W. 1994. Alpine Swift (*Tachymarptis melba*) photographed on St. Lucia, Lesser Antilles - third record for the Western Hemisphere. *El Pitirre*. 7(3): 3

Campbell, R. W., Dawe, N. K., McTaggart-Cowan, I., Cooper, J. M., Kaiser, G. W., McNall G. C. E. 1990. *Birds of British Columbia*. Royal Brit. Columbia Mus., Victoria, B.C.

Castan, R. 1955. Le martinet pale gabes *Apus pallidus brehmorum* (Hartert). *Oiseau* 25: 172-182.

Chantler, P. J. 1990. Identification of Pallid Swift. *Birding World* 3: 168-171.

— 1993. Identification of Western Palearctic Swifts. *Dutch Birding* 15 (3): 97-135.

— 1995. Belly patch shape on Alpine Swift. *Brit. Birds* 88:52.

— (in press). Identification of Pale-rumped, Grey-rumped and Band-rumped Swifts. *Cotinga*.

— & Driessens, G. (in press). The identity of 'whiskered' swifts. *Birding World*.

Chapin, J. P. 1939. The Birds of the Belgian Congo. *Bull. Amer. Mus. Nat. Hist.* 75: 1-632.

Chapman, F. M. 1917. The distribution of bird life in Colombia. *Bull. Amer. Mus. Nat. Hist*. 36.

— 1919. Descriptions of proposed new birds from Peru, Bolivia, Brazil and Colombia. *Proc. Biol. Soc. Washington* 32: 253-255.

— 1929. New birds from Mt. Duida. *Amer. Mus. Novitates* No. 380: 11-13.

— 1931. Bird-life of Mts. Roraima and Duida. *Bull. Amer. Mus. Nat. Hist.* 63: 68-70.

Chapman, P. 1985. Cave-frequenting vertebrates in the Gunung Mulu National Park, Sarawak. *Journ. Sarawak Mus.* 34:55.

Cheng, Tso-hsin. 1987. *A synopsis of the Avifauna of China*. Science Press, Beijing and Paul Parey, Hamburg & Berlin.

Clancey, P. A. 1980. Miscellaneous notes on African birds, 58. The mainland Afrotropical subspecies of the Little Swift *Apus affinis* (Gray). *Durban Mus. Novitates* 12 (13): 151-156.

Coates, B. J. 1985. *The Birds of Papua New Guinea*, Vol. 1. Dove,

Alderley.

Collar, N. J., & Stuart, S. N. 1985. *Threatened Birds of Africa and related Islands*. The ICBP/IUCN Red Data Book. Part 1. Cambridge.

—, Gonzaga L. P., Krabbe, N., Madroño Nieto, A., Naranjo L. G., Parker III, T. A., & Wege, D. C. 1992. *Threatened Birds of the Americas*. The ICBP/IUCN Red Data Book. Part 2, Cambridge.

—, Crosby M. J., & Stattersfield, A. J. 1994. *Birds to Watch 2. The World List of Threatened Birds*. BirdLife International, Cambridge.

Collins, C. T. 1967. Comparative biology of two species of Swifts in Trinidad, West Indies. *Bull. Florida State Mus.* 11: 257-320.

— 1968a. Notes on the biology of Chapman's Swift *Chaetura chapmani* (Aves, Apodidae). *Amer. Mus. Novitates* No.2320: 1-15.

— 1968b. Distributional notes on some Neotropical swifts. *Bull. Brit. Orn. Cl.* 88 (8): 133-134.

— 1972. A new species of swift of the genus *Cypseloides* from northeastern South America. *Contrib. Sci. Nat. Hist. Mus. Los Angeles County* 229: 1-9.

— 1974. Survival rate of Chesnut-collared Swift. *Western Bird Bander.* 49 (3): 10-13.

— 1980. The biology of the Spot-fronted Swift in Venezuela. *American Birds* 34: 852-855.

— & Brooke, R. K. 1976. A review of the swifts of the genus *Hirundapus. Contrib. Sci. Nat. Hist. Mus. Los Angeles County* 282: 1-22.

— & Landy, M. J. 1968. Breeding of the Black Swift in Veracruz, Mexico. *Bull. S. Acad. Sci.* 67:226-268.

Colston P. R., & Curry-Lindahl, K. 1986. *The Birds of Mount Nimba, Liberia*. British Museum (Natural History), London.

Cranbrook, Earl of. 1984. Report on the birds' nest industry in the Baram district and at Niah, Sarawak. *Journ. Sarawak Mus.* 54.

Cramp, S. 1985. *The Birds of the Western Palearctic* Vol. 4. Oxford Univ. Press, Oxford.

Cucco, M., & Malacarne, G. 1987. Distribution and nest-hole selection in the breeding Pallid Swift. *Avocetta* 11: 57-61.

—, Bryant D. M., & Malacarne, G. 1993. Differences in diet of Common (*Apus apus*) and Pallid (*A. pallidus*) Swifts. *Avocetta* 17:131-138.

Cuello J., & Gerzenstein, E. 1962. Las aves del Uruguay. *Com. Zool. Mus. Montevideo* 6:191.

Darlington, P. J. 1931. Notes on the birds of the Rio Frio (near Santa Marta), Magdalena, Colombia. *Bull. Mus. Comp. Zool. Harvard* 71: 391-393.

Dean, W. R. J., & Jensen, R. A. C. 1974. The nest and eggs of Bradfield's Swift. *Ostrich* 45: 44.

Dean, A. R. 1994. Identification of Alpine Swift. *Brit. Birds* 87: 174-177.

Dee, T. J. 1986. *Endemic Birds of Madagascar*. ICBP Report. Cambridge.

Deignan, H. G. 1955. The races of the Swiftlet, *Collocalia brevirostris* (McClelland). *Bull. Brit. Orn. Cl.* 75: 116-117.

— 1956. Eastern races of the White-rumped Swift *Apus pacificus* (Latham). *Bull. Raffles Mus.* 27: 147-149.

Dementiev, G. P., & Gladkov, N. A. 1966. *The Birds of the Soviet Union*. Israel Program for Scientific Translation, Jerusalem.

Demetrio, L. 1993. Aclaracioens sobre la prescencia de Vencejo de Chimnea (*Chaetura pelagica*) en el valle de Calama. *Bol. UNORCH* 15:16-17.

Dexter, R. W. 1979. Fourteen year life history of a banded Chimney Swift. *Bird Banding* 50(1): 30-33.

Diamond, J. M. 1972. *Avifauna of the Eastern Highlands of New Guinea*. Publ. Nuttall Orn. Club, 12.

Dickinson, E. C. 1989a. A review of larger Philippine swiftlets of the genus *Collocalia. Forktail* 4: 19-53.

— 1989b. A review of smaller Philippine swiftlets of the genus *Collocalia. Forktail* 5: 23-34.

—, Kennedy, R. S., & Parkes, K. C. 1991. *The Birds of the Philippines*. BOU Check-list No.12. BOU, Tring, Herts.

Dickerman, R. W., & Phelps, W. H. 1982. Birds of Cerro Uritani. *Amer. Mus. Novitates*. No. 2732: 5.

Donelly, B. G. 1982. Cold-induced mortality in African Palm Swifts. *Honeyguide* 111/112: 15-17.

Dowsett, R. J., & Forbes-Watson, A. D. 1993. *Checklist of Birds of the Afrotropical and Malagasy Regions. Vol. 1. Species limits and distribution*. Tauraco Press, Liège.

Dubs, B. 1992. *Birds of Southwestern Brazil: Catalogue and Guide to the Birds of the Pantanal of Mato Grosso and its border areas*. Betrona.

Dupont, J. E. 1971. *Philippine Birds*. Delaware Museum of Natural History, Delaware.

Dyer, M., Gartshore, M. E., & Sharland, R. E. 1986. The birds of Nindam Forest Reserve, Kagoro, Nigeria. *Malimbus* 8: 2-20.

Eck, S. 1976. Die Vogel der Banggai-Inseln, insebesondere Pelengs, (Aves). *Zool. Abh., Staatl. Mus. Tierk. Dresden* 34: 53-100.

Edwards, E. 1972. *A Field Guide to the Birds of Mexico*. Sweet Briar, Virginia.

— 1989. *A Field Guide to the Birds of Mexico*. Sweet Briar, Virginia.

Eisenmann, E., & Lehmann, F. C. 1962. A new species of swift of the genus *Cypseloides* from Colombia. *Amer. Mus. Novitates* No. 2117: 1-16.

Engbring, J. 1988. *Field Guide to the Birds of Palau*. Conservation Office, Koror, Palau.

Erickson, R. A., Morlan, J., & Robertson, D. 1989. First record of the White-collared Swift in California. *Western Birds* 20: 25-31.

Finlayson, J. F. 1979. *The ecology and behaviour of closely related species in Gibraltar with special reference to swifts and warblers*. Unpublished D Phil Thesis, Oxford Univ.

Fischer, R. B. 1958. The breeding biology of the Chimney Swift (Linnaeus). *New York State Museum and Science Service Bulletin* No. 368.

Friedmann, H. 1948. Birds collected by the National Geographic Society's Expeditions to northern Brazil and southern Venezuela. *Proc. U.S. Nat. Mus.* 97.

Fry, C. H., Keith, S., & Urban, E. K. 1988. *The Birds of Africa*. Vol. 3. Academic Press, London.

Fjeldså, J., & Krabbe, N. (1990). *Birds of the High Andes*. Zoological Museum, Univ. of Copenhagen and Apollo Books, Svendborg.

Garnt, K., & Dunn, J. 1981. *Birds of Southern California*. Los Angeles Audubon Soc., Los Angeles.

Gilliard, E.T. 1941. Birds of the Mt. Auyan-tepui, Venezuela. *Bull. Amer. Mus. Nat. Hist.* 77: 469.

Glutz von Blotzheim, U. N., & Bauer, K. M. 1980. *Handbuch der Vögel Mitteleuropas*. Vol 7. Akademische Verlagsgesellschaft, Wiesbaden.

Godfrey, W. E. 1966. *The Birds of Canada*. National Museums of Canada, Ottawa.

Goodman, S. M., & Meininger, P. L. 1989. *The Birds of Egypt*. Oxford Univ. Press, Oxford.

Gosler, A. 1991. *The Hamlyn Photographic Guide to the Birds of the World*. Reed, Hong Kong.

Griscom L. 1932. The distribution of bird-life in Guatemala. *Bull. Amer. Mus. Nat. Hist.* 64:1-439.

Hachisuka, M. 1934. *The Birds of the Philippine Islands with notes on the Mammal Fauna, 2 (part 3)*. H. F. & G. Witherby, London.

Hadden, D. 1975. Observations (in and around the Tari District). *PNG Bird Society Newsletter* 115: 8.

— 1981. *Birds of the North Solomons*. Wau Ecology Inst., Wau.

Haverschmidt, F. 1968. *Birds of Surinam*. Oliver and Boyd, Edinburgh and London.

Harvey, W. G. 1981. Pallid Swift: new to Britain and Ireland. *Brit. Birds* 74: 170-178.

Hazevoet, C. J. 1994. Species concepts and systematics. *Dutch Birding.* 16 (3): 111-115.

— 1995. *The Birds of the Cape Verde Islands*. BOU Check-list No. 13. BOU, Tring, Herts.

Heim de Balsac, H., & Brosset, A. 1964. Le martinet *Chaetura melanopygia* Chapin, au Gabon. *Alauda* 32: 241-244.

Higgins, N., Lewis, A., & Morris, P., 1989. West Indoneisa. Unpublished report.

— & — 1989. Sumatra, Java and Bali. Unpublished report.

Hilty, S. L., & Brown, W. L. 1986. *A Guide to the Birds of Co-*

lombia. Princeton Univ. Press, Princeton.

Holyoak, D. T. 1974. *Collocalia sawtelli* sp. nov. *Bull. Brit. Orn. Cl.* 94: 146-147.

— & Thibault, J. C. 1978. Notes on the biology and systematics of Polynesian swiftlets, *Aerodramus*. *Bull. Brit. Orn. Cl.* 98 (2): 59-65.

Hoogerwerf, A. 1965. Some remarks about the Crested Tree-Swift *Hemiprocne longipennis* (Rafinesque). *Treubia* 26 (4): 244-249.

Hopcraft, C. J. 1974. Observations on the breeding of the Horus Swift, *Apus horus*. *Witwatersrand Bird Club News.* 87: 6-7.

Howard, R., & Moore, A. 1991. A complete checklist of the Birds of the World. 2nd edition. Academic Press, London.

Howell, S. N. G. 1993a. Photo spot: White-naped and White-collared Swifts. *Euphonia* 2: 66-68.

— 1993b. More comments on White-fronted Swift. *Euphonia* 2:100-101.

— & Webb, S. 1995. *A Guide to the Birds of Mexico and Northern Central America.* Oxford Univ. Press, Oxford.

Ihering, H. von. 1900. Catalogo critico-comparativo do Ninhos e ovos das Aves do Brazil. *Rev. Mus. Paulista.* 4: 254-255.

Inskipp, C., & Inskipp, T. 1991. *A Guide to the Birds of Nepal.* 2nd edition. Christopher Helm, London.

— & — 1994. Birds recorded during a visit to Bhutan in spring 1993. *Forktail* 9: 121-143.

Jenkins, J. M. 1983. The native forest birds of Guam. *AOU Ornithol. Monogr.* 31: 1-61.

Johnson, A. W. 1967. *The Birds of Chile and adjacent areas of Argentina, Bolivia and Peru.* Platt Estab, Graficos, Buenos Aires.

Johnson, D. 1994. From the field. *Bull. Oriental Bird Cl.* 19: 65.

Kennedy, P. 1986. Pallid Swift occupying House Martin's nest. *Brit. Birds* 79: 339-340.

Kennerley, P. 1991. Three records of swiftlets of undetermined species in Hong Kong. *Hong Kong Bird Report 1990:* 194-197.

Kepler, C. B. 1971. First Puerto Rican record of the Antillean Palm Swift. *Wilson Bull.* 83 (3): 309-310.

— 1972. Notes on the ecology of Puerto Rican swifts, including the first record of the White-collared Swift *Streptoprocne zonaris. Ibis* 114: 541-543.

Kiff, L. F., Marin, A. M., Sibley, F. C., Matheus, J. C., & Schmitt, N. J. 1989. Notes on the nests and eggs of some Ecuadorian birds. *Bull. Brit. Orn. Cl.* 109 (1): 25-31.

King, B. 1987. The Waterfall Swift *Hydrochous gigas. Bull. Brit. Orn. Cl.* 107 (1): 36-37.

— & Dickinson, E. C. 1975. *A Field Guide to the Birds of Southeast Asia.* Collins, London.

Knox, A. 1994a. Species and subspecies. *British Birds.* 87: 51-58.

— 1994b. Lumping and splitting. *Brit. Birds.* 87: 149-159.

Lacey, E. 1910. Swifts eating drones of the hive bee. *Brit. Birds* 3: 263.

Lack, D. 1956a. The species of *Apus. Ibis* 98: 34-62.

— 1956b. A review of the genera and nesting habits of swifts. *Auk* 73: 1-32.

— 1956c. *Swifts in a Tower.* Chapman and Hall, London.

Land, H. C. 1970. *Birds of Guatemala.* Livingston, Wynnewood, Pennsylvania.

Langham, N. 1980. Breeding biology of the Edible-nest Switlet *Aerodramus fuciphagus. Ibis* 122; 447-461.

Langrand, O. 1990. *Guide to the Birds of Madagascar.* Yale Univ. Press, New Haven & London..

Lavauden, 1937. Supplement. A. Milne-Edwards & A. Grandidier. *Histoire physique, naturelle et politique de Madagascar, 12. Oiseaux.* Societe d'Editions Geographiques, Maritimes et Coloniales, Paris.

Lekagul, B., & Round, P. 1991. *A Guide to the Birds of Thailand.* Saha Karn Bhaet, Bangkok.

Lockwood, G., Lockwood, M. P., & Macdonald, M. A. 1980. Chapin's Spinetail Swift *Telacanthura melanopygia* in Ghana. *Bull. Brit. Orn. Cl.* 100: 162-164.

Loutit, R. 1980. Bradfield's Swift *Apus bradfieldi* feeding on bees. *Madoqua* 12: 125.

MacKinnon, J. 1988. *Field Guide to the Birds of Java and Bali.* Gadjah Mada Univ. Press, Yogyakarta.

— & Phillipps, K. 1993. *A Field Guide to the Birds of Borneo, Sumatra, Java and Bali.* Oxford Univ. Press, Oxford.

Mackworth-Praed, C. W., & Grant, C. H. B. 1953. On the affinities of *Apus somalicus* (Stephenson Clarke). *Bull. Brit. Orn. Cl.* 73: 105.

— & — 1970. *Birds of Eastern and North Eastern Africa.* Longman, London.

Mallory, E. 1994. Noteworthy bird sightings: JSSEUR Expedition. *Belize Audubon Soc. Newsletter.* 26 (3) 12.

Marin, A. M. 1993. Patterns of distribution of swifts in the Andes of Ecuador. *Avocetta* 17:117-123.

— & Stiles, F. G. 1992. On the biology of five species of Swifts (Apodidae, Cypseloidinae) in Costa Rica. *Western Foundation of Vertebrate Zoology.* 4 (5): 286-351.

— & — 1993. Notes on the biology of the Spot-fronted Swift. *Condor* 95: 479-483.

— , Carrion, B. J. M., & Sibley, F. C .1992. New distributional records for Ecuadorian birds. *Ornitologia Neotropical* 3: 27-34.

van Marle, J. G., & Voous, K. H. (1988). *The Birds of Sumatra.* BOU Check-list No. 10. BOU, Tring, Herts.

Marshall, J. T. 1949. The Endemic Avifauna of Saipan, Tinian, Guam and Palau. *Condor* 51: 209-210.

Mathews, G. M. 1918. *The Birds of Australia.* H. F. & G. Witherby, London.

Mayr, E. 1931. A systematic list of the birds of Rennell Island with Descriptions of New Species and Subspecies: Birds collected during the Whitney South Sea Expedition XIII. *Amer. Mus. Novitates.* No. 486:1-29.

— 1935. Descriptions of Twenty-five New Species and Subspecies: Birds collected during the Whitney South Sea Expedition XXX. *Amer. Mus. Novitates* No. 820: 1-6.

— 1937. Notes on New Guinea birds. I. Birds collected during the Whitney South Sea Expedition XXXIII. *Amer. Mus. Novitates.* No. 915: 1-19.

— 1944. The birds of Timor and Sumba. *Bull. Amer. Mus. Nat. Hist.* 83: 123-194.

— & van Deusen, H. M. 1956. The birds of Goodenough Island, Papua: Results of the Archbold Expeditions No. 74. *Amer. Mus. Novitates* No. 1792: 1-8.

— & Vuilleumier, F. 1983. New species of birds described from 1966-1975. *J. Ornithol.* 124: 217-232.

McKean, J. L. 1967. Sight records of Swiftlets. *Emu* 67: 98.

Mearns, E. A. 1909. Birds collected at Guam Island. *Proc. U.S. Nat. Mus.* 36: 476.

Medway, Lord, 1961. The identity of *Collocalia fuciphaga* (Thunberg). *Ibis* 103: 625-626.

— 1962. The swiftlets of Java and their relationships. *J. Bombay Nat. Hist. Soc.* 59: 146-153.

— 1963. The antiquity of trade in edible birds' nests. *Federation Mus. Journal* VIII: 36-47.

— 1966. Field characters as a guide to the specific relations of swiftlets. *Proc. Linn. Soc. Lond.* 177: 151-172.

— 1975. The nest of *Collocalia v. vanikorensis*, and taxonomic implications. *Emu* 75: 154-155.

— & Pye, J. D. 1977. Echolocation and the systematics of swiftlets Pp. 225-238 in B. Stonehouse & C. Perrins, eds. *Evolutionary Ecology.* Macmillan, London.

— & Wells, D. 1976. *The Birds of the Malay Peninsula.* Vol. 5. H. F. & G. Witherby, London.

Mees, G. F. 1973. The status of two species of migrant swifts in Java and Sumatra. *Zoologische Mededelingen* 46 (15): 197-205.

Meininger P. L., Duiven, P., Marteijn, E. C. L., & van Spanje, T. M. 1990. Notable bird observations from Mauritania. *Malimbus* 12:19-24.

Meyer de Schauensee, R. 1982. *A Guide to the Birds of South America.* 2nd edition. Academy of Natural Sciences, Philadelphia.

— & Phelps, W. H. 1978. *A Guide to the Birds of Venezuela.* Princeton Univ. Press, Princeton.

Mitchell, M. H. 1957. *Observations on birds of southeastern Brazil.* Univ. Toronto Press, Toronto.

Monroe, B. L. 1968. *A Distributional Survey of the Birds of Honduras.* Orthithol. Monograph No. 7.

Moreau, R. E. 1941. A contribution to the breeding biology of the Palm Swift, *Cypselus parvus. Journ. East Afr. Uganda Nat. Hist. Soc.* 15: 154-170.

Mortimer, J. 1975. Letter. *Bokmakierie.* 27: 88.

Morse, R. A., & Laigo, M. 1969. The Philippine Spine-tailed Swift *Chaetura dubia* McGregor as a honey-bee predator. *Philippine Entomol.* 1:138-143.

Mountfort, G. 1988. *Rare Birds of the World.* Collins, London.

de Naurois, R. 1972. Morphologie et position systematique du martinet *Apus affinis* au Banc d'Arguin (Mauritenie). *Oiseau et R.F.O.* 42:195-197.

— 1985. *Chaetura thomensis* Hartert 1900 endemique des îles de São Tomé et Príncipe. *Alauda* 53: 209-222.

Narosky, T., & Yzurieta, D. 1987. *Birds of Argentina and Uruguay: A Field Guide.* Vasquez Mazzini, Buenos Aires.

National Geographic Society 1987. *Field Guide to the Birds of North America.* Second edition. National Geographic Soc., Washington.

Navarro, A. G., Peterson, A. T., Escalante, B. P. & Benitez, H. 1992. *Cypseloides storeri,* A new species of swift from Mexico. *Wilson Bull.* 104 (1): 55-64.

—, Benitez, H., Sanchez, B. V., Garcia, R. S., & Santana, C. E. 1993. The White-faced Swift in Jalisco, Mexico. Short Communications. *Wilson Bull.* 105 (2):366-367.

Neumann, O. 1937. *Hemiprocne comata stresemanni,* subsp. nov. *Bull. Brit. Orn. Cl.* 52: 151.

Oberholser, H. C. 1906. A monograph of the genus *Collocalia. Proc. Acad. Nat. Sci. Philadelphia.* 58:177-212.

— 1912. A revision of the forms of the Edible-nest Swiftlet *Collocalia fuciphaga* (Thunberg). *Proc. U.S. Nat. Mus.* 42:11-20.

Ogilvie-Grant W. R. 1895. On the birds of the Philippine Islands - part V. The highlands of the province of Lepanto, north Luzon. *Ibis* (7)1:433-472.

Olrog, C. C. 1979. Nueva lista de la avifauna argentina. *Opera Lilloana* 27: 137-139.

Orr, R, T, 1963. Comments on the classification of swifts. *Proc. XIII Internat. Ornithol. Congress* : 126-134.

Osmaston, B. B. 1906. Notes on Andaman birds with accounts of the nidification of several species where nest and eggs have not been hitherto described. *J. Bombay Nat. Hist. Soc.* 17 : 486.

Parker, T. A., & Remsen, J. V. 1987. Fifty-two Amazonian bird species new to Bolivia. *Bull. Brit. Orn. Cl.* 107: 94-107.

Parker, M. 1990. Pacific Swift: new to the Western Palearctic. *Brit. Birds* 83: 43-46.

Parkes, K. C. 1993. Taxonomic notes on the White-collared Swift (*Streptoprocne zonaris*). *Avocetta* 17:95-100.

Penny, M. 1974. *Birds of the Seychelles and the Outlying Islands.* Collins, London.

Perrins, C. 1971. Age of first breeding and adult survival rates in the Swift. *Bird Study* 18: 61-70.

Peters, J. L. 1940. *Check-list of birds of the World.* Harvard Univ. Press, Cambridge. Vol. 4: 220-259.

Peterson, A. T., & Navarro, A. G. 1993. Systematic Studies of Mexican Birds - a Response. Point/Counterpoint - more on swifts. *Euphonia* 2: 98-99.

Phillips, A. R. 1962. Notas sistematicas solore avec Mexicanas. *An. Inst. Biol. Univ. Mexico* 32: 333-381.

Pizzey, G. 1980. *A Field Guide to the Birds of Australia.* Collins, Sydney.

Pratt, H. D. 1986. A review of the English and Scientific nomenclature of Cave Swiftlets (*Aerodramus*). *Elepaio* 46: 119-125.

—, Engbring, J., Bruner, P. L., & Berrett, D. G. 1987. *A Field Guide to the Birds of Hawaii and the Tropical Pacific.* Princeton Univ. Press, Princeton.

Prigogine, A. 1985. Recently recognised bird species in the Afrotropical region - a critical review. *Proc. Int. Symp. Afr. Vert., Zool. Forsch. Mus. A. Koenig, Bonn*: 91-114.

Procter, J. 1972. The nest and the identity of the Seychelles Swiftlet *Collocalia. Ibis* 114: 272-273.

Pyle, P., & Engbring, J. 1985. Checklist of the birds of Micro-

nesia. *Elepaio* 45: 57-68.

Ralph, C. J., & Sakai, H. F. 1979. Forest bird and fruit bat populations and their conservation in Micronesia: notes on a survey. *Elepaio* 45: 20-26.

Rand, A. L. 1936. Distribution and Habits of Madagascar Birds. *Bull. Amer. Mus. Nat. Hist.* 72: 413-415.

— 1941. New and interesting birds from New Guinea: Results of the Archbold expeditions No. 32. *Amer. Mus. Novitates.* No. 1102:1-15.

— 1942. Birds of the 1936-1937 New Guinea Expedition: Results of the Archbold expeditions No. 42. *Bull. Amer. Mus. Nat. Hist.* 79:289-366.

— & Gilliard, E. T. 1967. *Handbook of New Guinea Birds.* Weidenfeld and Nicholson, London.

Reboratti, J. 1918. Nidos y huevos de vencejos. *Hornero* 1:193.

Ridgely, R. S. 1976. *Birds of the Republic of Panama.* Princeton Univ. Press, Princeton.

— & Gwynne, J. A. 1989. *A Field Guide to the Birds of Panama with Costa Rica, Nicaragua and Honduras.* Second Edition. Princeton Univ. Press, Princeton.

Ridgway, R. 1910. Diagnoses of new forms of Micropodidae and Trochilidae. *Proc. Biol. Soc. Washington* XXIII: 53-54.

— 1911. Birds of Middle and North America. Part 5. *Bull. U.S. Nat. Mus.* 50: 1-859.

Riley, J. H. 1933. A new swift of the genus *Reinarda* from Venezuela. *Proc. Biol. Soc. Washington* 46: 39-40.

— 1927. New forms of birds in Northeast Borneo. *Proc. Biol. Soc. Washington* 40: 140-141.

Ripley, S. D. 1982. *A synopsis of the Birds of India and Pakistan together with those of Nepal, Bhutan, Bangladesh and Sri Lanka.* 2nd edition. Bombay Nat. Hist. Soc., Bombay.

— & Rabor, D. S. 1958. Notes on a collection of birds from Mindoro Island, Philippines. *Peabody Mus. Nat. Hist. Yale Univ. Bull.* 13 : 1-83.

Roberson, D. 1980. *Rare Birds of the West Coast of North America.* Woodcock, California.

Roberts, T. 1991. *The Birds of Pakistan.* Vol. 1. Oxford Univ. Press, Oxford.

Robinson, H. C. & Kloss, C. B. 1924. Birds from West Sumatra. *J. Fed. Malay. States. Mus.* 11:189-347.

Robson, C. R., Eames, J. C., Wolstencroft, J. A., Nguyen Cu & Truong Van La. 1989. Recent records of birds from Viet Nam. *Forktail* 5: 71-97.

de Roo, A. E. M. 1966. Age characteristics in adult and subadult Swifts, *Apus a. apus* (L.), based on interrupted and delayed wing moult. *Gievralk* 56: 113-131.

— 1968. Taxonomic notes on Swifts, with description of a new genus (Aves: Apodidae). *Rev. Zool. Bot. Afr.* LXXVII, 3-4: 412-417.

— 1970. A new race of the African Black Swift *Apus barbatus* (Sclater) from the Republic of Cameroon (Aves: Apodidae). *Rev. Zool. Bot. Afr.* LXXXI, 1-2: 156-160.

Rothschild, Lord. 1931. *Cypseloides fumigatus major,* subsp. nov. *Bull. Brit. Orn. Cl.* 52:36.

Round, P. 1988. *Resident Forest Birds in Thailand: Their Status and Conservation.* ICBP, Cambridge.

Rowley, J. S., & Orr, R. T. 1962. The nesting of the White-naped Swift. *Condor* 64 (5): 361-366.

— & — 1965. Nesting and feeding habits of the White-collared Swift. *Condor* 67 (6): 449-456.

— 1966. Breeding Birds of the Sierra Madre del Sur, Oaxaca, Mexico. *Proc. Western Found. Vert. Zool.* 1: 107-204.

Russell, S. M. 1964. *A distributional study of the birds of British Honduras.* Ornithol. Monograph No. 1.

Ryan, P. G., & Rose, B. 1985. Bradfield's Swift using cliffs and palm trees in Namibia. *Ostrich* 56: 218.

Rydzewski, W. 1978. The longevity of ringed birds. *Ring* 96-7: 218-262.

Salomonsen, F. 1962. Whitehead's Swiftlet (*Collocalia whiteheadi* Ogilvie-Grant) in New Guinea and Melanesia. *Noona Dan Papers* 3: 509-512.

— 1976. The main problems concerning avian evolution on islands. *Proc. 16th Internat. Ornithol. Congress:* 585-602.

— 1983. Revision of the Melanesian swiftlets (Apodes, Aves)

and their conspecific forms in the Indo-Australasian and Polynesian regions. *Biol. Skr. Dan. Vidensk. Selsk.* 23 (5): 1-112.

Salvin, O. 1863. Descriptions of thirteen new species of birds discovered in Central America by Frederick Godman and Osbert Salvin. *Proc. Zool. Soc. London* 190: 186-192.

Schmitz, E. 1905. *Orn. Monatsber.* 13: 197-201.

Schodde, R., & McKean, J. L. 1972. Comment on alleged sight record of Uniform Swiftlet from Northern Queensland. *Emu* 72: 116.

—, van Tets, G. F., Champion, C. R., & Hope, G. S. 1975. Observations on birds at glacial altitudes on the Carstensz Massif, Western New Guinea. *Emu* 75: 65-72.

Scott, A. J. 1979. A Scarce Swift (*Schoutedenapus myoptilus*) on the Nyika Plateau. *Bull. Zamb. Orn. Soc.* 11: 46-47.

Selander, R. K. 1955. Great Swallow-tailed Swift in Michoacan, Mexico. *Condor* 57 (2): 123-125.

Serle, W., & Morel, G. 1977. *A Field Guide to the Birds of West Africa.* Collins, London.

Short, L. L. 1975. A zoogeographic analysis of the South American Chaco Avifauna. *Bull. Amer. Mus. Nat. Hist.* 154.

Sibley, C. G., & Ahlquist, J. E. 1990. *Phylogeny and Classification of Birds.* Yale Univ. Press, New Haven & London.

— & Monroe, B. L. 1990. *Distribution and Taxonomy of Birds of the World.* Yale Univ. Press, New Haven & London.

— & — 1993. *A Supplement to Distribution and Taxonomy of Birds of the World.* Yale Univ. Press, New Haven & London.

Sick, H. 1948a. The nesting of *Reinarda squamata* (Cassin). *Auk* 65: 169-173.

— 1948b. The nesting of *Chaetura andrei meridionalis. Auk* 65: 515-520.

— 1958. Distribution and nests of *Panyptila cayennensis* in Brazil. *Auk* 75: 217-220.

— 1959. Notes on the biology of two Brazilian Swifts, *Chaetura andrei* and *Chaetura cinereiventris. Auk* 76: 470-477.

— 1991. Distribution of the Biscutate Swift *Streptoprocne biscutata. Bull. Brit. Orn. Cl.* 111 (1): 38-40.

— 1993. *Birds in Brazil: a natural history.* Princeton Univ. Press, Princeton.

Silva, 1975. Observations (Manus, Mussau amnd Emira Islands). *PNG Bird Soc. Newsletter* 112:4-6.

Skutch, A. F. 1935. Helpers at the nest. *Auk* 52: 257-273.

Smith, A. P. 1977. Observations of birds in Brunei. *Journ. Sarawak Mus.* 25:235-269.

Smythies, B. E. 1953. *The Birds of Burma,* 2nd Edition. Oliver and Boyd, Edinburgh and London.

— 1981. *The Birds of Borneo.* 3rd Edition. Sabah Society and Malayan Nature Society, Kuala Lumpur.

Snow, D. W. 1962. Notes on the biology of some Trinidad swifts. *Zoologica* 47: 129-139.

— 1978. *An Atlas of Speciation in African Non-Passerine Birds.* Brit. Mus. (Nat. Hist.), London.

Snyder, D. E. 1966. *The Birds of Guyana.* Peabody Mus., Salem, Mass.

Somadikarta, S. 1967. A recharacterization of *Collocalia papuensis* Rand, the Three-toed Swiftlet. *Proc. U.S. Nat. Mus.* 124: 1-8.

— 1968. The Giant Swiftlet *Collocalia gigas* Hartert and Butler. *Auk* 85 (4): 549-559.

— 1975. On the two new subspecies of Crested Tree-Swift from Peleng Island and Sula Island. (Aves: Hemiprocnidae) *Treubia* 28:119-127.

— 1986. *Collocalia linchi* Horsfield & Moore - a revision. *Bull. Brit. Orn. Cl.* 106 (1): 32-40.

Stadler, H., & Schmitt, C. 1917. Die Rufe der Mauersegler.*Verh. orn. Ges. Bayern* 13: 152-157.

Staub, F. 1976. *Birds of the Mascarenes and St. Brandon.* Mauritius Or. Normale Enterprises, Port Louis.

Steyn, P., & Brooke, R. K. 1971. Cold-induced mortality of birds in Rhodesia during November 1968. *Ostrich* 8: 271-282.

Stiles, F. G., & Skutch, A. F. 1989. *A Guide to the Birds of Costa Rica.* Cornell, New York.

Stresemann, E. 1912. Ornithologische Miszellen aus dem Indoaustralischen Gebiet. *Novit. Zool.* 19: 311-351.

— 1914. Die Vogel von Seran (Ceram). *Novit. Zool.* 21:25-153.

— 1926. Beitrage zur Ornithologie der Indo-australischen Regio. II. Bruchstucke einer Revision der Salanganen (*Collocalia*). II. *Mitt. Zool. Mus. Berlin.* 12: 349-354.

— 1931. Notes on the systematics and distribution of some swiftlets (*Collocalia*) of Malaysia and adjacent subregions. *Bull. Raffles Mus.* 6: 83-101.

— 1932. Vorlauges uber die ornithologischen Ergenbnisse der Expedition Heinrich 1930-1932. VII. Zur Ornithologie von Sudost-Celebes. *Ornith. Monatsber.* 40:104-115.

— 1940. Die Vogel von Celebes. Teil II. Systematik und Biologie. *J. Ornithol.* 88: 1-135.

— & Stresemann, V 1966. Die Mauser der Vogel. *J. Ornithol.* 107 (sonderheft).

Sutton, G. M. 1928. A new Swift from Venezuela. *Auk* 45 (2): 135-136.

— 1941. A new race of *Chaetura vauxi* from Tamaulipas. *Wilson Bull.* 53 (4): 231-233.

— & Phelps, W. H. 1948. Richmond's Swift in Venezuela. *Occ. Papers Mus. Zool.* 505: 1-6.

Tarburton, M. K. 1987. The population status, longevity and mortality of the White-rumped Swiftlet in Fiji. *Corella* 11: 97-110.

— 1993. Determinants of clutch size in the tropics, with reference to the White-rumped Swiftlet. *Avocetta* 17:163-175.

Thibault, J. C. 1975. *Birds of Tahiti.* Editions du Pacifique, Papeete.

Todd, W. E. C. 1916. Preliminary diagnoses of fifteen Neotropical birds. *Proc. Biol. Soc. Washington* 29: 95-98.

— 1937. Two new swifts of the genus *Chaetura. Proc. Biol. Soc. Washington* 50: 183-184.

Turner, A. 1989. *A Handbook to the Swallows and Martins of the World.* Christopher Helm, Beckenham.

Vaurie, C. 1965. *The Birds of the Palearctic Fauna.* Vol. 2. H. F. & G. Witherby, London.

Wells, D. R. 1975. The Moss-nest Swiftlet *Collocalia vanikorensis* Quoy and Gaimard in Sumatra. *Ardea* 63: 148-151.

— 1992. Night migration at Fraser's Hill, Peninsular Malaysia. *Bull. Oriental Bird Cl.* 16: 21-23.

Wetmore, A. 1926. Odservations on the birds of Argentina, Paraguay, Uruguay and Chile. *Bull. U.S. Nat. Mus.* 133: 232-234.

— 1957. Species limitations in certain groups of the swift genus *Chaetura. Auk* 74: 383-385.

— 1968. Additions to the list of birds recorded from Colombia. *Wilson Bull.* 80: 325-326.

White, C. M. N. 1960. Notes from the Cape Verde islands. *Ibis* 102: 138-139.

— & Bruce, M. D. (1986). *The Birds of Wallacea.* BOU Checklist No. 7. BOU, Tring, Herts.

Whitacre, D. F. 1989. Conditional use of nest structures by White-naped and White-collared Swifts. *Condor* 91: 813-825.

Wildash, P. 1968. *Birds of South Vietnam.*Tuttle, Rutland, Vermont.

Williams, M. D. (ed.) 1986. Report on the Cambridge Ornithological Expedition to China 1985.

Williams, J. 1960. Sabine's Spinetail (*Chaetura sabini* Gray). Nature Notes. *Uganda Nat. Hist. Soc.* 23: 306.

— 1980. *A Field Guide to the Birds of East Africa.* Collins, Glasgow.

Williams, L. P. 1986. Chimney Swift: new to the Western Palearctic. *Brit. Birds* 79: 423-426.

Woods, R. W. 1982. *Falkland Island Birds.* Nelson, Oswestry.

Xian, Y. H., & Zhang, H. Y. 1983. A new record of Chinese bird - Edible-nest Swiftlet *Collocalia fuciphaga* (Thunberg), from Dazhou Dao, Hainan Island. *Acta Zootaxonomica Sinica* 125.

Yamashina, Y. 1982. *Birds in Japan: a field guide.* 3rd edition. Shubin, Tokyo.

Yeatman, L. J. 1976. *Atlas des oiseaux nicheurs de France.* Societe Ornithologique de France, Paris.

Zehnter, L. 1980. Besuch einer Brutstatte des einfarbigen seglers (*Apus unicolor* Jard.). *Arch Naturges.* 56:189-220.

Zimmer, J. T. 1945. A new swift from Central and South America. *Auk* 62: 586-592.

— 1953. Studies of Peruvian Birds. No. 64 The Swifts: Family Apodidae. *Amer. Mus. Novitates* No.1609: 1-19.

— & Phelps, W. H. 1952. New birds from Venezuela. *Amer. Mus. Novitates* No. 1544: 1-4.

INDEX OF SCIENTIFIC AND ENGLISH NAMES

Species are listed by their English vernacular name (e.g. Alpine Swift), together with alternative names where relevant, and by their scientific names. Specific scientific names are followed by the generic name as used in this book (e.g. *melba, Tachymarptis*) and subspecific names are followed by both the specific and generic names (e.g. *bakeri, Tachymarptis melba*). In addition, genera are listed separately and alternative generic names are also given with a reference to the genus in which they have been placed in this work (e.g. *Aerodramus*, see *Collocalia*).

Numbers in italic type refer to the main systematic entry and those in bold type refer to plate numbers.

232

Mauritius Swiftlet
  see Mascarene Swiftlet
maxima, Collocalia **6/7** 48, 50, *133*
maxima, Collocalia maxima 134
maximus, Tachtmarptis melba 190
mayottensis, Apus barbatus 203
Mayr's Swiftlet **9** 30, 54, 119, *125*
mearnsi, Collocalia **8** 52, *117*
Mearnsia **12** 14, *140*
melanopygia, Telacanthura **10** 56, *147*
melba, Tachymarptis **20** 76, *188*
melba, Tachymarptis melba 189
mendeni, Hemiprocne longipennis 221
meridionalis, Chaetura andrei 66, *173*
mexicana, Streptoprocne zonaris 42, *104*
micans, Collocalia fuciphaga 136
Micropanyptila
  see *Tachornis*
minuta, Collocalia esculenta 112
misimae, Collocalia esculenta 111
Moluccan Swiftlet **8** 52, *118*
moluccarum, Collocalia vanikorensis 128
montivagus, Aeronautes **17** 70, *175*
montivagus, Aeronautes montivagus 176
Mossy Swiftlet
  see Mossy-nest Swiftlet
Mossy-nest Swiftlet **7** 50, *125*, 133, 135
Mottled Spinetail **10** 56, *145*, 152, 217
Mottled Swift **20** 76, 80, *190*
Mottled-throated Spinetail
  see Mottled Spinetail
Mountain Swift
  see White-tipped Swift
Mountain Swiftlet **9** 54, *118*, 121, 127
Moustached Treeswift **24** 84, *222*
myochrous, Cypsiurus parvus 185
myoptilus, Schoutedenapus **19** 74, 80, *137*
myoptilus, Schoutedenapus myoptilus 138
mystacea, Hemiprocne **24** 84, *222*
mystacea, Hemiprocne mystacea 222

nakamurai, Hemiprocne comata 84, *224*
Naked-legged Swiftlet
  see Bare-legged Swiftlet
natalis, Collocalia esculenta 111
natunae, Collocalia salangana 126
Neafrapus **10** 14, *150*
Needle-tailed Swift
  see White-throated Needletail
neglecta, Collocalia esculenta 111
Neotropical Palm Swift
  see Fork-tailed Palm Swift
Nephocetes
  see *Cypseloides*
New Guinea Needletail
  see Papuan Spinetail
New Guinea Swiftlet
  see Bare-legged Swiftlet
niansae, Apus **22** 80, 193, *198*
niansae, Apus niansae 199
niger, Cypseloides **1/2** 38, 40, *93*
niger, Cypseloides niger 95
nigrior, Aeronautes saxatalis 70, *175*
nigrodorsalis, Tachornis furcata 179
nipalensis, Apus **12/23** 60, 82, *212*
nipalensis, Apus nipalensis 213

nitens, Collocalia esculenta 44, *111*
noonaedanae, Collocalia spodiopygius 120
Northern Needletail
  see White-throated Needletail
novaeguineae, Mearnsia **12** 60, *141*
novaeguineae, Mearnsia novaeguineae 141
nubifuga, Tachymarptis melba 76, *190*
nudipes, Hirundapus caudacutus 62, *155*
nuditarsus, Collocalia **9** 54, *124*
numforensis, Collocalia esculenta 111
Nyanza Swift **22** 80, *198*, 202

oberholseri, Collocalia esculenta 111
occidentalis, Chaetura cinereiventris 164
ochropygia, Chaetura vauxi 169
ocista, Collocalia leucophaeus 133
ocypetes, Chaetura brachyura 66, *172*
ocyptera, Hemiprocne longipennis 221
Old World Palm Swift
  see African Palm Swift
oreobates, Apus barbatus 203
orientalis, Collocalia **9** 54, *125*
orientalis, Collocalia orientalis 125
origenis, Collocalia whiteheadi 124

Pacific Swift **20** 76, 155, *206*, 208
Pacific White-rumped Swiftlet
  see White-rumped Swiftlet
pacificus, Apus **20** 15, 76, *206*
pacificus, Apus pacificus 207
Palau Swiftlet **5** 46, *129*
palawanensis, Collocalia vanikorensis 52, *128*
Pale-rumped Swift **16** 23, 68, 161, 163, 164, *165*
pallens, Collocalia vanikorensis 54, *128*, 129
Pallid Swift **21** 21, 29, 78, 80, 194, *199*, 202
pallidifrons, Streptoprocne zonaris 104
pallidor, Cypsiurus balasiensis 187
pallidus, Apus **21** 15, 78, 80, 194, *199*
pallidus, Apus pallidus 201
Panyptila **18** 15, *181*
Papuan Spinetail **12** 60, *141*
Papuan Swiftlet **9** 12, 54, 124, *136*
Papuanapus
  see *Mearnsia*
papuensis, Collocalia **9** 54, *136*
parvulus, Aeronautes andecolus 70, *177*
parvus, Cypsiurus **19** 74, *184*
parvus, Cypsiurus parvus 185
pekinensis, Apus apus 78, *196*
pelagica, Chaetura **14** 64, *166*
pelewensis, Collocalia **5** 46, *129*
pellos, Collocalia brevirostris 123
perlonga, Hemiprocne longipennis 84, *221*
perneglecta, Collocalia esculenta 112
perplexa, Collocalia fuciphaga 136
peruvianus, Aeronautes andecolus 177
phaeopygos, Chaetura cinereiventris 68, *164*
phelpsi, Cypseloides **1** 11, 38, 40, *92*
Philippine Grey Swiftlet **8** 13, 52, *117*, 126
Philippine Needletail
  see Philippine Spinetail
Philippine Spinetail **12** 60, *140*
Philippine Spine-tailed Swift
  see Philippine Spinetail
Philippine Swiftlet
  see Glossy Swiftlet and Philippine Grey Swiftlet
phoenicobia, Tachornis **18** 15, 72, *178*